Environment and So

Advances in Research

Introducing Berghahn's *Advances in Research*

Advances in Research

Advances in Research responds to the growing need for a rigorous, in-depth review of current work in the social sciences and humanities and seeks to stimulate advanced research and action on particularly critical issues of today.

The annuals provide an overview of the latest developments in their area from an interdisciplinary perspective and encourage international communication and exchange among all relevant disciplines.

Now available
Religion and Society
Environment and Society

Forthcoming
Conflict and Society
Museum Worlds

Environment and Society

Advances in Research

Contents

Volume 1 ■ 2010

Introduction

A New Journal for Contemporary Environmental Challenges

Social scientists have been writing about the relationships between people and their surroundings for as long as there has been social scientific inquiry. Fields such as anthropology, economics, history, human geography, law, political science, psychology, and sociology all have long and rich histories of contributing to and pioneering socio-environmental analysis. However, the past 20 years have seen a proliferation of scholarship in the social sciences that is focused on environmental issues. This is due, in part, to changes in our environment that have profound implications for the future of both human society and the environment. It is also due, in part, to the ways in which environmental practitioners have portrayed the causes of these changes. In the 1970s, social scientists, concerned with the ways in which the causes of environmental changes were being attributed to some peoples and not others, felt that their knowledge of social processes and social systems could shed light on these issues (see Blaikie and Brookfield 1987). They thought that the methods and theories of the social sciences could and should be brought to bear on questions about contemporary environmental changes. Climate change, the water crisis, deforestation, desertification, biodiversity loss, the energy crisis, nascent resource wars, environmental refugees, and environmental justice are just some of the many compelling challenges facing society today that were identified by these early scholars as sites in need of social scientific analysis.

Most social scientists approach environmental issues by asking why a phenomenon is taking place and then trace the answers out across time and space (cf. Vayda 1983). More often than not, they also contextualize the answers they find within larger conversations, debates, and theories in and across disciplines and thus attend to cultural, historical, political, and economic contexts. In asking questions and then contextualizing the answers within social scientific theory and discourse, they sometimes write in ways that are difficult for outsiders to decipher. This is also the case in the natural sciences. Within atmospheric science, biology, chemistry, ecology, geology, geophysics, hydrology, oceanography, physical geography, physics, and soil science, scholars have developed their own vernaculars and systems of contextualization. These competing vernaculars often make it seem as if social and natural scientists cannot (and do not) communicate at all.

Environment and Society: Advances in Research is meant to address these key issues, among others. This new journal focuses on contemporary environmental topics that large numbers of social scientists in multiple disciplines write about. We provide peer-reviewed articles that analyze and summarize existing research and that offer suggestions for new directions in research. These articles are not steeped in disciplinary vernaculars, but they do incorporate contemporary theory. With this, they afford a window into social science without editing out the insights brought to the empirical world through rigorous engagements with theory. The journal also offers reviews of books concerned with environmental issues that are of significance for both the social and

Environment and Society: Advances in Research 1 (2010): 1–3 © Berghahn Books
doi:10.3167/ares.2010.010101

natural sciences. The editors see the journal as a site for dialogue across disciplines within the social sciences and as a resource to which natural scientists, policy makers, environmental practitioners, and activists can turn in order to understand better the work that social scientists do.

This first volume of *Environment and Society: Advances in Research* focuses on social scientific analyses of climate-related topics. The first article, by Noel Castree, is titled "Neoliberalism and the Biophysical Environment: A Synthesis and Evaluation of the Research." Many social scientists working on the environment in general and climate-related issues in particular frame their work in terms of the socio-economic philosophy that has come to be known as 'neoliberalism'. Castree explains this philosophy and shows how the policy ideas and practices that emerge from it affect both environments and societies. He then reviews the ways in which social scientists have attended to neoliberalism and environmental issues for the past decade. Castree's article is about neoliberalism, but it can also be read as an example of the ways in which social scientists work to understand the confluence of the various processes mentioned above and then build social theory from these understandings. 'Neoliberalism', 'neoliberalization', and 'eco-neoliberalism' are terms in the social scientific vernacular. In reviewing a substantial literature, Castro unpacks these concepts in a clear and compelling way, opening them up to wider transdisciplinary understanding.

In "Neoliberal Water Management: Trends, Limitations, Reformulations," Kathryn Furlong frames her analysis of water policy and management in terms of neoliberalization. Furlong describes the claims of neoliberal reformers and of their opponents in Bolivia, her long-term site of research, situating them in a much broader discussion of the environmental aspects of the philosophical tenets that drive neoliberal reforms. She shows how that philosophy is mobilized, not only when it comes to addressing water shortage issues, but also when other matters, such as climate, are dealt with.

In his article, "Controversies in Climate Change Economics," Robert Eastwood explains how economists understand and evaluate the most pressing climate-related issues. He demonstrates how the economic and scientific uncertainty associated with climate creates a policy context that requires innovative measures in assessment and policy crafting. Eastwood also examines other economic reviews of climate change and climate change policy, illustrating how different approaches to climate policy analysis exist within economics. Part of what is compelling about Eastwood's article is the way in which he clearly documents the issues that are important to economists and how those issues then drive the analytic tools that they bring to the entire climate debate.

Richard Ladle and Paul Jepson's article, "Origins, Uses, and Transformation of Extinction Rhetoric," examines the role of extinction in climate-related discussion and illustrates how social scientists contextualize the questions they ask in wider frames. The authors trace the history of academic and political discussions about extinction in order to show how these discourses influence policy and practice. They carefully review studies concerned with extinction from multiple disciplines, and through this they provide suggestions as to how conservation practitioners and policy crafters might draw on this wide literature in order to create more robust approaches to issue of extinction.

The next two articles, "Climate Changing Small Islands: Considering Social Science and the Production of Island Vulnerability and Opportunity," by Amelia Moore, and "Adaptation—Genuine and Spurious: Demystifying Adaptation Processes in Relation to Climate Change," by Thomas Thornton and Nadia Manasfi, provide literature analyses that work to contextualize case studies. In doing so, they demonstrate how contextualization makes the social scientific analysis of empirical evidence richer. In her review of research literature, Moore examines the processes by which small islands have been incorporated into climate-related debates. She then contextualizes state-specific discussions about climate change science and policy, vulnerability indexing, and sustainable economic growth in the Bahamas within this literature. Thornton and Manasfi review the social scientific literature on adaptation. After carefully assessing whether adaptation to climate change can be planned and managed, they show how contemporary governance frameworks fail to address the growing pressure that climate stress is exacting on indigenous peoples in the far north of Alaska and Canada and in the Arctic Circle. Both of these articles show the current and potential human aspects of the climatic-related economics discussed by other authors.

In their article "Climate Change Resilience and Adaptation: Perspectives from a Century of Water Resources Development," Clive Agnew and Philip Woodhouse argue that much of the debate about how societies adapt to climate change is in fact a discussion about water management. They offer a careful analysis of the literature on resilience and adaptation while exploring the reasons for failure and success in adapting to water scarcity and degradation in different circumstances. Their research is particularly pertinent for dialogues about coping with climate change.

Finally, in the book reviews section, we provide critiques of 12 environmentally focused social scientific books published between 2007 and 2010. Through this feature of the journal, we hope to show how publications in the social sciences speak to issues of broad concern across disciplines. The reviews in this volume examine environmental conservation practice, science and policy, long-term global environmental changes, the medical effects of pollution, the economic aspects of disasters, urban ecology movements, the relationship between ecological and social change, environmental history, and the environmental politics of economic development.

Paige West, Dan Brockington, Jamon Halvaksz, and Michael Cepek

REFERENCES

Blaikie, Piers M., and H. C. Brookfield. 1987. *Land Degradation and Society*. London: Methuen.

Vayda, Andrew P. 1983. "Progressive Contextualization: Methods for Research in Human Ecology." *Human Ecology* 11 (3): 265–281.

ARTICLES

Neoliberalism and the Biophysical Environment

A Synthesis and Evaluation of the Research

Noel Castree

ABSTRACT: This article both synthesizes and critically evaluates a now large, multi-disciplinary body of published research that examines the neoliberalization of environmental regulation, management, and governance. Since the late 1970s, neoliberal ideas and ideals have gradually made their way into the domain of environmental policy as part of a wider change in the global political economy. While the volume of empirical research is now such that we can draw some conclusions about this policy shift, the fact that the research has evolved piecemeal across so many different disciplines has made identifying points of similarity and difference in the findings more difficult. After clarifying what neoliberalism is and explaining why the term 'neoliberalization' is preferable, the article analyzes the principal components and enumerates the social and environmental effects of this multifaceted process. By offering a comprehensive and probing survey of the salient literature, I hope not only to codify the existing research but also to guide future critical inquiries into neoliberal environmental policy.

KEYWORDS: biophysical world, deregulation, environmental policy, governance, neoliberalism, neoliberalization

In this article I will examine the relationship between neoliberalism and the biophysical world. 'Neoliberalism' is very much a critics' term, an oppositional badge as much as an analytical concept.[1] For those who choose to use it, the word describes a worldview fleshed out over the last 30 years at the sub-national, national, and global scales. Rarely invoked before 2000, it is now part of the *lingua franca* of left-wing social scientists and activists. Researchers in development studies, sociology, area studies, anthropology, labor studies, political science, cultural studies, human geography, philosophy, environmental studies, international relations, education policy studies, and the radical fringes of the economics profession have—over the last decade—sought to define neoliberalism, identify its modes of operation, track its mutations, pinpoint its effects, and describe various modes of opposition to it. Where political activists have often used the term for polemical purposes, academic researchers have attempted to mount a rather cooler challenge to neoliberalism on both evidential and moral-political grounds. There have been

Environment and Society: Advances in Research 1 (2010): 5–45 © Berghahn Books
doi:10.3167/ares.2010.010102

two main camps, theoretically speaking: namely, a political-economic one, which is broadly neo-Marxist, and a Foucauldian one—with some overlap between the two (e.g., Lockwood and Davidson 2009).[2] Together, their members have produced a now sizable literature comprising monographs, co-authored books, edited collections, and peer-reviewed articles like this one. In surveying this literature, I will be focusing on those parts of it where the analytical attention has principally addressed neoliberal environmental use and management. I have selected only those publications in which the term 'neoliberalism' is—in my view—central to the arguments and research findings presented.[3]

The original architects of neoliberal thinking, such as Milton Friedman and Friedrich von Hayek, said little about environmental issues or natural resource management. However, since the mid-1980s, many neoliberal values and principles have steadily made their way into the domain of environmental policy,[4] conceived here in the widest possible sense to include situations where any element of the biophysical world is a major focus of policy makers' attention at any spatio-temporal scale. Accordingly, a previous trickle of research into neoliberal environmental governance has given way to something of a torrent over the last five years or so. As we will see, this research comprises predominantly single-site case studies, presented in rich empirical detail. It also covers the full spectrum of environmental and resource policy domains, from water to fisheries to farming and beyond. Up to the present, this research has grown organically in a rather disparate fashion, with some elements of disciplinary crossover.[5] In order to steer its future development, it is useful to take stock of the insights that the research offers to date, to draw wider lessons about neoliberal environmental governance, and to assess the way that critics have investigated it so far.

To summarize, this article will operate at two levels toward two rather different ends. First, it will parse the insights of numerous studies in order to clarify how neoliberal environmental governance has operated to date and with what effects. Secondly, it will treat the authors of these studies as a loose 'epistemic community' whose worldview and research practices do not simply hold a mirror up to a neoliberal world existing 'out there'. I identify broadly with its members' value set and normative agendas, but I also believe that this community may want to reflect critically on its own habits of thought and analysis. Having written along these lines for a human geography readership (Castree 2008a, 2008b), this article is directed at a wider and more intellectually mixed audience, spanning several social science disciplines (as befits the remit of this journal).[6] I hope that those who are new to the subject, as well as those familiar with it, will profit similarly from reading this comprehensive review.[7] As ever with review articles, there is the risk that I am imposing a false order on the literature I survey. I will thus try to be scrupulous about justifying my various 'moves' as I proceed, and I invite readers to assess these moves critically.

The article is organized as follows. I begin in a rather obvious place by seeking to define and delimit the term 'neoliberalism'. Here I summarize the insights of critical social scientists who do not themselves study environmental management but whose writings have influenced those who do. Then I introduce the research literature on the neoliberalization of nature in general terms. This done, the next two sections distill the substantive insights of this largely case study–based literature, focusing on the process of reregulation and its socio-ecological outcomes, respectively. I then further refine these insights into a small set of provisional 'take-home' lessons. The final main section presents a sympathetic critique of the literature, wherein I raise several key questions about the published research analyzed earlier. While this article is extensive, I believe that its length is justified. Considering the number of publications reviewed, as well as the scope of my 'take' on them, a shorter survey would sacrifice much detail and many key points—especially for those new to this literature.

Before I get down to business, I should make one final introductory observation. In light of the recent financial crisis and current global economic recession, some have suggested that the 'neoliberal era' has come to an end—or at least the beginning of the end. If true, this would imply no future for the sort of research surveyed here and would make this article very untimely indeed. Needless to say, I reject this suggestion (cf. Brand and Sekler 2009). Periodizing history is notoriously tricky, as is the identification of temporal 'breaks'. Craig and Cotterell (2007: 510) note that "the various conjunctural factors bundled together [by analysts] to constitute 'neoliberalism' (and any plausible period shift in it) are quite diverse in nature, so that first of all comparing the relative weight or importance of any them to an overall periodizing assessment is analytically fraught." Moreover, even supposing that we *have* been living through a historic period sufficiently homogeneous to be called neoliberal, experience tells us that there are rarely punctual transitions between one putative era and another. The traces of the recent past will inevitably continue to affect both the present and the short- to medium-term future (see Brenner et al. 2010). As *New Left Review* editor Susan Watkins (2010: 14) notes, "The widely proclaimed end of neo-liberalism looks more and more like the continuation of its agenda by other means." I doubt, in other words, that the term 'neoliberalism' will disappear from the vocabulary of social scientists (or political activists) any time soon. If it does, we will probably be employing new words to capture its meanings and to describe many of its real-world objects (Clarke 2010). Therefore, throughout this article I will talk of neoliberalism in the present tense, presuming that the term and the things that it names retain their currency for the time being.[8] The question then becomes not *whether* we should (still) use the term, but rather *how*.

What Is 'Neoliberalism'?

Conceptual Issues

The research into neoliberal environmental governance is in one sense parasitic on a wider theoretical and empirical literature in which environmental issues are not strongly thematized. This broader literature mostly predates the research being surveyed in this article, and this antecedence explains why it has proven formative for many who interrogate neoliberal environmental policy. There has, in my view, been a tendency for the latter to borrow definitions and insights from the former rather than to rework and question them. That does not make their work entirely derivative—far from it, in fact. As we will see later, the primarily empirical (rather than theoretical) character of recent research into neoliberal environmental governance is its major strength and contribution. Along with Adam Tickell, Nik Theodore, and Neil Brenner, Jamie Peck has done much to shape social scientific conceptions of neoliberalism. Not long ago, Peck (2006: 731) observed that "there remains a paucity of 'grounded' work on the specific and concrete routines, practices, networks, and structures through which the neoliberal project has been constructed and sustained." This is no longer true, and the recently published research into neoliberalism and the environment can take considerable credit for filling the empirical gap that Peck identified. Even so, in conceptual terms much of this research has tended to work with 'off the peg' definitions of neoliberalism developed by others (such as Peck himself). What is more, these definitions have been used selectively and partially, depending on the case.

Note that I use the word 'definitions' in the plural. The reason that economic sociologist Stephanie Mudge (2008) so recently felt compelled to ask "What is neoliberalism?" is because of the diversity of meanings and applications in play. Despite the term's current popularity among a cohort of left-wing social scientists, the more familiar it has become, the less consensus there appears to be about what it means. James Ferguson (2010: 170) observes that "there is [now] huge variation in the way the word 'neoliberalism' is used in contemporary scholarship." And yet, perhaps curiously, most researchers who employ the term typically proceed as if the meaning is clear and maps onto a definite set of values, ideas, and/or practices that exist in the wider world. Thus far, there has been little sustained discussion about the term's ambiguities and complexities—a conversation that is now certainly overdue.

So what is going on here? One view is that neoliberalism is an unusually complex word (like 'globalization' or 'nature') and as such signifies a range of related meanings that can be applied to a plethora of real-world referents. Seen from this perspective, as long as the term's meaning is clear for each context of application, there is no especial problem with it signifying several things and having myriad objects of empirical reference through strong 'family ties'. Another view is that we should relieve the term of some of its current denotations, not least because we have other well-established words for them, such as privatization, commodification, the free market, the Washington Consensus, and structural adjustment. By doing so, we would restrict—and so render more precise—the meaning of those situations or things we still prefer to describe as neoliberal (see Boas and Gans-Morse 2009). Still another perspective is that we are now beyond the point of no return: so various and confusing are the meanings of neoliberalism that it has become as 'chaotic' a concept as globalization was after a decade of debate and use (circa 1990–2000). For instance, in their recent review, anthropologists Catherine Kingfisher and Jeff Maskovsky (2008: 123) confessed a temptation "to abandon the term altogether"—and they are not alone (see Barnett 2010).

For now, I will adopt the first of these viewpoints (although I will come back to the other two toward the end of this article). In other words, I will presume that while the term 'neoliberalism' is polysemic and refers to a plurality of material and discursive things, there are nonetheless a set of fairly stable, circumscribed, but also related meanings in use that are applied in a relatively consistent way by academic analysts. In effect, this is the viewpoint adopted by those social scientists investigating neoliberalism and the biophysical world (as we will see later). Each researcher has defined neoliberalism in a certain recognized manner and then undertaken an empirical investigation of environmental governance with this particular definition in place. The assumption is that there is a meaningful similarity (or even relationship) between otherwise different and separate studies.

The Meanings of Neoliberalism

As some readers will know, the term 'neoliberalism' was coined by a group of economists and legal scholars based in Freiburg between World War I and World War II. Subsequently, it was used quite briefly by those now thought to be neoliberalism's principal intellectuals, that is, the already mentioned Hayek and Friedman, even though neither man entirely accepted the Freiburg (or 'ordoliberal') model of a state-managed 'market society'. As I intimated above, those individuals and institutions that the critics choose to call 'neoliberals' did not/do not use the term as a self-descriptor and rarely ever have. After a smattering of appearances during the 1990s (e.g., Barry et al. 1996; Fraser 1993; Gowan 1995; Tickell and Peck 1995), these critics started to invoke the term with increasing frequency from the turn of the millennium. My own reading of the now voluminous academic literature suggests that, for this interdisciplinary epistemic community of social scientists, the term 'neoliberalism' describes one or more of the following three related things: first, a worldview (i.e., a body of normative principles, goals, and aspirations amounting to a philosophy of life, or something close to one); second, a policy discourse (i.e., a set of specific values, norms, ambitions, and associated policy proposals professed by those who control, or realistically seek to control, the formal apparatuses of government); and, third, a set of practical policy measures (i.e., concrete regulations and procedures that make both the worldview and the policy discourse evident in some tangible way). As a shorthand, we can think in terms of 'three p's': philosophy, program, and practice.[9] I will now discuss each of neoliberalism's three aspects in turn.

1. *Neoliberalism as worldview (philosophy).* Although Hayek, Friedman, and others of their ilk rarely used the term, their neoliberalism, according to David Harvey (2007: 24), "took the political ideals of individual liberty and freedom as sacrosanct." However, there are many different ways in which to define and engender liberty and freedom.[10] The neoliberal ideals articulated in books such as *The Road to Serfdom* (Hayek 1944), *The Constitution of Liberty* (Hayek 1960), and *Capitalism and Freedom* (Friedman 1962) accented two things. First, the state's role was to maximize the independence of both real and institutional-juridical individuals: anything less would be anti-liberal, a travesty of 'true freedom'. Governments and bureaucracies, it was argued, should refrain from imposing collective agendas and otherwise 'interfering' in the lives of people. In this sense, the early neoliberals strongly emphasized not only the rights of individuals but also their responsibility to make their own way in the world. Second, neoliberalism's founding thinkers saw money-mediated markets as the best mechanism for coordinating among the diverse needs and wants of ostensibly free people. This is because markets were seen as highly 'intelligent' and 'efficient'. Price signals, it was claimed, enable disparate providers and users of goods and service to achieve many of their desires, given whatever restraints of resource availability happen to be in place for those involved. Indeed, market competition was seen as bringing the best out of entrepreneurs, while delivering value for money to intermediate and end consumers. This belief segued into a critique of 'state failure', that is, the idea that government bureaucracies are administratively cumbersome and economically wasteful vehicles for service delivery.[11]

The neoliberal worldview, as summarized above, is not—despite appearances—fixated on economic liberty alone, although it is emphasized very strongly indeed. Political and civil liberties are featured as well: electoral democracy was the early neoliberals' favored political system, and freedom of expression (within or without the market) was also seen as fundamentally important. This wide-ranging doctrine began to take shape during World War II and its immediate aftermath. Fashioned in reaction to the totalitarian impulses of fascism and communism, it was also presented as an alternative to the new Keynesian welfare-state paradigm, which licensed state intervention in the market and in many aspects of citizens' daily lives. In the former respect, neoliberal thinking was at one with the *zeitgeist*, but this was not so in the latter respect. Commenting on *The Road to Serfdom*, Peck (2008: 5–6) says that "the book may have been a best-seller, but it was practically an act of self-immolation for Hayek-the-economist." Writing in the same year of its publication, Hayek's contemporary Karl Polanyi (1944: 142) declared that "our age will be credited with having seen the end of the self-regulating market." In advocating an alternative to Keynesianism, the neoliberal worldview remained well outside the perimeter of cognitive and moral 'common sense' until the economic and political crises of the 1970s.

2. *Neoliberalism as policy discourse (program).* For over three decades, neoliberal ideas were incubated within a select group of university economics departments, foundations, and think tanks. As the excellent histories recounted by Peck (2008) and Mirowski and Plehwe (2008) show so well, the neoliberal 'thought collective' was transnational, composed of groups in both Europe and the US, with a good deal of interchange between them—not least because of the efforts of the Mont Pelerin Society (formed by Hayek in 1947 and based in Chicago).[12] The collective's eventual success was hardly predictable during the immediate post-1945 period.

Lacking political influence and much academic credibility, neoliberal thinkers initially produced a combination of general manifestos and fundamental works of theory, only later turning their attention to policy programs in live settings. In the latter respect, Pinochet's Chile was a key opportunity: a group of Friedman-trained economists were invited to remake the country's political and moral economy almost overnight (while being safeguarded by a military dictatorship). It was the first of several opportunities provided by domestic crises of one sort or another; however, without the groundwork laid by the prolonged efforts of the Mont Pelerinians, these opportunities could not have been exploited. As is well known, by the late 1970s, New Zealand, the United Kingdom, and the United States were the other three countries where some version of the neoliberal worldview was taken seriously by political-economic elites. It gained a subsequent hearing in many other countries—especially after the late-1980s 'revolutions' in Eastern Europe—and also in a range of US-dominated global institutions, such as the World Bank and the International Monetary Fund (IMF).

As a policy discourse, the neoliberal worldview is typically understood by critics to include the following seven proposals for significant society-wide change. Note that national governments are the major institutions tasked with delivering these changes. Like all political philosophies and programs, neoliberalism is necessarily a state-led project.

1. *Privatization*: assigning clear, legally enforceable private property rights to hitherto unowned, state-owned, or communally owned aspects of the social, cultural, and/or natural worlds.[13]
2. *Marketization*: rendering alienable and exchangeable things that might not previously have been subject to a market logic expressed through commodity transactions within and between nation-states measured in monetary terms.[14]
3. *State roll-back or deregulation*: withdrawing or diminishing state intervention in certain areas of social, cultural, and environmental life in order to enable firms and consumers to exercise 'freedom of choice'; creating new quasi-state or state-sanctioned 'non-political' actors to take on functions that states themselves could, in theory or in practice, otherwise perform; and contracting private or third sector bodies to deliver some state services through a process of competitive bidding or through partnership agreements.
4. *Market-friendly reregulation*: reconfiguring the state so as to extend the frontiers of privatization and marketization. The state in its various forms becomes more a 'market manager' and less a 'provider' to the citizenry or to 'special interests' therein: it intervenes *for* the market economy, not, as it were, *in* it. This entails fiscal discipline, a focus on supply-side investments, entrepreneur- and consumer-friendly tax policies, firm-friendly labor market policies, and measures to enable 'free' movements of money capital, as well as other less 'fluid' commodities.
5. *Use of market proxies in the residual state sector*: making the remaining state functions and services more market-like in their operation through the use of measures such as internal markets, cost recovery, and budget capping. This amounts to embedding an ethos and the practices of 'commercialization' into state services.[15]
6. *Strong encouragement of 'flanking mechanisms' in civil society*: using state-led measures to promote the growth of (a) robust informal and social economies, and (b) voluntary, charitable, non-profit, and community groups, all preferably well-funded and professionalized. Together, these mechanisms are intended to fill the vacuum created by the absence/diminution of direct state support in the social and environmental domains. They could be interpreted as a 'shadow state' that is emergent organically, once prodded by the state.
7. *Creation of 'free', 'self-sufficient', and self-governing individuals and communities*: cultivating an ethic among persons, other juridical units, and communities that emphasizes less

(and ultimately limited) reliance on public services and state agencies for life's necessities. This ethic extends to those operating within the state apparatus itself. It equates to a 'responsibilization' of all producers, consumers, citizens, families, communities, institutions, and spatial units of governance, as well as their 'autonomization'.[16]

Clearly, there is a dialectical relationship between neoliberalism as a worldview and as a program. To the extent that the program gains traction and exerts influence within any given context, it not only lends legitimacy to the worldview but poses a barrier to its future removal by critics.

3. *Neoliberalism as a set of actionable policy measures (practice).* For any program to translate into actual policies, its advocates first need to legitimate their worldview. As Clarke (2004: 34) reminds us, "Politics is both a regime of truth (ruling out the irrational, marginal or unworkable …) and a field in which … some discourses [are allowed] to contend." Because the neoliberal worldview and policy program are so wide-ranging and ambitious, it is no surprise that they can be—and have been—translated into a plurality of concrete policy measures. The following are some common practices of neoliberalism in action that have been identified by analysts (even if these measures have frequently been more honored in the breach):

- *Macro-economic policies* that place controls on government borrowing, keep inflation low, place constraints on domestic money supply, keep taxation levels low, allow exchange rates to float, and allow interest rates to be determined by the market (or at least not to be determined by the government).
- *Industrial and business policies* that (a) remove selective subsidies, trade barriers, investment barriers, and ownership barriers, and (b) incentivize innovation, competition, and entrepreneurial risk taking. In effect, these policies widen and intensify commercial competition.
- *Labor market policies* that remove collectivist 'obstacles' to competition and reward, such as wage controls and trade union membership.
- *Education and training policies* that focus on the supply side and encourage individuals to build their 'human capital', be adaptable, and commit to 'lifelong learning'.
- *Managing, monitoring, and auditing measures* that—whether in the private, state, or third sector—focus hard on setting targets, establishing benchmarks, measuring performance, penalizing failure, and rewarding success. In the state sector, these measures have been inspired, variously, by approaches known as 'new public management', 'transaction cost theory', and 'principal-agent theory'.
- *Social policies* that are oriented to 'workfare' not 'welfare' and that offer state support only to the very needy or chronically disadvantaged. This entails a remoralization of the poor and the 'excluded', so as to 'responsibilize' them for their livelihoods, their successes, and their 'failures'. They are thus exposed to the various risks of life and to living without much assistance from society or the state.
- *Law and order policies* that take an uncompromising approach to rule breakers, 'troublemakers', and those who otherwise cause social disruption and infringe upon the rights of others.
- *Civil rights policies* that encourage free speech, freedom of information, lifestyle choice, privacy rights, and freedom of assembly—as long as the rule of law is observed.
- *Governance policies* that, in a range of policy areas, democratize and devolve decision making by empowering a wide range of actors outside the formal sphere of government.[17] Empowerment is about making decisions and dealing with their consequences, for good or ill.

I will deal with environmental and natural resource policies later in the article, but for now, suffice it to say that these policies aim to achieve given ends—such as environmental protection or resource conservation—in efficient and competitive ways by variously privatizing, marketizing, and de-statizing a range of biophysical goods and ecosystem services.

Again, it almost goes without saying that neoliberal policies are linked recursively to program and philosophy. For instance, Peck and Tickell (2002) have argued that the 'roll-back' policies of a 'hard' neoliberal program have, in countries like Britain and the US, given way to 'roll-out' policies that seek to embed core neoliberal principles and values as norms governing everyday life and living. A key point to make here is that a wide range of different concrete policy measures—alone and in combination—will 'deliver' various of the seven elements of the neoliberal policy proposals detailed above. There is no universal or perfect 'one-to-one' mapping of these elements onto discrete policy measures.

Neoliberalism or Neoliberalizations?

This three-part disaggregation of neoliberalism usefully clarifies the term's complex meanings and myriad referents. It may help readers to understand better what different researchers are referring to when they describe something as 'neoliberal'. Clearly, one should not elide philosophy, program, and practice, even though they are necessarily related. Mudge (2008), following sociologist Pierre Bourdieu's concept of different discursive-institutional 'fields', uses a 'horizontal' metaphor to understand the 'three p's'. For her, the philosophy is located in the 'intellectual field' (dominated by university economics departments, think tanks, and foundations), the program in the 'political field' (political parties, professional politicians, and elected leaders), and the practice in the 'bureaucratic field' (civil servants, administrators, managers appointed by elected politicians, nominated firms, sanctioned NGOs and charities, etc.). Obviously, the fields greatly overlap and mutually condition one another—but how?

Here, one or two commentators have come unstuck. For instance, in an uncharacteristically ill-judged (and much cited) observation, *New Left Review* editor Perry Anderson (2000: 7) once said that "neo-liberalism as a set of principles rules undivided across the globe: the most successful ideology in world history." Similarly, another observer, otherwise attentive to the uneven development of neoliberalism, has sometimes described it using blanket metaphors: "Neoliberal[ism] has in effect swept across the world like a vast tidal wave of institutional reform and discursive adjustment … no place can claim total immunity (with the exception of a few states such as North Korea)" (Harvey 2007: 23). Statements like these paint a picture of 'hegemonic neoliberalism', thus spatializing at the global scale the notion of a supposedly coherent period or era (with the oft-used term 'neoliberal globalization' performing the same function). Such statements imply a one-way, unadulterated relationship between philosophy, program, and practice in which a peculiarly homogeneous geography of neoliberalism (with a capital 'N') writ large is imagined—a sort of spreading ink blot that has its origins in the academic field. However, very few empirical analysts of neoliberalism see it in such simplistic terms, which is why the process term 'neoliberalization' has been favored since it was coined nearly a decade ago by Peck and Tickell (2002). Neoliberalization describes an ongoing, unfinished process of proposing, revising, testing, applying, and further altering neoliberal ideas and policies. As Neil Brenner and Nik Theodore (2002) have argued, 'actually existing neoliberalism' is not the same as the neoliberal philosophy. Indeed, Harvey (2005: 19) regards the latter as utopian, as a rhetorical cloak used to describe and justify the messy pragmatics of programs and policies in practice. It is thus ironic that he inadvertently bolsters the utopian rhetoric by discussing neoliberalism with a capital 'N'.

These arguments suggest that what is referred to as neoliberalism in the singular is, in reality, a complex historical-geographical formation that is marked by unevenness and variety as much as it is by similarity—that is to say, it is a set of interconnected local, regional, and national neoliberalizations (in the plural). Neoliberal ideas may well have 'gone global' from the mid-1980s, courtesy of the US and its influence on the World Bank and the IMF. But this has not resulted in a tidy process of downward and outward diffusion from neoliberalism's North Atlantic heartlands. Instead, there has been path dependency, contingent couplings, unplanned adaptations, organic mutations, and a good deal of social resistance to 'new liberal' policies. Varying combinations of coercion, consent, contestation, and compromise describe the spatio-temporal evolution of neoliberal projects in different parts of the world. In some cases, their reach is wide and deep; in others, it is not. Peck (2006: 732) summarizes well the research agenda that follows from this: "If neoliberalism can only exist in hybrid, in a kind of parasitic relation to the social formations that provide its hosts, then there is … considerable work to do in mapping varieties and transmutations of the project" (see fig. 1).[18]

Figure 1: From neoliberalism to neoliberalization

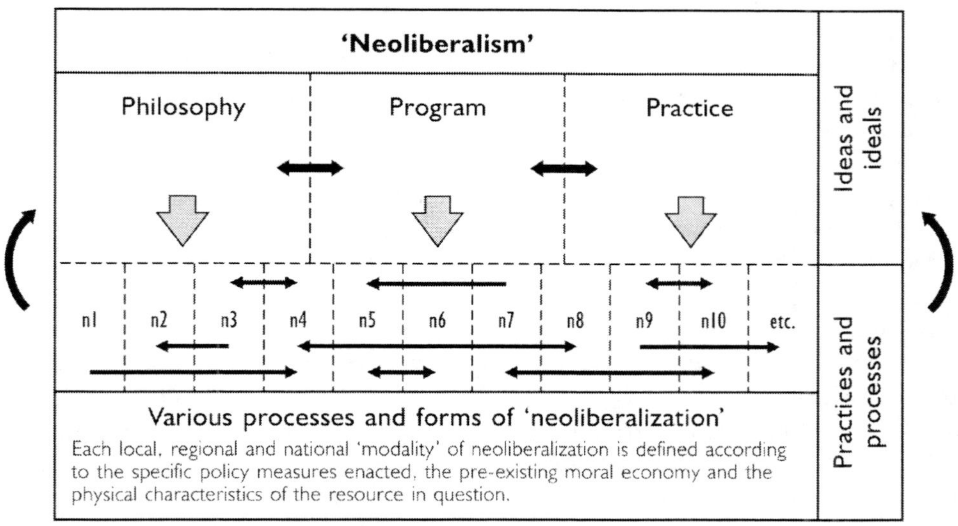

Note: In order to ensure that 'philosophy' and 'program' are not perceived by readers to 'determine' 'practice', the last has been placed in the top part of the figure. However, perhaps somewhat confusingly, the pragmatics of practice are signified in the lower half of the figure. This is confusing only if readers divorce practice from philosophy and program. In actuality, they are inseparable; hence, the dotted lines used in the figure above.

Neoliberal Environments: Introducing the Topic and the Published Research

As already stated, critical social scientists interested in environmental management, natural resource use, and related issues came late to the discourse of neoliberalism, when compared with most of those whose works I have cited previously. Even so, they had been discussing at least some of the same phenomena by way of other terms, such as 'free market environmentalism' (see, e.g., Eckersley 1993). Since roughly 2000, they have linked these terms to the concept of neoliberalism or, for various reasons, have eschewed the former and used the latter as an analytical framing device in their research.

Neoliberal Environmental Policy: History and Declared Benefits

Environmental and natural resource policies that deliver one or more elements of the seven neoliberal policy proposals have been implemented in a wide range of contexts and locations. The question arises: given that neoliberal ideas originally made little or no reference to natural resources or environmental issues, why did they find expression in water, forestry, and fisheries management (to take just three examples) in many places, regions, and countries? The published research literature suggests that there are five answers to this question, even though the complete history has yet to be detailed.

First, Steven Bernstein (2002) points to the entirely contingent coupling of environmentalism and neoliberalism during the 1930s—what David Driesen (2008) refers to as a 'shotgun wedding'. The global rise of environmental concern through the 1960s and 1970s, Bernstein argues, coincided with the success of neoliberalism in the Anglo-American world and, via the World Bank, the IMF, and the World Trade Organization, at the global level through the 1990s (see also Hartwick and Peet 2003). Yet the post-war origins of environmentalism can, of course, be traced back to the Fordist-Keynesian, pre-neoliberal period, one that ended with widespread talk of an 'environmental crisis'. Second, environmentalism aside, the idea of 'green' development took hold in several global institutions from the late 1980s. The idea was that in the global South there were many 'unpriced' and often unowned biophysical 'assets' that could, if inserted into global markets, create revenue streams that would be able to support much-needed socio-economic development. These assets to be traded overseas included everything from rare and beautiful animal species to plantation trees and mineral resources. Third, in some (but by no means all) domains of environmental management and natural resource policy, neoliberal ideas had already been aired quite separately from the broad manifestos authored by the likes of Friedman and Hayek. For instance, during the 1950s, economist H. Scott Gordon (1954) had suggested a proto-neoliberal solution to overfishing in which private property and markets played a major role, while the famous essays authored by Garrett Hardin (1968, 1974) more than a decade later popularized similar ideas for all 'open access' resources. The term 'neoliberal' was not used by Gordon or Hardin, but their arguments were consistent with the neoliberal philosophy. Fourth, because many natural resources (e.g., water and forests) had been managed by state bodies as public services or national assets following World War II, it was inevitable that neoliberal politicians, such as Margaret Thatcher, would seek to manage these resources using the same political-economic rationality utilized in all other areas of economic and social policy. Fifth, in the US a small group of think tanks and foundations worked very hard from the late 1970s to popularize 'green neoliberalism' (Goldman 2005) as a way of responding to the concerns of the green movement without capitulating to 'command and control' solutions. This later had effects on American environmental policy arguments on the world stage. In short, there was no single reason why neoliberal ideas seeped into the arena of environmental policy. It was an uneven process, both temporally and spatially, and the three fields identified by Mudge (i.e., intellectual, political, and bureaucratic) were all involved in various ways and to varying degrees.

With the benefit of hindsight, it is clear that the growth of environmental and resource economics as an applied field of academic research was very important in codifying an environmental version of neoliberalism as, variously, a worldview, a policy discourse, and a set of practical policy measures. This sub-discipline's prodigious growth in universities from the early 1980s was a reflection of and response to the five developments recounted above and, in turn, gave neoliberal ideas further impetus in the environmental domain. What is also clear in hindsight is that the international policy networks and epistemic communities, which ensured that neoliberal ideas 'traveled' in areas such as social and labor market policy, were also effective in disseminating green neoliberalism and green developmentalism far and wide. The first United Nations Conference on Environment and Development (informally, the Earth Summit), held in 1992, was a key event in this regard because the now famous Framework Convention on Climate Change and the Convention on Biological Diversity both embodied neoliberal principles—and they did so at a global level. However, this is not to say that these principles infused environmental policy everywhere and equally.[19]

The neoliberalization of environmental management and resource use necessarily occurs against the background of pre-existing political and moral economies. It must also grapple with the biophysical specificities and peculiarities of particular resources, ecosystems, and environmental assets. In short, the world is never a *tabula rasa*, waiting to be freshly inscribed by omniscient elites. Seen in this light, neoliberalization must overcome, or at least adapt to, challenges and barriers residing in both the human and non-human domains. In any given case, neoliberal ideas, policies, and practices must hybridize—more or less successfully—with what is already there.

The socio-economic and cultural obstacles to the neoliberalization of nature are, in theory at least, more tractable than the biophysical ones. Although 'nature'—in the sense of the environment in general and natural resources in particular—is undoubtedly a social construction at one level, it also possesses material properties that any governance regime or policy measure must work with (or around).[20] By contrast, those stakeholders who stand to lose or gain from the neoliberalization of nature in any given case can, in principle, be persuaded about its merits or otherwise be obliged to live with it. The various arguments in favor of neoliberal environmental policies, made by their various supporters, include the following:

- open access resources can be protected, once private property rights and prices are assigned to them, with the result that the 'tragedy of the commons' (Hardin 1968) can be avoided;
- currently unowned or unpriced portions of the biophysical world can yield a profit for existing or aspiring entrepreneurs, either through conservation/protection/remediation or through productive use;
- environmental goods that are not currently valued economically, or are not in some way subsidized by the state, equate to 'lost income', 'unrealized value', or 'mispriced/underpriced assets' for nature and people;
- the private sector can manage natural resources and environmental services so as to deliver value for money for consumers or citizens relative to state bodies and other non-market actors;
- the introduction of competition and pricing into the management of the biophysical world can boost both management standards and environmental outcomes;
- instilling commercial principles into state bodies can make them more efficient managers and deliverers of resources and environmental amenities and/or services;
- the off-loading of some state responsibility for the quantity and quality of environmental goods and services to civil society actors not only empowers those actors but also allows for tailored, creative, and non-bureaucratic approaches to resource governance; and
- empowering consumers, citizens, firms, and other juridical units to take responsibility for their environmental impact respects the rights of the individual (real, communal, or corporate).

We might synthesize these various arguments as follows. Depending on the case, neoliberal environmental policy can (so its advocates claim) deliver benefits summarized in the acronym GEDDS (growth/efficiency/development/democracy/sustainability). In more detail, this acronym stands for economic growth (through privatizing and marketizing biophysical resources, goods, and services); economic efficiency (the best environmental outcome per unit cost, where outcomes are described in various management standards and codes); economic and social development for marginal or low-income communities (as new revenue streams kick in, underpinned by secure property rights in land and other resources); democracy (by destatizing responsibility for, and the benefits of, environmental management); and environmental sustainability (by making conservation, preservation, restoration, and renewable use into profitable activities). I am aware that this is something of a contrivance on my part because I am grouping arguments made by different advocates who operate in different fields of environmental thinking and policy. In any given case, all five criteria would not be relevant. However, to the extent that these arguments are advanced selectively to support the policies analyzed by the critics whose work I turn to in the next main section, I think GEDDS is a useful heuristic when seeking to map the diverse effects of these various policies, as detailed later in this article.

Researching Neoliberal Environmental Policy

Most analysts of 'neoliberal nature' are not so sanguine about the purported benefits of market-led environmental governance. The first purposeful and collective use of the idea of neoliberalism to investigate environmental questions occurred in 2004 and 2005, when the journals *Geoforum* and *Capitalism, Nature, Socialism* both devoted whole issues to the subject.[21] This led to the edited book *Neoliberal Environments* (Heynen et al. 2007), inspired special issues of the journals *Antipode* (Mansfield 2008) and *Conservation and Society* (Igoe and Brockington 2007), prompted a special section of *Geoforum* (Guthman 2008a), and triggered a steady stream of empirical studies in several disciplines that use neoliberalism as a framing concept or analytical lens. These studies focus on the full spectrum of environmental and natural resource uses, from extraction (e.g., mining or agriculture) to conservation, and from green policies to those in which nature is simply something to be exploited for profit. Neoliberal environmental policies are neither intrinsically 'anti-ecological' nor always hard-wired to the sustainability agenda: it very much depends.

As I said earlier, the signature feature of this recent literature is its commitment to case study research. Although some of it is synoptic and general (see, e.g., Buscher et al., forthcoming; Guthman 2007; Roberts 2008), for the most part it comprises in-depth analyses of neoliberal environmental policies in specific places, regions, and countries. To recall Peck and Tickell's term, this literature is thus interested in various neoliberalizations. What is more, and broadly speaking, it takes a political-economic perspective on neoliberalization rather than a Foucauldian one.[22] The latter, sometimes called a 'governmentality' approach to neoliberalization, has inquired into the 'technologies of government' or the 'rationalities of rule', not just administrative ones, but also those rhetorical and discursive technologies that have generated new subject positions and identities in the wider society. The key point has been that neoliberalism, despite its principles, has involved more (or as much) government, not less—especially outside the formal apparatuses of the state. It is 'rule at a distance' or by 'remote control' because of the way it reformats social norms so as to create—rather than simply activate—the supposedly latent and intrinsic capacities of individuals (be they mortals or corporations).[23]

Inspired by neo-Marxist, Polanyian, and Gramscian ideas, some political-economic perspectives differ from Foucauldian ones in that they see neoliberal statecraft as, variously, (1) a project of class domination hidden behind, or expressed in terms of, the rhetorical worldview; (2) a means to create new social inequalities and injustices or to intensify existing ones; and (3) a far-from-smooth process that is often generative of social resistance framed in class, community, gender, or other terms (which power elites within or without the state then have to manage in some way—what Polanyi termed 'the double movement'). I might add that some of this Marxian-Polanyian-Gramscian literature has analyzed neoliberalism in light of recent theoretical studies on capitalism-ecology relations in general, including the works of 'eco-Marxists' such as James O'Connor. A motif of these eco-Marxists, unsurprisingly, is contradiction—not just the classic first contradiction internal to capitalism as identified by Marx, that is, between the forces and relations of production, but also the second contradiction between a growth-oriented capitalism and the finite biophysical world upon which that growth depends. In this light, three questions arise. First, does the neoliberalization of nature constitute a widening or deepening of class-based social power?[24] Second, does neoliberalism simply perpetuate, mitigate, or possibly even overcome the 'ecological contradictions' that are characteristic of capitalism to date?[25] Third, how successful have neoliberals been in framing their policies discursively in order to gain the support of various affected constituencies? When combined, the answers to these questions speak to the organic relationship between issues of social justice and of environmental justice.

These theoretical-political commonalities in the research literature notwithstanding, the empirical inquiries are far from easy to parse and synthesize so that a broader understanding of 'neoliberal nature' can be achieved. Why is this so? Five reasons help to explain this difficulty. First, the published studies are now relatively numerous and are scattered across a wide range of journals in a variety of academic fields (e.g., human geography, rural studies, planning, anthropology, agrarian studies). Second, these studies together cover a wide range of environmental policy areas (e.g., water resources, forestry, mining, fisheries). Third, while some studies are on a local scale, others examine national or international policy measures. Fourth, the geographical cases that have been selected are highly diverse and span the developing and developed worlds. Fifth, specific studies have focused on different aspects of neoliberalization. For instance, geographer Julie Guthman's (2007) excellent research on voluntary food labels combines a focus on one specific policy instrument (devised in the policy field) with a wider discussion of how 'neoliberal consumers' are created (an aim of the neoliberal worldview writ large). Due to the scope of what neoliberalism is (or is understood to be), few studies would be able to consider all of its elements in one go. For these five reasons, a major effort is required on the part of readers of this literature to detect the proverbial signals in the noise. The empirical studies into 'neoliberal nature' have emerged bit by bit and, in many cases, have not been cross-referenced by their authors to all the relevant published literature. Additionally, these studies have recently increased in number in a short space of time.[26]

In what follows, I will attempt to locate the findings of these separate studies on a broader cognitive map without, I hope, imposing a false sense of order or unity on them. The value of this exercise is, I trust, obvious. If neoliberalism is as widespread and influential as its critics have claimed, we need to examine it holistically and comprehensively, rather than fixating on one or a few cases and examples, as if they could tell us the whole story. Because the published studies have proceeded according to no common template or method, it is timely to create a greater sense of analytical order when deciphering their potentially disparate results and insights. Inevitably, some of the empirical (and conceptual) richness of the research I describe will be sacrificed in the process of presenting my synopsis.[27]

Neoliberalizing Nature 1: Processes of Reregulation

Above I summarized the purported benefits of neoliberal environment policies with reference to the acronym GEDDS. In toto, the arguments made by advocates here can be said to constitute the neoliberal worldview in the environmental domain. In what real life situations have they been translated into practice? I will address this question systematically, according to the seven neoliberal policy proposals presented earlier. Note that no one author's study speaks to all of these points, yet when taken together, different studies do speak to these points as a whole. I will then, in the next section, describe the outcomes of the neoliberalization of nature, organizing these according to social and environmental impacts and then the GEDDS template. This, it seems to me, is far more incisive than discussing policy change on a sector-by-sector basis (e.g., agriculture, fisheries, forestry, etc.). It also allows us to see how sector-specific reforms are part of a broader transformation of economy, society, and environment. Together, specific neoliberalizations of environmental and natural resource use amount to the neoliberalization of nature as a whole.

1. *Privatizing and propertizing nature.* Assigning rights of ownership in, or use of, the biophysical world can occur in three situations: (1) where state bodies relinquish or 'loan' their sovereign rights, (2) where a recognized or established rights regime outside the state realm is fundamentally altered by policy makers, and (3) where no recognized (or enforceable) rights currently exist.

Karen Bakker (2003, 2005) and Loftus and McDonald (2001) provide examples of the first situation. Bakker's detailed studies of water and sewerage services in post-1989 England and Wales show how a few large private sector firms took direct control of the hydrological infrastructure. Similarly, Loftus and McDonald describe Argentine President Carlos Menem's 1989 Administrative Reform Law and focus on the privatization of water delivery and sewerage services in Buenos Aires through the 1990s. This reform concentrated water management rights in the hands of one large company, Aguas Argentinas, an entity established by a consortium of private European water companies with the assistance of the World Bank.

Studies of the second situation identified above include those by Thomas Perreault, Diana Davis, Gavin Bridge, and Becky Mansfield. Perreault (2005) focuses on the legal enclosure of Bolivia's water resources in the 1990s and its gas resources too (Perreault 2006). In the former case, central state control of the resource was relatively weak prior to privatization. Davis (2006) describes the several laws passed in Morocco in the 1990s pertaining to unfenced, rural farmland and to dryland agriculture. These laws have involved enclosing the environmental commons or communal lands and creating large parcels of agricultural land, with ownership being concentrated in a new set of well-capitalized farmer-operators. Bridge (2002) charts the enclosure of land in Guyana by small and medium domestic investors and by large overseas investors, who together have purchased prospecting and extraction gold mining rights from the national state. After the 1989 liberalization of mining laws, this poor and indebted country saw a 15-fold increase in the area given over to mining permits or claims by 1994. Gold mining in the country has historically been dominated by small and medium domestic operators. An early 1990s amendment to national mining legislation was designed to encourage large overseas investors to enter the Guyanese gold industry. But the amendment was designed so that overseas interests had to have domestic partners—a way of trying to make external investment pay dividends for certain Guyanese nationals. In contrast to Bridge's land-based study, Mansfield's (2004a, 2004b, 2007b) excellent essays examine the enclosure during the 1990s of the US portion of an open access fishery in the North Pacific Ocean.[28] This enclosure has excluded new fishery entrants and, indeed, reduced the number of existing ones.

With respect to the third situation identified above, geographer Scott Prudham (2007) analyzes a 2004 Canadian Supreme Court decision to reject a Monsanto patent claim covering genetically modified canola. Prudham focuses on the legal complications involved in trying to abstract discursively parts of nature from their social and environmental integument in order to be presented, by companies such as Monsanto, as putatively 'autonomous inventions'. This case is linked to ongoing attempts by many other pharmaceutical, biotech, and agro-foods companies to privatize previously unowned (and unownable) elements of the biophysical world. Relatedly, there are Morgan Robertson's (2004, 2006, 2007) fascinating studies of the wetland banking and water quality credit markets in the US during the 1990s and the 2000s. His essays examine how the right to destroy and create wetlands was invented and institutionalized, along with the right to produce dirty and clean water. In both cases the rights were new, creating a property regime where one had not actually existed before. The same applies to the Californian conservation easements investigated by Amy Morris (2008).

2. *Marketizing biophysical resources, goods, and services.* Rights of ownership and the use of nature do not necessitate the marketization of biophysical resources, services, or assets (Bakker 2005). However, for neoliberal policy makers the two are umbilically connected. Once property rights are assigned and legally 'real', the assets owned or accessed should, neoliberals argue, generate a stream of revenue. Therefore, each of the studies cited in the previous three paragraphs necessarily includes mention of marketization. For instance, Mansfield's research explores how a transferable quota system was introduced—one that allowed fishermen and other stakeholders (e.g., coastal indigenous peoples in Alaska) to establish a price for the annual right to harvest a given amount of fish. Additional studies in which marketization is a central theme include the following. Fraser Sugden (2009) reports on the Nepal government's Agriculture Perspective Plan, which was initiated in 1995. This plan aimed to get subsistence farmers in rural areas to commercialize their operations so as to earn exchange values from food sales to domestic and overseas markets. In other words, the plan sought to replace a long-standing peasant way of life with a capitalist one, characterized by numerous agrarian entrepreneurs vying for market share. Likewise, Cristobal Kay's (2002) superb analysis of Chile's neoliberal agrarian transformation in the 1990s includes a discussion of attempts to bring peasant farmers into the country's capitalist economy by trading their produce overseas or domestically. This 'second modernization' of Chilean agriculture is also the focus of Warwick Murray's (2002) wide-ranging study of rural 'reconversion'.

A striking case of marketization is presented by James McCarthy (2004) in his analysis of how large firms within the North American Free Trade Agreement (NAFTA) region use the investor protections written into this compact. Whereas Robertson (2004, 2006, 2007) and Morris (2008) both show marketization to be an attempt to mitigate environmental harm, McCarthy tells a different story. In 1993, Metaclad, a US waste disposal company, bought and subsequently developed an inactive toxic waste dump in the Mexican state of San Luis Potosí. The local government halted Metaclad's activities in 1995 on health and safety grounds. Metaclad subsequently began a proceeding against the Mexican government under Chapter 11 of NAFTA, citing the 'regulatory takings' concept, which presumes that investors are entitled to earnings lost through the actions of others. In 2000, a NAFTA tribunal found in favor of Metaclad, ordering Mexico to pay $16.7 million for outlays and lost revenue. McCarthy calls this the 'primitive accumulation' of the conditions of production, because revenues are earned from the biophysical world through compensation for *not* undertaking extractive activities. In this case, the right of firms to make money regardless of the human or environmental cost is taken as sacrosanct, in McCarthy's view.

3. *State roll-back or deregulation.* Clearly, the withdrawal (or decrease) of state control over environmental goods, ecological services, and natural resources has not been universal during the last 30 years. In many countries, the level of state involvement has been minimal from the beginning (e.g., in developing countries), while in others it has been considerable (notably in Western capitalist democracies, former communist states, and former 'developmental states' in the global South). Only in these latter cases is roll-back a significant and tangible process,

often driven by a perceived need of state bodies to save money and reduce public borrowing or taxation. For instance, Harold Perkins (2009) reports on a sharp reduction in local government expenditure on environmental amenities in the famously 'green city' of Milwaukee through the 1990s and 2000s (see also Heynen and Perkins 2005). Also in the US, Paul Robbins and April Luginbuhl (2005) show that in California, Colorado, Nevada, New Mexico, Oklahoma, Oregon, Utah, and Washington there has been a transfer of fiscal and management responsibility for wild game from states to (mostly) private landowners, such as farmers, according to their land acreage and the habitats involved. As already mentioned, Bakker (2003, 2005) relates how the post-1989 central government off-loaded responsibility for water supply and sewerage infrastructure in England and Wales. Also in the Anglophone world, Brad Coombes (2003) reports on New Zealand's 1991 Resource Management Act and its effects on habitat husbandry in the Auckland region. After the act was passed, central government devolved many of its strategic planning powers, in the process weakening the influence of professional planners on land use decision making. In turn, this left the protection of rare or otherwise valuable habitats on privately owned land parcels to a combination of voluntary agreements and market-based instruments. More dramatically, Prudham's (2004) study of a serious water poisoning incident in Walkerton, Ontario, describes the context as being a contracting out of provincial government responsibility for water testing as part of Ontario Premier Mike Harris's 'common sense' agenda in the 1990s.

Western capitalist democracies aside, Maria Rodrigues (2003) recounts how one of Brazil's post-1945 parastatals, Companhia Vale do Rio Doce (a conglomerate of 50 firms operating in the mining, transportation, and forestry sectors), was sold off to private investors in 1997. This involved a massive withdrawal of national state authority, although it generated an equally massive infusion of funds into the public purse. Also in South America, Jessica Budds (2004) focuses on the Pinochet government's reversal of previous government policy on water rights and water delivery so that the hydrosocial cycle was de-statized.

4. *Market-friendly reregulation.* The processes of privatization, marketization, and deregulation do not, of course, mean that national governments somehow play a minor role in neoliberalization. On the contrary, their active involvement is imperative, both as lawmakers and as direct (or indirect) regulators of actors inhabiting the private and civil society domains. This involvement can redefine market relations across the board and affect all market participants, not least by creating new markets altogether or by significantly altering existing ones. Typically, the regulatory environment has shifted from formal, state-centered government to more dispersed, extra-state forms of governance.

Studies of market-friendly reregulation by national governments abound. Once again, I need to mention Bakker (2003), whose book *An Uncooperative Commodity* details the extraordinary lengths to which technocrats operating under British Conservative governments went to create a market in the water supply. Bill Pritchard (2005a, 2005b) and Cocklin et al. (2006) relate how post-1980s Australian governments lent their full support to free trade policies in the agricultural arena, creating a 'hyper-competitive' farming sector oriented to global export markets. This contrasts with the EU, which used the idea of multi-functionality to protect certain of its farmers and rural communities from the negative effects of *laissez-faire* (see Dibden et al. 2009). Relatedly, Clive Potter (2006) reports on how a globally powerful discourse of free trade in the agricultural sector, which cast a long 'discursive shadow', has heavily conditioned national attempts to exempt some rural spaces from competitive, productivist agriculture.

Budds's (2004) already cited study of Chile's 1981 Water Code shows how it created an expanded market in water rights and thus in water itself. Focusing on nearby Peru, Jeffrey Bury

(2004, 2005) explains how successive governments opened up their country's mineral resources to overseas investors through the 1990s, echoing Bridge's (2002) study. Arielle Levine (2007) describes how Zanzibar's Environmental Management for Sustainable Development Act of 1996 devolved responsibility for protected areas to private parties. These parties can include local communities, as detailed by Igoe and Croucher (2007) for Tanzania and by McCarthy (2006) for British Columbia. In her account of Madagascar's turn to ecotourism as a major source of overseas income and a way of protecting special or biodiverse sites, Rosaleen Duffy (2008) places emphasis on the wide array of actors who have been invited to neoliberalize access to the island's flora and fauna. These actors cross-cut both geographical scales and the public, private, and third sectors. Together they show how a national government has been influenced by, or has actively enrolled, quasi-state institutions (such as the World Bank), well-funded environmental NGOs, and many others besides. This theme of the state-sanctioned turn away from government to governance is emphasized by Perreault (2005, 2006) in his already mentioned accounts of the privatization of rights to water and gas resources in 1990s Bolivia. This privatization entailed a radically altered access regime and a newly centralized regulatory system dominated by national 'quangos'.[29] Relatedly, Buscher and Dressler (2007) show that national states in the global South have, since the early 1990s, designated more cross-border areas in order to encourage the emergence of privately (or communally) managed conservation spaces.

Prudham and Morris (2006) also look at a quango, this one in Canada. They scrutinize a national review of genetically modified (GM) foods regulation by the Canadian Biotechnology Advisory Committee (CBAC) between 1999 and 2004. GM foods have, of course, been heavily criticized on environmental and health grounds by organizations such as Greenpeace and the UK's Soil Association. Prudham and Morris show that, from the start, the CBAC—a supposedly impartial expert body advising the Canadian government—was already committed to creating a market for GM foods and spent a good deal of its resources trying to persuade the Canadian public that this was a good thing. Finally, there is Julie Guthman's (2007) study of voluntary food labels that purport to connect farmers and shoppers. These schemes, she argues, create an ethical market in socio-environmental 'goods' and 'bads' in place of direct national state regulation of farming and food prices. They thus put a monetary value on morality in competitive markets, making exchange values a vehicle for expressing consumers' environmental and social values.

5. *Use of market proxies in the residual state sector*. In relation to the non-human world or society, the use of market proxies entails making the remaining state-run activities as economically efficient as possible. 'Efficiency' here means that, whether delivering goods and services to citizens or acting as a regulator, state institutions operate as if they were private sector firms subject to a competitive environment. Where 'artificial' competition cannot be created among state institutions for practical reasons, other measures can be used, such as rules insisting on full cost recovery, balanced budgets, and high standards of service provision.

In the biophysical domain, one service area where even many neoliberal states find it difficult to 'let go' relates to basic natural resources, such as water and oil. These resources are either universally required by all citizens or have strategic importance for a given country. Laila Smith's (2004) study of water provision in Cape Town is a case in point. In an empirically rich essay, Smith shows how the Cape Town government introduced water demand management through metering and pricing, the outsourcing of some water services as a cost recovery strategy, and water cutoffs for non-paying citizens. This, she concludes, was done in the name of economic efficiency and a market-defined notion of citizens' rights and privileges. Relatedly, Priya Sangameswaran's (2008) study of water resource governance in the Indian state of Maharashtra suggests that attempts to commercialize a parastatal service provider—Maharashtra

Jeevan Pradhikaran—are radically altering its *modus operandi*. However, aside from Smith's and Sangameswaran's studies, there are currently few others that interrogate the neoliberalization of nature in the residual state sector.

6. *Strong encouragement of 'flanking mechanisms' in civil society*. The gaps left in environmental provision by state bodies can be plugged by various civil society actors, operating either outside or within the market. The Milwaukee research conducted by Perkins (2009) provides an example of the former, as does Ryan Holifield's study (2004). Perkins recounts how three volunteer organizations stepped into the vacuum created by the Milwaukee municipal government's partial withdrawal from maintenance of public parks and trees. Holifield examines how so-called environmental justice communities (EJCs) were, during the Clinton presidency, brought into the regulatory practices of the US's Environmental Protection Agency (EPA). In neoliberal terms, EJCs are those communities suffering the ecological 'externalities' of economic activity—that is, they are victims of 'market failure', including the inability to gain financial compensation for toxic waste being dumped nearby or to ensure that the offending firm(s) will pay the clean-up costs. Holifield shows how, under Clinton, EJCs—whose politics over the last 30 years have typically been radical left, linked as they are with the civil rights movement of the 1960s—were given much more visibility within the EPA's remit to remediate toxic sites. Although EJCs are not, Holifield argues, products of neoliberalism, they have been brought within its logics of property and payment by the EPA. This contrasts with the previous situation in which EJCs were typically ignored by state officials unless they fought hard and loud enough to get their grievances heard. In a recent study of agricultural biosafety in Costa Rica, Thomas Pearson (2009) recounts the appearance of 'informal' auditors in civil society who are called on to check the work of state-sanctioned private auditors. In contrast to Holifield's study, Pearson shows that these auditors were not, in fact, brought forth intentionally through state encouragement. They appeared because of a lack of trust in the private sector auditors.

As I intimated above when discussing Guthman's (2007) research into voluntary food labels, civil society actors have been increasingly enjoined to take on a regulatory role outside the state apparatus through the moralization of certain commodity markets. Paige West (2010) and Lovell et al. (2009) examine similar cases where commodity exchange becomes a substitute for state management of environmental and labor conditions. West looks at how Papua New Guinean specialty coffee is marketed to US consumers, whose dollars purport to deliver decent wages and to improve environmental conditions on the other side of the world. She shows how the subject effects of marketing—to the extent that they exist—individualize consumers and systematically misrepresent coffee producers' lives in neo-colonial, romanticized narratives of peasants and family farmers. The equally contrived ethical narratives-cum-discourses thrown at purchasers of voluntary carbon offsets by offset retailers is described by Lovell et al.

Consumers are, of course, positioned at the end of 'commodity chains'. But these commodities are often produced and marketed according to standards that are prescribed by non-state actors and are adhered to voluntarily by certain firms and commercial outfits. Dan Klooster (2010) details the globally widespread adoption of Forest Stewardship Council (FSC) certification standards for wood product emanating from plantation forests. These voluntary schemes have the value—so argue their advocates—of creating global standards without the need for complicated, cross-jurisdictional administration by national regulators.

7. *Creation of 'free', 'self-sufficient', self-governing, and entrepreneurial individuals and communities*. Most markets involve a wide array of actors who are often separated in space and time and whose relations are scrutinized by commodity chain analysts. Because all markets are

'embedded', their operation affects other diverse constituencies. We can thus imagine neoliberal discourse having to work in several different arenas so as suitably to 'interpellate' commodity producers, sales, and marketing intermediaries, commodity consumers, regulatory actors, and other stakeholders affected by the neoliberalization of nature. These various actors may inhabit the same national space. But then again, they may not.

The already mentioned studies by Sugden, Murray, Kay, and St. Martin focus squarely on those who derive a livelihood directly from natural resources (e.g., land or fish). Sudgen (2009) points to the failure of neoliberal agrarian reform in the Nepali lowlands: existing social identities and relations have not been dislodged discursively or practically. By contrast, although identities and subject positions are not examined, Murray's (2002) research into agrarian transformation under Chile's center-left Concertación coalition governments demonstrates how small, medium, and larger farmers embraced neoliberal policy, if not always willingly. This, too, is the message of Kay (2002) in a similar study of Chilean agriculture. St. Martin (2007) examines how a neoliberal fisheries discourse promoted by policy makers articulates with a non-capitalist political and moral economy specific to New England fisheries. It does not, in his view, erase this existing economy, but it does threaten to unsettle it considerably, even as fishermen seek to maintain their traditions (see also St. Martin 2006).

Other key research that focuses on commodity production includes works by Peter Wilshusen, Wendy Wolford, Becky Mansfield, and Gabriela Valdivia. Wilshusen's (2010) field research in southeastern Mexico in the state of Quintana Roo relates how rural communities have creatively worked with, and around, neoliberal discourses and policies emanating from the national state. These communities, he shows, have accommodated neoliberalization by blending its favored subject positions and rationalities with those of a moral economy based on collectivist ideas of sharing and mutual aid. Wolford's (2007) investigations of land reform in northeastern Brazil demonstrate how neoliberal discourse has been adopted enthusiastically by members of the populist Movement for Landless Workers. However, the price for their support, she argues, is that many now feel obliged to participate in a competitive food economy (based on exchange values), rather than a peasant one. In an essay on indigenous Alaskan peoples' involvement in a transferable quota fisheries scheme, Mansfield (2007b) reports a similar enthusiasm for neoliberalism among an otherwise excluded social group. The rights and revenues enjoyed by First Nations Alaskans, she shows, are consistent with their sense of themselves as a distinct community with historical and cultural claims to a share of the fishery and with a need for money to address poverty and to underpin socio-economic development. Relatedly, Valdivia (2005) examines the way that neoliberal reform measures in the Ecuadorian Amazon have affected indigenous peoples' discourse with respect to land and other resources. She reports a creative, complex engagement with neoliberal principles and ideas that allows indigenous peoples to adapt their claims and agendas to changing political-economic circumstances. Of course, in many cases involving the neoliberalization of nature, there is little or no need to 'cultivate' suitable producer identities—notably, in situations where large private corporations are invited to invest in land, water, forests, and fisheries in order to make money.

Perreault's studies of Bolivia's water and gas wars cross-cut people's identities as citizens and commodity consumers. Although the reasons are very different from those recounted in Sugden's (2009) Nepal study, Perreault (2005, 2006) shows how neoliberal rhetoric failed to prevent serious public protests over the reform of water and sewerage services and the management of natural gas reserves. By contrast, and as earlier mentioned, in her more consumer-oriented study, West (2010) suggests that neoliberal subjectivity 'works' for specialty coffee sellers in the US, but not because consumers really understand (or even care about) the ethical issues or the salient facts. Meanwhile, Guthman (2008b) maintains that voluntary labeling schemes for organic products have taken hold among food activists and many left-wing consumers because these measures appear to deliver some sort of genuine socio-ecological justice at the production end of the commodity chain. As she explains elsewhere (Guthman 2008c), this appearance is made possible in part by the active interpellation of individuals into the policy norms and moral goals of neoliberal discourse—norms and goals that are, as it were, the only 'realistic' ones on offer.

Neoliberalizing Nature 2: Social and Environmental Effects

In the previous section I focused on the multifaceted process of neoliberalizing human engagements with the non-human world. Depending on the study cited, more than one facet of the process has been the focus of analytical attention. Although the precise details vary, I have described these facets under one or more of the seven neoliberal policy proposals. This reveals the sheer breadth of projects that are intended to neoliberalize environmental use and resource management. These undertakings range from the reform of ownership rights to engendering new forms of subjectivity among resource users. So far so good. But what have been the effects of the neoliberalization of nature in various times and places, according to the research published to date? In posing this question, the answer to which I have only hinted at in the previous section, I am well aware that separating process from outcome is, ontologically speaking, artificial, since arguably the latter is part of the former and not some sort of stable endpoint. Even so, the distinction has a heuristic value in that it describes the (albeit eventually changeable) outcomes of process at a given moment in time. Indeed, several published studies have focused specifically on these outcomes, as we will see below. I will further distinguish between the social and environmental effects of policy measures, although, again, in reality they are intertwined. Needless to say, the precise socio-environmental effects of neoliberal policies vary in their details according to their particular type and the spatio-temporal scale we are interested in. It almost goes without saying too that all effects are relative with regard to who (or what) is affected, in what way, and to what degree.[30]

Rather than summarize the findings of all the relevant studies, I will discuss some of the indicative ones and simply tabulate the many others that I could mention if I had more space (see table 1). Several contributions highlight problems of ordinary people's exclusion from, or reduced access to, environments and resources upon which their livelihoods or well-being depends. For instance, consider Davis's already mentioned research in Morocco and Sugden's research in Nepal. Davis (2006) shows how peasant pastoralists were alienated from grazing territory in the name of large-scale, irrigated dryland farming by private landholders. Sugden (2009), meanwhile, confirms that agricultural reform has done little to release the very poorest farmers from disadvantageous feudal and rentier relationships. Budd's (2004) research in Chile reveals that, subsequent to neoliberal water reform, large-scale farmers have gained greater access to rural water compared with peasant cultivators. In her study of conservation policies in the Virgin Islands, Crystal Fortwangler (2007) points to the increase in real estate prices outside the conservation zone, which makes land ownership more difficult for local people. Lisa Grandia (2007) reports on the exclusion of Guatemalan peasants from land given over to new commercial tree farms, while Igoe and Croucher (2007) make a similar observation about rural Tanzanians who live adjacent to a new wildlife management zone. Finally, Smith's (2004) Cape Town research shows the socially regressive effects of water pricing under a new cost recovery regime in the late 1990s—an arrangement that prioritized economic equity over social equity.

By contrast, a minority of other studies are more equivocal than those just mentioned. In his detailed examination of how several rural households in the Cajamarca region of Peru have

Table 1: Social and environmental outcomes of the neoliberalization of nature, as specified by the authors named

Location	Biophysical Resource	Issue	Socio-economic Outcomes	Environmental Outcomes	Author/s
British Columbia	Forests	Forest management	Democratization of management for small percentage of forest area and new income streams	N/A*	McCarthy (2006)
North America	Forests	Forest management	Co-optation of community groups to neoliberal ideas plus new income streams	N/A	McCarthy (2006)
England and Wales	Water	Water and sewage management	Increase socio-spatial inequity in relative water costs and cutoffs	Improved national water quality; regional water shortages	Bakker (2001, 2003)
Cajamarca, Peru	Gold	Gold mining	Less access to produced and human capital resources, and greater access to social capital and natural resources among rural households; reworking of land tenure to favor private owners	Intensified land use by house-holders in areas adjacent to mining	Bury (2004, 2005)
Southeastern Mexico	Forests	Forest management	Community adaptation to neoliberal norms	N/A	Wilshusen (2010)
Bolivia	Water	Urban water management	Serious public protest	N/A	Nickson and Vargas (2002)
Morocco	Land	Land ownership	Enclosure of the grazing commons and exclusion of many pastoralists	N/A	Davis (2006)
Nepal lowlands	Land	Agricultural production and trade	Failure of agrarian plan to improve the livelihoods of many peasant farmers	N/A	Sugden (2009)
Northeastern Brazil	Land	Agricultural livelihoods	Land given to the landless, but on condition that they now 'work' the land and conform to the norms of the agrarian elite	N/A	Wolford (2005, 2007)
Brazil's Amazonia	Forest ecosystem	Environmental protection from the externalities of private mining and metals production	Indigenous communities offered less support under the post-public regime	Environmental improvements locally, but wider environmental externalities	Rodrigues (2003)
Ecuador	Land and oil	Rescripting of indigenous identities	Complex reworking of identities to both accommodate and challenge neoliberal reform	N/A	Valdivia (2005)
Cochabamba, Bolivia	Water	Dam project	Regional resistance to damming	N/A	Laurie and Marvin (1999)

(Continued)

Table 1: Social and environmental outcomes of the neoliberalization of nature, as specified by the authors named (*cont.*)

Location	Biophysical Resource	Issue	Socio-economic Outcomes	Environmental Outcomes	Author/s
US	Hazardous waste	Environmental justice procedures	Attempted co-optation of community activism by federal state and depoliticization of community claims about environmental injustice	N/A	Holifield (2004)
Walkerton, Ontario	Water	Water quality testing	Poisoning of public	Drinking water pollution	Prudham (2004)
NAFTA	Pollutants	Right of corporations to pollute the commons	Taxpayers having to pay firms for 'regulatory takings'	Actual or potential point pollution of the commons	McCarthy (2004)
Chicago area	Wetlands	Wetland mitigation scheme	Creation of new profit opportunities for firms; partial subsumption of ecological science to abstractions imposed by monetary valuation	Loss of wetlands and creation of 'equivalents' elsewhere	Robertson (2004, 2006, 2007)
Cape Town	Water	Water service delivery	Attempted depoliticization of water distribution issues; increased socio-spatial inequity in customer charges and service delivery	N/A	Smith (2004)
Milwaukie and global atmosphere	Trees	Management of urban forests	Decline of public forest area; concentration of urban trees on private land; increased urban energy use	Increased city level of hydrological and temperature ranges; minor loss of CO_2 absorption globally	Heynen and Perkins (2005)
Chile	Land	Land ownership and agricultural production	Increased rural income inequality and poverty among farmers and farm workers	N/A	Murray (2002)
Chile	Land	Land ownership and agricultural production	Proletarianization of peasants, land ownership concentration, and economic success for relatively few farmers	N/A	Kay (2002)
Buenos Aires	Water	Water service delivery	Increased water network coverage; price increases for consumers; layoffs of water sector workers	Inadequate sewage treatment capacity	Loftus and McDonald (2001)
Rural Guatemala	Forests	Carbon offset tree farm projects	Displacement of rural peasants	Reduction in forest biodiversity	Grandia (2007)
Rural Tanzania	Land	Conservation easements	Inadequate compensation to local residents for loss of access to land; exclusions from land use	N/A	Igoe and Croucher (2007)

Location	Biophysical Resource	Issue	Socio-economic Outcomes	Environmental Outcomes	Author/s
California	Land	Food certification and labeling schemes	Reduced farm worker representation in decisions about farm labor practices	N/A	Brown and Getz (2008)
California	Land	Pesticides on farmland	Visible and effective forms of activism against pesticide drift	Off-farm pesticide drift	Harrison (2008)
California	Land	Food politics	Co-optation of normative goals of a sustainable production project to neoliberal norms	N/A	Guthman (2008c)
Bolivia	Rural water resources	Water management	Popular protests against water reform based on an existing moral economy	N/A	Perreault (2008)
Mexico	Rural water resources	Water management	Democratization of water governance but no gains in the efficiency, equity, or sustainability of water use	N/A	Wilder and Lankao (2006)
Southern Africa	Land-based ecosystems	Transfrontier conservation	Neoliberal discourse depoliticizes conservation issues, attenuates existing community-based conservation discourse, and as yet has delivered few tangible development gains	Little evidence of improved levels or geographical spread of environmental conservation	Buscher (2010a, 2010b)
Botswana and Thailand	Elephants	Ecotourism	Income streams and jobs from elephant tourism	Productive use for elephants that might otherwise be neglected or culled	Duffy and Moore (2010)
Various locations	Plantation forests	Environmental and social standards	Attenuation of some, but not all, elements of FSC standards	Attenuation of some, but not all, elements of FSC standards	Klooster (2010)
Southern Africa	Land-based ecosystems	Transfrontier conservation	Private sector benefits while communities benefit less than previously	N/A	Buscher and Dressler (2007)
California	Commercial farmland for crop production	Management of permissible seed and nursery stock	Corporate attempts to remove local ordinances banning genetically modified crops failed	N/A	Roff (2008)
South Island, New Zealand	Commercial sheep farming	Farming practices: intensive-productivist or organic?	Changing consumer demands for meat may produce post-productivist sheep farming but do not guarantee it	Some moves toward a more organic style of farming, with beneficial effects for sheep and farmers	Haggerty et al. (2009)

*N/A (not applicable) means that environmental outcomes were not a focus of the research.

Note: Only resolutely empirical studies are included in this table, rather than ostensibly conceptual ones or those including empirical 'vignettes'.

been affected by overseas mining investment, Bury (2004, 2005) concludes that many families now enjoy increased access to both produced and human capital, although several have fewer opportunities to acquire both natural and social capital. Somewhat differently, Perkins's (2009) analysis of third sector involvement in the maintenance of Milwaukee's green spaces shows that such involvement has proven personally empowering for many citizens.

Of course, the implementation of neoliberal policy in the environmental domain has resulted in several spectacular and well-publicized reactions within civil society. Andrew Nickson and Claudia Vargas (2002) report on the failure of the Cochabamba water concession in Bolivia: in 2000 the private company Aguas del Tunari had its contract canceled after just six months. Several factors—low rates of economic growth, persistent poverty, and low or stagnant wages among large segments of the population— came into play. The sharp increase in water tariffs in 2000, allied to a loss of water use rights by many stakeholders, contributed to the national unrest that resulted in the cancellation. Similar large-scale dissent was repeated following the privatization of gas resources in 2003 (Perreault 2006). Less dramatically, Jill Harrison (2008) shows that neoliberalized agriculture in California has led to increased pesticide drift, eliciting vocal expressions of agro-food activism in civil society. Still, in other cases neoliberal environmental policy has persisted despite manifest problems. This is (or was) true for water governance in Buenos Aires. Loftus and McDonald (2001) demonstrate that even though consumer prices increased, many workers were laid off in the water sector, and advances in new sewerage connections and treatment were slow, the citizenry tolerated the policy shift—grudgingly or otherwise.

On the environmental side, neoliberal environmental policy seems to have had mixed effects. There was the shocking water poisoning case in Walkerton, Ontario (Prudham 2004). Equally alarming was the attempt of two American firms to seek monetary compensation from state bodies for *not* polluting the commons in Mexico and California (McCarthy 2004). Less dramatically, Robbins and Luginbuhl (2005) examine recent attempts to create 'game farms' in some Midwestern and Western US states. These initiatives, they show, go against a long US history that considers wildlife just that—wild. They also represent attempts to make wildlife pay for the benefit of a select group of ranchers and hunters. Robbins and Luginbuhl find little evidence to support the contention that private wildlife managers are delivering effective animal and habitat husbandry as compared to a previous generation of state officials.

Coombes's (2003) study of bush lot subdivision in New Zealand (conceived as a new way to protect indigenous habitats on private land) suggests that this approach is too ad hoc. In addition, strategic interventions in order to ensure proper levels of ecological protection are proscribed. Nik Heynen and Harold Perkins (2005) find that local government divestment from the maintenance of urban tree cover has led to a noticeable loss of ecological services (e.g., provision of shade) and that private landholders cannot, left to their own devices, compensate adequately for the reduction in tree numbers. A more mixed picture is painted by Rodrigues (2003). Recently, privatized iron ore and manganese operations in the Brazilian Amazon, she shows, are much more effective at protecting the local environment than their state-run predecessor. But, she argues, they remain ineffective at tracking environmental externalities outside the areas immediately adjacent to the mines. Bakker (2003) is more positive about the environmental effects of neoliberal reform. Her study of water privatization in England and Wales shows both a reduction in leakage rates (due to a huge investment in infrastructural renewal) and a notable increase in the quality of drinking water.

Bakker, like Bury, is among the few analysts who has sought to record both social and environmental effects of neoliberal reform in some detail. Although she acknowledges the environmental benefits of the shift away from direct state management of water resources in England and Wales, she also points to new social inequities in water service delivery that were evident in the early 1990s. Specifically, lower income households were subject to water cutoffs on a scale previously unmatched, leading to a response by concerned regulators (Bakker 2001).

What do these social and environmental effects tell us about the neoliberalization of nature when mapped against the criteria of GEDDS? To start with, it is clear that certain kinds of growth, efficiency, development, democracy, and sustainability have, in fact, been delivered by market-led environmental policies. However, the key—and obvious—point is that each of these terms has a contested meaning. The lack of universal or agreed upon definitions means that analysts of the neoliberalization of nature, as well as the many people directly affected by it on the ground, have good reason to question the 'success' of the project. In various situations, economic growth has disproportionately benefited private sector actors; economic efficiency has been achieved at the expense of social equity and justice; a very particular kind of development, one that does not reflect the full range of development thinking, has been achieved; democracy has been neutered; and sustainability has been realized, but only to the extent that it is consistent with the peculiarities of private property rights and market pricing.

Key Points and Lessons to Be Learned

Important as it is to pay close attention to the insights afforded by each of the empirical studies cited in the last two sections, it would be easy to get bogged down in the details. What, then, are the principal take-home lessons from the literature reviewed above, quite aside from the suggestion that the realization of GEDDS is open to question? In my view, there are six lessons to be learned, and I have identified them on the basis that they are evident in several separately conducted studies (although by no means in all). In each case I present an illustrative example or two. Most of the key points below apply to neoliberal statecraft more generally, regardless of whether or not we are discussing environmental management, regulation, or governance.

1. *Markets in environmental goods, services, and assets typically require considerable state intervention.* One of the shibboleths of the neoliberal worldview is that a reduction of state intervention in economy and society is both desirable and possible. However, as Polanyi (1944: 141) realized over 60 years ago, "Laissez-faire [is] planned; planning [is] not." In one of her several investigations into the privatization of open access fisheries, Mansfield (2004a) makes the same point with compelling detail. She examines the implementation of the 1998 American Fisheries Act (AFA) as it affected the Alaska pollock population, the world's largest single-species fishery, which is of great economic and social importance to Alaskans. The AFA led to sweeping changes in the organization of this open access fishery, introducing a catch quota system among a restricted group of relevant parties in order to prevent overharvesting on environmental and economic grounds. Mansfield highlights the remarkable degree of state regulation that was required to privatize and marketize the pollock fishery, as the AFA had to be interpreted and enforced by the North Pacific Fishery Management Council. This involved the development of exceedingly detailed and complex regulations in order to (1) negotiate fair relations between inshore fisherman, offshore fisherman, fish processors, commercial fishers, independent fishers, and indigenous (Native American) communities; (2) sort out the AFA's relations to endangered species legislation (e.g., pollock are eaten by the Steller sea lion, a threatened species); and (3) determine the impact that pollock fishery rules would have on other Alaskan fisheries, such as crab. In short, Mansfield shows that the neoliberalization of the pollock fishery involved as much (if not more) state regulation than previously. The conclusion appears paradoxical: the market is both created and regulated by the state. It suggests that, in Mansfield's view, neoliberalization

is not about the state stepping aside. Rather, the state has changed its role to ensure actively that markets work where they might founder, if left to arise spontaneously.[31]

2. *Markets in environmental goods, services, and assets must carefully adapt to biophysical obstacles in order to avoid being inhibited by them.* Throughout the 1990s, many social scientists were wont to say that nature is a social construction. However, research into the neoliberalization of nature shows that policy makers must pay very close attention indeed to the biophysical specificities of their intended targets; otherwise, their best laid plans might come to grief. In short, nature's material properties and affordances matter, and they can complicate the form and functioning of neoliberal policies. The title of Bakker's 2003 monograph, *An Uncooperative Commodity*, indicates as much. Because water is both heavy and bulky, and because the infrastructure needed to both purify and move it is very expensive, it has proven almost impossible for policy makers to create direct competition between private water companies on a regional scale in post-1989 England and Wales. As a result, policy makers and regulators have had to devise solutions that simulate direct competition in order to prevent water companies from abusing the de facto privileges that attach to natural monopolies. The resulting market is, necessarily, highly contrived and a far cry from the textbook model—an act of political will, if you like. This is why most water privatization schemes over the last 30 years have been at the level of one or another locality, city, or region, rather than genuinely national.[32]

3. *Markets in environmental goods, services, and assets must carefully adapt to the prevailing socio-cultural and political-economic context, unless they can reconfigure it.* All plans to neoliberalize environmental goods, services, and assets necessarily occur against the background of existing policies, established conventions, and prevailing economic interests. As I intimated earlier, this context must be either changed or (failing that) negotiated in order for neoliberal measures to realize their self-declared ambitions. In Polanyi's terms, if one wishes to establish a market economy, then one needs to alter the moral, cultural, and political climate—that is, one needs to create a market society in which neoliberal norms become the prevailing common sense of the day. This is why several analysts of the neoliberalization of nature have seen fit to use Antonio Gramsci's ideas when analyzing the way that neoliberal environmental policies have (or have not) taken hold. Consent, after all, must be constructed: it is never suddenly or spontaneously achieved.

An example of neoliberal environmental policies that encountered relatively little communal or public opposition is presented by Wolford (2005, 2007), who analyzes the mid-1990s national land reform program in Brazil. Led by President Fernando Cardoso, this program offered many landless rural workers the possibility of owning and working their own land, in the context of a long history of large landowners controlling a disproportionate amount of rural space. As Wolford explains, even though members of the Movement for Landless Workers had a very different understanding of the role of land and property than did agrarian elites, many supported Cardoso's reform program because it seemed to promise a degree of justice (via the language of 'rights') and the prospect of improved livelihoods. In other words, the language of reform could be made consistent with the moral economy of many landless workers.

In contrast, Perreault (2006, 2008) examines the reasons why neoliberal environmental policy failed the test of popular legitimacy in Bolivia. As indicated earlier, Perreault focuses on the now famous water and gas 'wars' of 2000 and 2003, respectively. In the former case especially, neoliberal reform triggered widespread protests (especially among peasant farmers) in which historic norms and cultural values were crystallized into a powerful anti-neoliberal discourse. Quite aside from the fact that the water reforms were antithetical to these established norms and values, both water and gas are viewed as basic resources that are crucial to people's well-being. In Bolivia, they are seen as national resources that should be shared fairly, not appropriated privately.

Clearly, in any given case the precise mix of factors determining the relative degree of traction and longevity of neoliberal environmental policies will vary. In cases where a fair bit of policy adaptation has occurred (so to speak, when some rooms have been rebuilt in order to spare the foundation), the gap between the ideals of the neoliberal worldview and the messy specifics of practice becomes manifest (see also Bakker 2005).

4. *Neoliberal environmental policy is often and in large measure constitutively 'impure'.* Earlier in this article, it was argued that neoliberalism is a spatio-temporally differentiated process—following Peck and Tickell (2002), a set of connected neoliberalizations (in the plural)—and not a globally homogeneous thing. Building on the third point above, some have suggested that this insight applies as much to environmental management, regulation, and governance as to anything else. Both Becky Mansfield (e.g., 2007a, 2007b) and James McCarthy (2006) have been especially vocal in this regard. As part of her extensive research into the enclosure of North Pacific fisheries, Mansfield has looked closely at how catch quotas are allocated in practice. As marketable rights to fish, the quotas have commanded considerable economic value since their creation in the late 1990s. They have included a so-called community development quota (CDQ) for the poor, mostly indigenous communities of western Alaska. Within the neoliberal logic of privatization and marketization, the CDQ addresses issues of both social justice, as an anti-poverty measure, and cultural justice, as a recognition of and partial redress for indigenous people's historic exclusion from land and water, following the European occupancy of North America. As Mansfield (2007b: 495) concludes, "What is interesting about the CDQ … is not that it is incoherent or inconsistent, but that … privatization [here] manages to bridge seemingly contradictory goals." This is done by enhancing personal (in this case, communal) freedom while redistributing wealth to favor the needy.

The wider significance, when the second and third points above are connected to this one, is that "neoliberalism … is something created in *practice*, and that through practice, it becomes varied, fractured, and even contradictory. In this sense, 'neoliberalism' is inherently geographical" (Mansfield 2004a: 580; emphasis added). In his analysis of community forestry projects, McCarthy (2006: 87) concurs: "Processes of neoliberalization never occur on blank slates, but rather hybridize with existing institutions, regionally and nationally specific policy ensembles, and so on in ways that always and inevitably produce unique, contingent variants." This explains why empirical research is so important and why blanket descriptions, explanations, and evaluations are likely to founder on the shoals of spatio-temporal difference.

5. *Neoliberal environmental policy frequently tends to disadvantage the poor and the powerless.* As detailed in the previous section, this fifth point is evident in Bakker's (2003) findings, Budd's (2007) research on Chilean peasant irrigators, Prudham's (2004) Walkerton study, McCarthy's (2004) analysis of regulatory takings, Smith's (2004) review of Cape Town water resources, and several other works as well, including Tad Mutersbaugh's (2003, 2005) investigations of Mexican small farmers. Buscher's (2010a, 2010b) research on transfrontier conservation likewise suggests that local communities tend to lose out. And where neoliberal policies appear to offer openings for the socially disadvantaged—as described in Wolford's (2005) research on Brazil—there is evidence that these opportunities are more apparent than real. However, as noted earlier, there are exceptions to this very rough rule, depending on how analysts choose to define and measure disadvantage.

6. *Neoliberal environmental policy produces environmental improvements as much as problems, and problems as much as improvements.* This point, which has been detailed in the previous sections, challenges the argument that applying neoliberal principles to natural resource management and environmental problems is usually an effective way to 'green' capitalism. Quite how fundamental that challenge is remains a subject for discussion. For critics, the question is whether the improvements in environmental quality and protection delivered by certain neoliberal policy measures could have been—or could be—delivered by alternative, non-neoliberal policy devices. Another issue is how consistent such measures are with the achievement of social justice. Regardless, it is clear that neoliberal environmental policies can deliver certain benefits for nature, depending on the circumstances. Bakker's research on water quality shows this, so too Duffy and Moore's (2010) research into elephant tourism.

What light do these six summary observations shed on the questions I posed earlier when discussing neoliberal environments and the research published on this topic? There, you will recall, I contrasted the political-economic approach to the neoliberalization of nature (favored by many of the authors whose work I have reviewed here) with a more Foucauldian approach. The former viewpoint focuses our attention on whether the neoliberalization of nature constitutes (1) a widening or deepening of class-based social power; (2) a perpetuation, mitigation, or even overcoming of the 'ecological contradictions' that are characteristic of capitalism to date; and/or (3) a 'winning over' of the various constituencies with a stake in the reform of resource and environmental regulation and use. These are grand questions, and the findings of the studies reviewed in this article cannot be satisfactorily summed up to provide robust answers to any of them. What we *can* say is that there is now plenty of evidence to suggest that neoliberal environmental policy respects the rights of owner-operators above those of other stakeholders; that it does something, but not nearly enough, to address the ecologically destructive and wasteful patterns of capital accumulation; and that it frequently co-opts (often unwilling) people into its rationalities. The neoliberalization of nature is a project that is far from complete and perhaps precarious.

Researching the Neoliberalization of Nature: Problems of Conceptualization, Theory, Method, and Evaluation

As the previous paragraph has intimated, I believe that a critical examination of the research literature on the neoliberalization of nature is warranted. There are, it seems to me, two forms that such a critique could take. The first would focus on some rather obvious and important limitations of the studies conducted so far. For instance, one might ask for more balance in the number of studies centering on the three fields (i.e., the academic, the political, and the bureaucratic) where neoliberal ideas and practices have environmental and resource content and implications. Alternatively, one might ask for a greater number of studies involving a wider range of locations with a better sectoral balance so that, for example, water resource issues are not favored over investigations of mining. Despite the large number of published studies into the neoliberalization of nature, there are arguably not nearly enough when compared to the sheer number and variety of policy experiments that could be investigated. A second form of critique would focus in more detail on some fundamental cognitive and normative questions when analyzing the way that researchers are conceiving and executing their research projects. It is this second approach that I want to take because I believe that there are several key issues that deserve serious attention, as they speak to the (perhaps unconscious) analytical habits of the loose, multi-disciplinary epistemic community whose research I have been reviewing.

The first key issue relates to how the word 'neoliberalism' is defined and conceptualized in any given empirical case. Early on in this article, I noted—but then bracketed—the lack of conceptual consensus among analysts about the meaning of the term. Opening the brackets, it is clear that different investigators have employed the word in a range of ways when examining environmental and resource issues. Some authors (e.g., Kay 2002; Murray 2002) use very generic or highly implicit definitions that are fairly devoid of conceptual substance. Others (e.g., McCarthy 2006) are conceptually precise, yet they define the term in a way that does not quite correspond with the equally precise definitions employed by still others (e.g., Robertson 2004). Throughout, there is no real agreement about which (or how many) of the seven elements of the neoliberal policy program need to be evident or implemented in any given case for us to describe it reasonably as neoliberal in degree or kind. This is unfortunate.[33] For example, should moves toward privatization alone be deemed instances of neoliberalism in action? And is what we call privatization in one place really the same as that in another? Surely, the significance of the seven neoliberal policy proposals listed earlier depends entirely on the context in which they are embedded and operate. In this sense, conceptual abstraction is a fraught process.

This last observation directs our attention to what 'context' means in any given case, and to the related question of how we recognize different modalities or variants of neoliberalism—that is to say, related but different neoliberalizations. As I have suggested, the sheer presence of one (or more) of the seven neoliberal policy proposals does not necessarily mean that it is a definitive element of the situation in which it is enacted. Mansfield (2007a, 2007b) rightly points to the constitutive 'impurity' of all neoliberal environmental measures (see also Sugden 2009). But the researchers whose work I have surveyed have yet to separate conceptual from empirical impurity. It is not surprising that there are detailed variations in the way that neoliberal environmental policies have been implemented in different times and places. The more exacting task is to identify conceptually the varieties of neoliberalism by abstraction from some of the concrete empirical details. Otherwise, we are left with empirical variation alone, meaning that each and every situation in which neoliberal policy measures are implemented is considered a specific and unique neoliberalization. This then greatly limits the potential for cross-case comparison and the identification of commonalities between different sets of cases. Indeed, it risks falling into the trap of 'idiography'—that is, the study of spatio-temporal difference for its own sake, at the expense of identifying common processes and outcomes across space and over time (see Castree 2005).[34]

Just as there is currently no uniform definition of neoliberalism among analysts, those researchers whose work I have reviewed here utilize a range of theoretical lenses when examining environmental and resource policy. Although all are critical political economists (in the general sense of the term), attempts to achieve greater theoretical consistency have been rather limited thus far. In discussing this second key issue, I use the word 'theory' in a conventional sense to denote a descriptive and explanatory framework that focuses researchers on what they presume to be the most salient processes, relationships, or issues in any given case. For instance, Mansfield has made ample use of Polanyi's concept of a 'fictitious commodity' in her fisheries research, while McCarthy deploys James O'Connor's Marxist notion of the 'underproduction of the conditions of production'. To cite one more case, my colleague Erik Swyngedouw (2005) uses David Harvey's concept of 'accumulation by dispossession' to discuss water resource privatization. The challenge, it seems to me, is to weave some of this otherwise disparate use of political-economic theory together into a more coherent framework, one that can sensitize us to the complex, contradictory, and dialectical dynamics of neoliberal environmental and resource governance in practice. Currently, what seems to happen is that any given researcher employs a theoretical insight or idea drawn from one of his or her favored thinkers. Thus, one rarely finds a research project that integrates critically the ideas of, for example, Gramsci *and* Polanyi, or O'Connor *and* Gramsci.[35]

A third key issue I wish to highlight is methodology. Here, by the word 'issue', I really mean a family of issues. Whether they focus on human-environment relations, or anything else, social scientists must typically deal with 'open systems' in which it is virtually impossible to exert 'experimental control'. Given this, they are faced with a large number of methodological choices and options that relate to the quality and quantity of evidence, the methods of data acquisition and analysis, and the scale (micro-, meso-, or macro-) of methodological resolution. Decisions must be made about what evidence and data to include and exclude, how to code and categorize this information, and which questions will (and will not) be asked. Inevitably, all research projects are tailored to the specific opportunities and constraints operative in any given case, even as analysts aim for rigor and systematicity. In respect of the literature reviewed in this article, what is striking is the multiplicity of ways in which the neoliberalization of nature has been investigated. In many cases, it is not even clear how—methodologically speaking—the research was conducted or why it was carried out in the way that it was. The question arises: does this reflect the specifics of the situations being investigated (i.e., necessary compromises and adjustments), or is it a contingent reflection of investigators' varied expertise, time availability, level of experience, energy, commitment, etc.? Regardless, the upshot is that it is very difficult indeed for readers of the research to identify methodological consistency between disparate studies. I am not so naive as to expect, or wish for, more researchers to use the same, fairly detailed methodological template. However, it is difficult to combine the insights offered by diverse studies when those studies have been conducted so very differently. It is not simply an issue of having to somehow correlate very different kinds of evidence; there is also the issue of how robust and comprehensive the evidence from different research projects really is.

One specific methodological concern worth highlighting relates to comparative research. Thus far, there has been virtually no attempt made to investigate two or more cases of the neoliberalization of nature simultaneously. This is a pity because, in theory at least, it could help us to determine with some precision what a 'variety' or 'modality' of 'actually existing neoliberalism' looks like (in reference to my earlier point about how analysts have defined neoliberalism). One could look at two or more situations where the same translocal (or transnational) policy measures have been implemented; or one could look at two or more cases of *sui generis* environmental policy that appear to be ostensibly similar—or very different—cases of neoliberal reform. McCarthy (2006) is almost alone in having tried to conduct a cross-case study and, in my view, has set a precedent that ought to inspire others (see also Bailey 2007b; Duffy and Moore 2010). One methodological virtue of comparative research is that it is incumbent upon the investigator to ensure a certain consistency in the questions posed, the methods used, and the evidence garnered.

Finally, with regard to a fourth key issue—evaluation—let me voice some normative concerns about the way that research into the neoliberalization of nature has thus far been conducted. Setting oneself up as a 'critic' of anything presumes not only that one has clear criteria against which the item being analyzed is measured, but also that one can flesh out and justify those criteria in a moral-ethical sense. As I have stated, the term 'neoliberalism' is very much one that is employed by the academic and activist arms of the political Left, which are typically in opposition to this policy. Those who have researched the neoliberalization of nature are thus in some sense skeptical about, or even opposed to, their objects of analysis. But on what grounds and in what ways? The answers to these questions are surprisingly difficult to decipher because, for the most part, the act of evaluation is left implicit by those whose work I have reviewed in the preceding pages.[36] In my view, they need to be made far more explicit. What is more, the basis on which any evaluation is undertaken needs to be fleshed out in reasonable detail and justified—no simple matter. As Andrew Sayer (1995) noted many years ago, social scientists typically devote far more of their energy to issues of theory and method than they do to issues of normative reasoning. That is certainly true of the work I have examined here.

The exceptions prove the proverbial rule. Mansfield's (2006) highly normative essay on the success/failure of North Pacific fisheries policy is a form of internal or 'immanent' critique. In other words, she holds the marked-based fisheries policy to its own evaluative criteria in order to provide a systematic assessment of how its performance to date should be judged. This is one example of what could (and arguably should) be a wider effort to evaluate neoliberal environmental policy by focusing on whether GEDDS (i.e., growth/efficiency/development/democracy/sustainability) is a myth or reality. But there is another form of evaluation that has not yet been employed in a formal and explicit way in research on the neoliberalization of nature—namely, 'external' critique. Here the critic evaluates the world using criteria and related moral-ethical arguments that are intentionally different from those contained within the object of analysis. Bakker's (2010b) new book, *Privatizing Water*, hints at this alternative form of evaluation. Her assessment of water resource markets rests, not uncritically, on the idea of water as a human right. Although the concept of 'rights' is central to the neoliberal worldview, Bakker reminds us that it is a far more expansive and complex idea than liberals would have us believe. It can speak to issues of social justice and resource redistribution to the needy and vulnerable, as much as to issues of individual sovereignty.

The utility of explicit and robust forms of normative assessment is clear. They are a potentially powerful weapon to use against the authors and advocates of neoliberal forms of environment management, regulation, and governance. But they also keep critics honest, obliging them to acknowledge those situations where the ostensible object of their animus can be credited with certain successes.

Conclusions

This article is a multi-disciplinary review of social scientific research into the neoliberalization of nature, possibly the most inclusive to date.[37] Reviews such as this one organize and codify research that is developing organically over time, in this case in more than one academic field. The published research is the raw material out of which the survey is actively fashioned. But the cognitive mapping exercise is by no means easy. In the present case, it is not possible to begin with a uniform understanding of neoliberalism that is evident within the literature (let alone a coherent theoretical approach to it), and then straightforwardly track the operation and effects of neoliberalism in the fields of environmental management and natural resource governance. Instead, the published literature has examined different aspects of the neoliberalization of nature in different times and places and at different scales of analytical resolution. As Diana Liverman and Silvina Vilas (2006: 358) note in their recent attempt to review systematically the research on this topic, "Most studies tend to be case specific and difficult to generalize." They also maintain that there has been a lack of "comparative research … set within a rigorous *a priori* framework." I have therefore tried to throw a rope around these disparate studies in the hope of making some sense out of them overall. In so doing, I have made an admittedly contrived distinction between the processes and outcomes of neoliberal policy reform in the environmental domain. Inevitably, I have also organized the insights of the studies into other analytical categories that may, in the end, be too neat and tidy to be either useful or plausible.

Let me close with what might seem to be a very self-serving observation, although it is meant sincerely and relates to Liverman and Vilas's candid reflection on their own attempt to author a comprehensive review. In my experience, surveys of established or new fields of research are sometimes regarded as 'lesser outputs', academically speaking, compared to works of original empirical research or fundamental contributions to philosophy, theory, or method. This is unfortunate. As I tried to show in the previous section, surveying a research field is not about piecing together a jigsaw whose parts happen to be scattered hither and thither, so that the bigger picture eventually becomes clear—even though this is often assumed to be the case. The fact that a plethora of researchers use the same keywords and core concepts does not mean that they are, upon close inspection, actually analyzing or evaluating the same thing, let alone in a theoretically or methodologically consistent or commensurable way. In the present case, one can suggest that the nominal commonality of the work I have surveyed belies a set of serious substantive differences that are variously conceptual, theoretical, methodological, and normative in kind. Some might say that these differences are productive—although this is surely not the case if the differences are so significant that they threaten to prevent meaningful advances in research.

What is to be done? Many decades ago, Thomas Kuhn, the physicist and historian of science, popularized the term 'paradigm' as a way of characterizing the working habits and outputs of a set of like-minded researchers. Perhaps those investigating the neoliberalization of nature have not been paradigmatic enough and need to be much more so in the future. Indeed, this suggestion might apply to a great deal of contemporary social science, which is resolutely post-paradigmatic for the most part (economics being a notable exception). Topics such as neoliberalism bring disparate researchers from various disciplines together to interrogate what is ostensibly the same thing. However, they also make plain the quantitative and qualitative variations in the ways that social scientific research is being conducted. Perhaps it is not possible to change this state of affairs. But, looking ahead, it would surely pay dividends to aim for greater analytical consistency among researchers operating in different academic disciplines. There is hard mental labor to be performed, but it has very practical—and not purely cerebral—implications. The sorts of policy measures that analysts of the neoliberalization of nature have been concerned with are rarely trivial in their effects, for good or ill. A more collaborative and less piecemeal effort by social scientists to examine these and future policy measures could have a positive and very material bearing on the well-being of people and of the non-human world. Ideally, their research would actively shape the thinking of the politicians and policy makers whose decisions significantly affect our lives in so many different ways.

ACKNOWLEDGMENTS

I am extremely grateful to Dan Brockington and Paige West for permitting me to write at such length on the neoliberalization of nature. I also wish to thank the four reviewers of this article, who took considerable time and trouble to offer me superb evaluations of the first draft. One of the two anonymous reviewers posed some especially sharp and challenging questions that I dearly wanted to respond to when revising the article. However, for reasons beyond my control, I had very little time to alter the article, and so the published version is—for better or worse—similar to the original one. This being said, I hope that I have now knitted together the article's various parts so as to address at least one concern expressed by the most critical of the four reviewers. Finally, I am very grateful to Shawn Kendrick for her peerless copyediting. As ever, I take full responsibility for the remaining various flaws.

NOEL CASTREE is Professor of Geography in the School of Environment and Development, Manchester University. His principal research interest is in the political economy of environmental change, part of a wider interest in the geographical dimensions of capital accumulation. He was editor of *Antipode: A Journal of Radical Geography* (2004–2009) and is currently managing editor of *Progress in Human Geography*. His forthcoming book, *Making Sense of Nature* (2011), is about the governance of expert knowledge, ranging from that about ice caps to understandings of the human genome. He is also co-editor of the forthcoming four-volume *Human Geography* (2012), part of the Sage Major Works in Social Science and the Humanities series.

NOTES

1. The term 'neoliberalism' initially gained popularity in left-wing circles in Latin America as a reaction to the market-led reform agenda of General Augusto Pinochet in Chile.
2. Those in the Foucauldian camp are sometimes given to calling neoliberalism 'advanced liberalism'. In the political-economic camp, Karl Polanyi's economic history is used by some in conjunction with Marx's late political-economic writings because of Polanyi's critique of the 'classic liberalism' of the nineteenth and early twentieth centuries. Others call upon Antonio Gramsci's ideas in order to assess whether and how neoliberal values and practices are embedded in everyday life outside the formal spheres of 'state' and 'economy'. Few authors whose work is reviewed here have used all of Marx, Polanyi, and Gramsci together in a substantive way. Typically, one or two of these theorists' ideas are used (usually selectively) in any given study. (I say more about this in the penultimate section of the article.) Although some authors in the political-economic fold have drawn upon other theorists—such as Jean Baudrillard and Jim O'Connor—Marx, Polanyi, and Gramsci currently seem to be the favored ones.
3. In other words—for better or worse—I have not included publications in which environmental use and management have been analyzed in the context of policy changes that some would describe as neoliberal, such as 'structural adjustment policy' in the global South in the 1980s and 1990s. Unless authors discuss neoliberalism explicitly, I have excluded their publications, even when these publications cover related matters such as the privatization of environmental goods and services.
4. In part, the fact that neoliberal values and principles have made their way into the domain of environmental policy reflects the nature of many environmental goods and services: they inevitably impinge upon, or are themselves affected by, ostensibly 'non-environmental' policy domains, such as international trade policy. But it also reflects two other things. First, many environmental goods and services are of great social and economic importance (water and sewerage being prime examples). It is no surprise, therefore, that neoliberal reformers were intent on altering the modes of delivery and governance of these services, for they were hardly marginal to any project of remaking the economy, state, and society *tout court*. Secondly, a set of worrying environmental problems already evident in the 1960s have subsequently grown in number, diversity, and seriousness. Neoliberal policy makers have sought to address these problems in ways consistent with their particular worldview. The terms 'free market environmentalism', 'liberal environmentalism', 'green capitalism', 'ecological modernization', 'green neoliberalism', and 'ecological capitalism' all capture, with rather different cognitive and normative valences, this explicit problem-solving agenda.
5. For instance, researchers in geography, anthropology, and development studies have drawn upon each other's research of late when thinking through new developments in nature conservation. See, for example, the recent special issue, titled "Capitalism and Conservation," of the journal *Antipode* 42 (3) (2010): 469–799.
6. There is also a good review by Himley (2008) in the online journal *Geography Compass*. I have three new student-oriented essays on neoliberalism and nature in the same journal (Castree 2010a, 2010b,

2010c). Because I am presuming little prior knowledge of at least some readers, this article inevitably recapitulates ideas presented in my back-to-back 2008 articles for *Environment and Planning A*. However, knowledgeable readers will, I hope, see the 'value added' in the present article compared to the two earlier ones—not least in my attempt to cast a wider net and encompass works published in several disciplines outside my 'home base' of geography.

7. Inevitably, despite my attempt to be thorough, I will have missed some things. For instance, a special issue of the journal *Review of Radical Political Economics* (42 [2] [2010]) on the political economy of water service privatization was published as this article went to press. As stated in note 3, absences like this raise the wider question of how far I should have cast my net when researching this article. My rule of thumb was to read only those studies in which the term 'neoliberalism' is used in a more-than-passing way. However, this means that numerous works that focus on topics such as markets and privatization without mentioning neoliberalism are not included in this review. Only studies that have gone through peer review and are published are referenced here; working papers and conference papers are excluded.
8. I am in good company. At a three-day conference titled "A Brief Environmental History of Neoliberalism," which was held on 6–8 May 2010 at Lund University in Sweden, many researchers presented their findings pertaining to this topic. To view the working papers from this conference, see http://www.worldecologyresearch.org/?p=1.
9. A leitmotif of all these aspects is more or less far-reaching change, which is why neoliberalism has been commonly described using action terms such as 'project', 'strategy', 'roll-back', 'roll-out', and 'regime change'. It is understood by the critics to pose a definite challenge to what has heretofore constituted 'common sense' discourse and practice in the domains of state, economy, and society. This strongly performative element leads Harvey (2007) to regard neoliberalism as a synonym for 'creative destruction'—that signature element of capitalist modernity in all its geographically and historically varied forms.
10. Modern liberal philosophy is a diverse and complex thing. Not all self-declared liberals could be fairly described as neoliberals in the sense meant by critical social scientists or left-wing political activists. What is more, the neoliberal worldview is rather more radical than that of the Freiburgers, who originally claimed the term as their own. Indeed, some of the latter considered the likes of Hayek to be 'paleo-liberals', a pejorative reference to their aspirations for a world of *laissez-faire* writ large.
11. In this latter respect, the neoliberal worldview has frequently been called 'market triumphalist', 'market extremist', or 'free market capitalist'. More than any other institution in modern society, the market is seen as the handmaiden of liberty and freedom, by virtue of Adam Smith's famous 'hidden hand'. This also dovetailed with a belief that markets should, in many cases, substitute for decisions made currently within the domain of politics. In other words, Friedman and those who shared his views argued that in many Western democracies the political sphere had, in some senses, become 'bloated' since World War II. They believed that this was the case not just in the administrative-technical sense ('the big state', reliant on too much tax income and borrowing), but in the sense that too many private issues were being made into matters of public concern ('the nanny state', intruding into matters that should rightly be decided by individuals, families, and communities).
12. Still in existence, the Mont Pelerin Society is an international organization composed of economists, philosophers, historians, intellectuals, business leaders, and others who champion classical liberalism. The society advocates free market economic policies and the values of an open society, with political freedoms and human rights as its foundation. See http://www.montpelerin.org.
13. In spatio-temporal terms, privatization typically amounts to geographical exclusion and denying current and future generations the use and/or benefits of a given portion of the biophysical world. All privatizations are, at one and the same time, acts of inclusion and preclusion, creating both entitlements and proscriptions.
14. It is worth noting that, for many commentators, it is the conjunction of privatization and marketization that defines commodification (i.e., the reshaping of goods and services into commodities). Needless to say, in practice both privatization and marketization take a number of concrete forms, as befits the particular commodities, firms, consumers, and other relevant parties in question.

15. In the West, the term 'new public management' has become a familiar descriptor for this process, denoting a management paradigm that emphasizes value for money, budget capping, cost recovery, and the avoidance of deficit spending whenever possible.
16. Three things make this policy discourse 'neo' (i.e., new) -liberal when compared with the so-called classical liberalism of Adam Smith. First, from the 1970s it was presented as an explicit critique of post-war 'managed capitalism', be it in parts of the former communist bloc, the Western social democracies, or the many 'developmental states' of the global South. Secondly, it takes a fairly dim view of the state, public goods, and common resources—except insofar as any of them can aid the cause of individual freedom or liberty. In Colin Crouch's (2004: 248) words, "Not only is the state seen as having no goals or *modi operandi* different from those of market actors, but it is seen to gain by subordinating its activities as much as possible to those of market actors." Finally, this discourse has traveled far and wide geographically, courtesy of various institutions and networks in which US neoliberals have played a highly active role (for more, see Peck 2010).
17. Because many of these empowered individuals not only are outside the formal sphere of government but also are unelected, many observers have regarded neoliberal policies as anti-democratic.
18. This research agenda is advocated by several others, including Castree (2005), Clarke (2004), England and Ward (2007), Kingfisher and Maskovsky (2008), Larner (2000, 2003), and Leitner et al. (2007). It amounts to tracking the temporal and geographical circulation, modification, hybridization, implementation, revision, and (in some cases) abandonment of neoliberal ideas. At its most ambitious, it involves determining the reciprocal links between, and conjoint effects of, neoliberalism as philosophy, program, and practice in any given case. This would enable us to identify varieties of neoliberalism and to understand whether, how, and why they have succeeded in their own less-than-homogeneous normative terms. It would also oblige us to acknowledge the fact that, in some cases, the mere presence of elements of philosophy, program, and practice does not, in itself, entitle us to conclude that these are defining elements. They may, in fact, be less than hegemonic in certain situations, since their presence in one or another policy domain does not make them definitive of an entire policy regime, let alone an entire social formation. As things stand, there is no consensus on how one identifies a 'variety' of neoliberalism, although several economic sociologists and international relations scholars have scarcely hesitated here—as if the complex ontological, epistemological, and methodological issues just highlighted were of little consequence. For now, I will also put aside these complexities as I move on, at long last, to discuss the neoliberalization of nature.
19. Before proceeding any further, we should make an important distinction between real and analytical consequence (which has only been implicit thus far in this article). When considering the neoliberalization of nature, we can discuss one or both of the following, depending on how thoroughly we wish our examination to be. First, there are those neoliberal policies that are not ostensibly about environmental goods and services (such as free trade policies) yet nonetheless have real biophysical impacts. Secondly, there are those neoliberal policies (such as water resources policies) that take environmental phenomena as the explicit object of attention. In what follows I will be surveying research that focuses on the latter for the most part, meaning that I am—in truth—omitting a key section of literature on the neoliberalization of nature. However, including this other literature would, at the least, double the length (and complexity) of this already long article.
20. These properties are relative rather than absolute, but they are real nonetheless. They can perhaps be ignored in the short term but not in the medium to long term, since inappropriate regulatory measures will eventually lead to visible and possibly harmful environmental problems.
21. The special issues referred to are "Neoliberal Nature and the Nature of Neoliberalism," *Geoforum* 35 (3) (2004), and "The Commodification of Nature," *Capitalism, Nature, Socialism* 16 (1) (2005).
22. Higgins and Lockie (2002), Sullivan (2006), and Fletcher (2010) are among the relatively few exceptions to this viewpoint. Some regard the two approaches to neoliberalism as complementary: see, for example, Lockwood and Davidson (2009).
23. I suspect that this Foucault-inspired research into neoliberalism and environmental governance will grow in importance, not least because Foucault's lectures on biopolitics were recently translated into English.

24. This matter of class-based social power has frequently arisen in respect to enclosures of environmental commons (i.e., community land and resources).
25. The theoretical literature authored by eco-Marxists such as James O'Connor, Ted Benton, Elmar Altvater, and John Bellamy Foster is undecided on this question of ecological contradictions.
26. For example, in 2010 several publications by Bram Buscher appeared in a range of journals, with more to come (see Buscher 2010a, 2010b; Buscher et al., forthcoming).
27. I hope, therefore, that this article will lead readers to a close study of the many publications that have been cited in it.
28. See also St. Martin's (2007) study of recent New England fisheries and Ibarra et al.'s (2000) analysis of Chilean, Mexican, and Peruvian fisheries management.
29. The acronym 'quango', coined and used primarily in the UK, signifies a quasi non-governmental organization that is either financed by the government or formally linked to it but operates at arm's length.
30. The relativity of effects invites complex and differentiated judgments about how they are to be registered and evaluated. It would be facile to suppose that all the effects of the neoliberalization of nature registered in the published research can somehow be added up and included in some sort of aggregate scorecard, although table 1 gives the impression that this is a permissible practice. For more on this, see Castree (2008b: section 4).
31. A similarly convincing case about the state and its involvement in markets is provided by Bakker (2003). See also Bailey (2007a) and Bell and Quiggin (2008).
32. See Mansfield (2004a, 2004b) and Robertson (2004) for similar arguments about the impact of nature's material properties.
33. According to Clive Barnett (2010), this lack of agreement is symptomatic of the role neoliberalism plays as a shibboleth for the Left, its value being more symbolic than analytical.
34. In her most recent article, Bakker (2010a) makes a serious attempt to provide a vocabulary for undertaking such single-site and comparative studies.
35. See Castree (2008a) for a rather abstract attempt to synthesize some of the different theoretical ideas that have been used in the literature, and Bakker (2009) for a constructive critique of it.
36. For instance, Jessica Budd's (2004) article on Chilean water reform makes use of the concept of social equity but nowhere details or defends this normative idea.
37. Leila Harris (2009) has recently published a useful review, focusing specifically on the gender dimensions of neoliberal environmental policy.

REFERENCES

Anderson, Perry. 2000. "Renewals." *New Left Review* 1 (January–February): 1–9.

Bailey, Ian. 2007a. "Neoliberalism, Climate Governance and the Scalar Politics of EU Emissions Trading." *Area* 39 (4): 431–442.

______. 2007b. "Market Environmentalism, New Environmental Policy Instruments, and Climate Policy in the UK and Germany." *Annals of the Association of American Geographers* 97 (3): 530–550.

Bakker, Karen. 2001. "Paying for Water." *Transactions of the Institute of British Geographers* 26 (2): 143–164.

______. 2003. *An Uncooperative Commodity: Privatizing Water in England and Wales.* Oxford: Oxford University Press.

______. 2005. "Neoliberalizing Nature? Market Environmentalism in Water Supply in England and Wales." *Annals of the Association of American Geographers* 95 (3): 542–565.

______. 2009. "Neoliberal Nature, Ecological Fixes and the Pitfalls of Comparative Research." *Environment and Planning A* 41 (8): 1781–1787.

______. 2010a. "The Limits of 'Neoliberal Nature': Debating Green Neoliberalism." *Progress in Human Geography* 34 (6): 715–735.

______. 2010b. *Privatizing Water.* Ithaca, NY: Cornell University Press.

Barnett, Clive. 2010. "Publics and Markets: What's Wrong with Neoliberalism?" Pp. 269–296 in *The Sage Handbook of Social Geography*, ed. Susan J. Smith, Rachel Pain, Sallie A. Marston, and John Paul Jones III. London: Sage.

Barry, Andrew, Thomas Osborne, and Nikolas Rose, eds. 1996. *Foucault and Political Reason: Liberalism, Neo-liberalism, and Rationalities of Government.* Chicago: University of Chicago Press.

Bell, Stephen, and John Quiggin. 2008. "The Limits of Markets: The Politics of Water Management in Rural Australia." *Environmental Politics* 17 (5): 712–729.

Bernstein, Steven. 2002. "Liberal Environmentalism and Global Environmental Governance." *Global Environmental Politics* 2 (3): 1–16.

Boas, Taylor C., and Jordan Gans-Morse. 2009. "Neoliberalism: From New Liberal Philosophy to Anti-Liberal Slogan." *Studies in Comparative International Development* 44 (2): 137–161.

Brand, Ulrich, and Nikola Sekler, eds. 2009. *Postneoliberalism: A Beginning Debate*. Special issue of *Development Dialogue* 51 (January): 5–211.

Brenner, Neil, and Nik Theodore. 2002. "Cities and the Geographies of 'Actually Existing Neoliberalism.'" *Antipode* 34 (3): 349–379.

Brenner, Neil, Jamie Peck, and Nik Theodore. 2010. "After Neoliberalization?" *Globalizations* 7 (3): 327–345.

Bridge, Gavin. 2002. "Grounding Globalization." *Economic Geography* 78 (3): 361–386.

Brown, Sandy, and Christy Getz. 2008. "Privatizing Farm Worker Justice: Regulating Labor through Voluntary Certification and Labeling." *Geoforum* 39 (3): 1184–1196.

Budds, Jessica. 2004. "Power, Nature and Neoliberalism: The Political Ecology of Water in Chile." *Singapore Journal of Tropical Geography* 25 (3): 322–342.

Bury, Jeffrey. 2004. "Livelihoods in Transition: Transnational Gold Mining Operations and Local Change in Cajamarca, Peru." *Geographical Journal* 170 (1): 78–91.

______. 2005. "Mining Mountains: Neoliberalism, Land Tenure, Livelihoods, and the New Peruvian Mining Industry in Cajamarca." *Environment and Planning A* 37 (2): 221–240.

Buscher, Bram. 2010a. "Seeking Telos in the 'Transfrontier': Neoliberalism and the Transcending of Community Conservation in Southern Africa." *Environment and Planning A* 42 (3): 644–660.

______. 2010b. "Anti-Politics as Political Strategy: Neoliberalism and Transfrontier Conservation in Southern Africa." *Development and Change* 41 (1): 29–51.

Buscher, Bram, and Wolfram Dressler. 2007. "Linking Neoprotectionism and Environmental Governance: On the Rapidly Increasing Tensions between Actors in the Environment-Development Nexus." *Conservation and Society* 5 (4): 586–611.

Buscher, B., D. Brockington, J. Igoe, K. Neves, and S. Sullivan. Forthcoming. "Towards a Consolidated Critique of Neoliberal Biodiversity Conservation." In revision for *Capitalism, Nature, Socialism*.

Castree, Noel. 2005. "The Epistemology of Particulars." *Geoforum* 36 (4): 541–544.

______. 2008a. "Neoliberalising Nature: The Logics of Deregulation and Reregulation." *Environment and Planning A* 40 (1): 131–152.

______. 2008b. "Neoliberalising Nature: Processes, Outcomes and Effects." *Environment and Planning A* 40 (1): 153–173.

______. 2010a. "Neoliberalism and the Biophysical Environment 1: What 'Neoliberalism' Is, and What Difference Nature Makes to It." *Geography Compass* 4 (12): 1725–1733.

______. 2010b. "Neoliberalism and the Biophysical Environment 2: Theorising the Neoliberalisation of Nature." *Geography Compass* 4 (12): 1734–1746.

______. 2010c. "Neoliberalism and the Biophysical Environment 3: Putting Theory into Practice." *Geography Compass* 4 (12): 1747–1766.

Clarke, John. 2004. "Dissolving the Public Realm?" *Journal of Social Policy* 33 (1): 27–48.

______. 2010. "After Neoliberalism?" *Cultural Studies* 24 (3): 375–394.

Cocklin, Chris, Jacqui Dibden, and Naomi Mautner. 2006. "From Market to Multifunctionality?" *Geographical Journal* 172 (3): 197–205.

Coombes, Brad. 2003. "Ecospatial Outcomes of Neoliberal Planning." *Environment and Planning B* 30 (2): 201–218.

Craig, David, and Gerry Cotterell. 2007. "Periodising Neoliberalism?" *Policy & Politics* 35 (3): 497–514.
Crouch, Colin. 2004. "Markets and States." Pp. 240–250 in *The Blackwell Companion to Political Sociology*, ed. Kate Nash and Alan Scott. Oxford: Blackwell.
Davis, Diana. 2006. "Neoliberalism, Environmentalism, and Agricultural Restructuring in Morocco." *Geographical Journal* 172 (2): 88–105.
Dibden, Jacqui, Clive Potter, and Chris Cocklin. 2009. "Contesting the Neoliberal Project for Agriculture." *Journal of Rural Studies* 25 (3): 299–308.
Driesen, David. 2008. "Sustainable Development and Market Liberalism's Shotgun Wedding." *Indiana Law Journal* 83 (1): 21–69.
Duffy, Rosaleen. 2008. "Neoliberalising Nature: Global Networks and Ecotourist Development in Madagascar." *Journal of Sustainable Tourism* 16 (3): 327–344.
Duffy, Rosaleen, and Lorraine Moore. 2010. "Neoliberalising Nature? Elephant-Back Tourism in Thailand and Botswana." *Antipode* 42 (3): 742–766.
Eckersley, Robyn. 1993. "Free Market Environmentalism: Friend or Foe?" *Environmental Politics* 2 (1): 1–19.
England, Kim, and Kevin Ward, eds. 2007. *Neoliberalization: States, Networks, Peoples*. Oxford: Wiley-Blackwell.
Ferguson, James. 2010. "The Uses of Neoliberalism." *Antipode* 41 (s1): 166–184.
Fletcher, Robert. 2010. "Neoliberal Environmentality: Towards a Poststructural Political Ecology of the Conservation Debate. *Conservation and Society* 8 (3): 171–181.
Fortwangler, Crystal. 2007. "Friends with Money: Private Support for a National Park in the US Virgin Islands." *Conservation and Society* 5 (4): 504–533.
Fraser, Nancy. 1993. "Clintonism, Welfare and the Anti-social Wage: The Emergence of a Neoliberal Political Imaginary." *Rethinking Marxism* 6 (1): 9–23.
Friedman, Milton. 1962. *Capitalism and Freedom*. Chicago: University of Chicago Press.
Goldman, Michael. 2005. *Imperial Nature*. New Haven, CT: Yale University Press.
Gordon, H. Scott. 1954. "The Economic Theory of a Common Property Resource: The Fishery." *Journal of Political Economy* 62: 124–142.
Gowan, Peter. 1995. "Neoliberal Theory and Practice for Eastern Europe." *New Left Review* 213: 3–60.
Grandia, Lisa. 2007. "Between Bolivar and Bureaucracy: The Mesoamerican Biological Corridor." *Conservation and Society* 5 (4): 478–503.
Guthman, Julie. 2007. "The Polanyian Way? Voluntary Food Labels as Neoliberal Governance." *Antipode* 39 (3): 456–478.
______, ed. 2008a. "Agro-Food Activism in California and the Politics of the Possible." Special section of *Geoforum* 39 (3): 1171–1260.
______. 2008b. "Neoliberalism and the Making of Food Politics in California." *Geoforum* 39 (3): 1171–1183.
______. 2008c. "Thinking Inside the Neoliberal Box: The Micro-politics of Agro-Food Philanthropy." *Geoforum* 39 (3): 1241–1253.
Haggerty, Julia, Hugh Campbell, and Carolyn Morris. 2009. "Keeping the Stress Off the Sheep? Agricultural Intensification, Neoliberalism, and 'Good' Farming in New Zealand." *Geoforum* 40 (5): 767–777.
Hardin, Garrett. 1968. "The Tragedy of the Commons." *Science* 162: 1243–1248.
______. 1974. "Living on a Lifeboat." *Bioscience* 24 (10): 561–568.
Harris, Leila. 2009. "Gender and Emergent Water Governance." *Gender, Place & Culture* 16 (4): 387–408.
Harrison, Jill. 2008. "Abandoned Bodies and Spaces of Sacrifice: Pesticide Drift Activism and the Contestation of Neoliberal Environmental Politics in California." *Geoforum* 39 (3): 1197–1214.
Hartwick, Elaine, and Richard Peet. 2003. "Neoliberalism and Nature: The Case of the WTO." *The Annals of the American Academy of Political & Social Science* 590 (1): 188–211.
Harvey, David. 2005. *A Brief History of Neoliberalism*. Oxford: Oxford University Press.
______. 2007. "Neoliberalism as Creative Destruction." *Annals of the American Academy of Political and Social Science* 610 (March): 22–44.
Hayek, Friedrich von. 1944. *The Road to Serfdom*. London: Routledge.

______. 1960. *The Constitution of Liberty*. Chicago: University of Chicago Press.

Heynen, Nik, and Harold Perkins. 2005. "Scalar Dialectics in Green: Urban Private Property and the Contradictions of the Neoliberalization of Nature." *Capitalism, Nature, Socialism* 16 (1): 99–113.

Heynen, Nik, James McCarthy, Scott Prudham, and Paul Robbins, eds. 2007. *Neoliberal Environments*. London: Routledge.

Higgins, Vaughn, and Stewart Lockie. 2002. "Re-discovering the Social: Neo-liberalism and Hybrid Practices of Governing in Rural Natural Resource Management." *Journal of Rural Studies* 18 (4): 419–428.

Himley, Matthew. 2008. "Geographies of Environmental Governance: The Nexus of Nature and Neoliberalism." *Geography Compass* 2 (2): 433–451.

Holifield, Ryan. 2004. "Neoliberalism and Environmental Justice in the US Environmental Protection Agency." *Geoforum* 35 (3): 285–297.

Ibarra, Alonso A., Chris Reid, and Andy Thorpe. 2000. "Neo-liberalism and Its Impact on Overfishing and Overcapitalisation in the Marine Fisheries of Chile, Mexico and Peru." *Food Policy* 25 (5): 599–622.

Igoe, Jim, and Dan Brockington. 2007. "Neoliberal Conservation: A Brief Introduction." *Conservation and Society* 5 (4): 432–449.

Igoe, Jim, and Beth Croucher. 2007. "Conservation, Commerce, and Communities: The Story of Community-Based Wildlife Management Areas in Tanzania's Northern Tourist Circuit." *Conservation and Society* 5 (4): 534–561.

Kay, Cristobal. 2002. "Chile's Neoliberal Agrarian Transformation and the Peasantry." *Journal of Agrarian Change* 2 (4): 464–501.

Kingfisher, Catherine, and Jeff Maskovsky. 2008. "The Limits of Neoliberalism." *Critique of Anthropology* 28 (2): 115–126.

Klooster, Dan. 2010. "Standardizing Sustainable Development? The Forest Stewardship Council's Plantation Policy Review Process as Neoliberal Environmental Governance." *Geoforum* 41 (1): 117–129.

Larner, Wendy. 2000. "Neoliberalism: Policy, Ideology, Governmentality." *Studies in Political Economy* 63 (Autumn): 5–25.

______. 2003. "Neoliberalism?" *Environment and Planning D: Society and Space* 21: 509–512.

Laurie, Nina, and Simon Marvin. 1999. "Globalisation, Neoliberalism, and Negotiated Development in the Andes: Water Projects and Regional Identity in Cochabamba, Bolivia." *Environment and Planning A* 31 (8): 1401–1415.

Leitner, Helga, Jamie Peck, and Eric Sheppard, eds. 2007. *Contesting Neoliberalism*. New York: Guilford.

Levine, Arielle. 2007. "Staying Afloat: State Agencies, Local Communities, and International Involvement in Marine Protected Area Management in Zanzibar, Tanzania." *Conservation and Society* 5 (4): 562–587.

Liverman, Diana M., and Silvina Vilas. 2006. "Neoliberalism and the Environment in Latin America." *Annual Review of Environment and Resources* 31: 327–363.

Lockwood, Michael, and Julie Davidson. 2009. "Environmental Governance and the Hybrid Regime of Australian Natural Resource Management." *Geoforum* 41 (3): 388–398.

Loftus, Alexander, and David McDonald. 2001. "Of Liquid Dreams: A Political Ecology of Water Privatization in Buenos Aires." *Environment & Urbanization* 13 (2): 179–199.

Lovell, Heather, Harriet Bulkeley, and Diana Liverman. 2009. "Carbon Offsetting: Sustaining Consumption?" *Environment and Planning A* 41 (10): 2357–2379.

Mansfield, Becky. 2004a. "Rules of Privatization: Contradictions of Neoliberal Regulation of North Pacific Fisheries." *Annals of the Association of American Geographers* 94 (3): 565–584.

______. 2004b. "Neoliberalism in the Oceans: 'Rationalization', Property Rights and the Commons Question." *Geoforum* 35 (3): 313–326.

______. 2006. "Assessing Market-Based Environmental Policy Using a Case Study of North Pacific Fisheries." *Global Environmental Change* 16 (1): 29–39.

______. 2007a. "Articulation between Neoliberal and State-Oriented Environmental Regulation: Fisheries Privatization and Endangered Species Protection." *Environment and Planning A* 39 (12): 1926–1942.

______. 2007b. "Property, Markets, and Dispossession: The Western Alaska Community Development Quota as Neoliberalism, Social Justice, Both, and Neither." *Antipode* 39 (3): 479–499.

______, ed. 2008. *Privatization: Property and the Remaking of Society-Nature Relations*. Oxford: Wiley-Blackwell. (First published as *Antipode* 39 (3).)

McCarthy, James. 2004. "Privatizing Conditions of Production: Trade Agreements as Neoliberal Environmental Governance." *Geoforum* 35 (3): 327–341.

______. 2006. "Neoliberalism and the Politics of Alternatives." *Annals of the Association of American Geographers* 96 (1): 84–104.

Mirowski, Philip, and Dieter Plehwe eds. 2008. *The Road from Mont Pèlerin: The Making of the Neoliberal Thought Collective*. Cambridge, MA: Harvard University Press.

Morris, Amy. 2008. "Easing Conservation? Conservation Easements, Public Accountability and Neoliberalism." *Geoforum* 39 (3): 1215–1227.

Mudge, Stephanie. 2008. "What Is Neoliberalism?" *Socio-Economic Review* 6 (4): 703–731.

Murray, Warwick. 2002. "The Neoliberal Inheritance: Agrarian Policy and Rural Differentiation in Democratic Chile." *Bulletin of Latin American Research* 21 (3): 425–441.

Mutersbaugh, Tad. 2003. "Ethical Trade and Certified Organic Coffee: The Implications of Agricultural Product Certification for Mexican Producer Households and Villages." *Transnational Law and Contemporary Problems* 12 (1): 89–107.

______. 2005. "Fighting Standards with Standards: Harmonization, Rents, and Social Accountability in Certified Agrofood Networks." *Environment and Planning A* 37 (11): 2033–2051.

Nickson, Andrew, and Claudia Vargas. 2002. "The Limitations of Water Regulation." *Bulletin of Latin American Research* 21 (1): 128–149.

Pearson, Thomas. 2009. "On the Trail of Living Modified Organisms: Environmentalism within and against Neoliberal Order." *Cultural Anthropology* 24 (4): 712–745.

Peck, Jamie. 2006. "Countering Neoliberalism." *Urban Geography* 27 (8): 729–733.

______. 2008. "Remaking Laissez-Faire." *Progress in Human Geography* 32 (1): 3–43.

______. 2010. *Constructions of Neoliberal Reason*. Oxford: Oxford University Press.

Peck, Jamie, and Adam Tickell. 2002. "Neoliberalizing Space." *Antipode* 34 (3): 380–404.

Perkins, Harold. 2009. "Out of the (Green) Shadow? Neoliberal Hegemony through the Market Logic of Shared Urban Governance." *Geoforum* 28 (3): 395–405.

Perreault, Thomas. 2005. "State Restructuring and the Scale Politics of Rural Water Governance in Bolivia." *Environment and Planning A* 37 (2): 263–284.

______. 2006. "From the *Guerra del Agua* to the *Guerra del Gas*: Resource Governance, Neoliberalism and Popular Protest in Bolivia." *Antipode* 38 (1): 150–172.

______. 2008. "Custom and Contradiction: Rural Water Governance and the Politics of *Usos y Costrumbres* in Bolivia's Irrigators' Movement." *Annals of the Association of American Geographers* 98 (4): 834–854.

Polanyi, Karl. 1944. *The Great Transformation*. New York: Farrar & Rinehart.

Potter, Clive. 2006. "Competing Narratives for the Future of EU Agriculture." *Geographical Journal* 172 (3): 190–196.

Pritchard, Bill. 2005a. "Implementing and Maintaining Neoliberal Agriculture in Australia—Part I." *Journal of the Sociology of Agriculture* 13 (1): 1–12.

______. 2005b. "Implementing and Maintaining Neoliberal Agriculture in Australia —Part II." *Journal of the Sociology of Agriculture* 13 (2): 1–14.

Prudham, Scott. 2004. "Poisoning the Well: Neoliberalism and the Contamination of Municipal Water in Walkerton, Ontario." *Geoforum* 35 (3): 343–359.

______. 2007. "The Fictions of Autonomous Invention: Accumulation by Dispossession, Commodification and life patents in Canada." *Antipode* 39 (3): 406–429.

Prudham, Scott, and Angela Morris. 2006. "Making the Market 'Safe' for GM Foods." *Studies in Political Economy* 78 (Autumn): 145–175.

Robbins, Paul, and April Luginbuhl. 2005. "The Last Enclosure." *Capitalism, Nature, Socialism* 16 (1): 45–61.

Roberts, Adrienne. 2008. "Privatizing Social Reproduction: The Primitive Accumulation of Water in an Era of Neoliberalism." *Antipode* 40 (4): 535–560.
Robertson, Morgan. 2004. "The Neoliberalization of Ecosystem Services." *Geoforum* 35 (3): 361–374.
______. 2006. "The Nature That Capital Can See: Science, State and Market in the Commodification of Ecosystem Services." *Environment and Planning D: Society and Space* 24 (3): 367–387.
______. 2007. "Discovering Price in All the Wrong Places: The Work of Commodity Definition and Price under Neoliberal Environmental Policy." *Antipode* 39 (3): 500–526.
Rodrigues, Maria G. M. 2003. "Privatization and Socioenvironmental Conditions in Brazil's Amazonia: Political Challenges to Neoliberal Principles." *Journal of Environment & Development* 12 (2): 205–238.
Roff, Robin J. 2008. "Preempting to Nothing: Neoliberalism and the Fight to De/Re-regulate Agricultural Biotechnology." *Geoforum* 39 (6): 1423–1438.
Sangameswaran, Priya. 2008. "Neoliberalism and Water Reforms in Western India." *Geoforum* 40 (2): 228–238.
Sayer, Andrew. 1995. *Radical Political Economy*. Oxford: Blackwell.
Smith, Laila. 2004. "The Murky Waters of the Second Wave of Neoliberalism: Corporatization as a Service Delivery Model in Cape Town." *Geoforum* 35 (3): 375–393.
St. Martin, Kevin. 2006. "The Impact of 'Community' on Fisheries Management in the US Northeast." *Geoforum* 37 (2): 169–184.
______. 2007. "The Difference That Class Makes: Neoliberalization and Non-capitalism in the Fishing Industry of New England." *Antipode* 39 (3): 527–549.
Sugden, Fraser. 2009. "Neoliberalism, Markets and Class Structures in the Nepali Lowlands." *Geoforum* 40 (5): 634–644.
Sullivan, Sian. 2006. "The Elephant in the Room? Problematizing 'New' (Neoliberal) Biodiversity Conservation." *Forum for Development Studies* 33 (1): 105–135.
Swyngedouw, Erik. 2005. "Dispossessing H_2O: The Contested Terrain of Water Privatization." *Capitalism, Nature, Socialism* 16 (1): 81–98.
Tickell, Adam, and Jamie Peck. 1995. "Social Regulation after Fordism: Regulation Theory, Neo-liberalism and the Global-Local Nexus." *Economy and Society* 24 (3): 357–386.
Valdivia, Gabriela. 2005. "On Indigeneity, Change, and Representation in the Northeastern Ecuadorian Amazon." *Environment and Planning A* 37 (2): 285–302.
Watkins, Susan. 2010. "Blue Labour?" *New Left Review* 63 (May–June): 5–15.
West, Paige. 2010. "Making the Market: Specialty Coffee, Generational Pitches, and Papua New Guinea." *Antipode* 42 (3): 690–718.
Wilder, Margaret, and Patricia R. Lankao. 2006. "Paradoxes of Decentralization: Water Reform and Social Implications in Mexico." *World Development* 34 (11): 1977–1995.
Wilshusen, Peter. 2010. "The Receiving End of Reform: Everyday Responses to Neoliberalisation in Southeastern Mexico." *Antipode* 42 (3): 767–798.
Wolford, Wendy. 2005. "Agrarian Moral Economies and Neoliberalism in Brazil: Competing Worldviews and the State in the Struggle for Land." *Environment and Planning A* 37 (2): 241–261.
_____. 2007. "Land Reform in the Time of Neoliberalism." *Antipode* 39 (3): 550–570.

Neoliberal Water Management

Trends, Limitations, Reformulations

Kathryn Furlong

ABSTRACT: The impact of neoliberal policy reform on water management has been a topic of significant debate since the mid-1980s. On one side, a number of organizations have generated an abundant literature in support of neoliberal reforms to solve a range of water governance challenges. To improve water efficiency, allocation, and management, supporters have advocated the introduction and/or strengthening of market mechanisms, private sector ownership and operation, and business-like administration. Other individuals and groups have responded critically to the prescribed reforms, which rarely delivered the predicted results or became fully actualized. This article endeavors to articulate the varying sets of claims, to analyze the trends, to test them against their forecasted benefits, and to examine certain prominent proposals for reforming the reforms. The water sector experience with neoliberalization reveals several sets of contradictions within the neoliberal program, and these are discussed in the final section of the article.

KEYWORDS: development, governance, neoliberalism, privatization, reform, regulation, water supply

In 1999, the Bolivian government granted a 40-year concession for water supply in the city of Cochabamba to Aguas del Tunari, an international consortium led by Bechtel. The concession encompassed all water sources within the catchment, including the aquifer and collectively built wells. In some areas, prices skyrocketed by as much as 200 percent, and in January 2000, thousands of people took to the streets in protest. One died in the standoff, and a 90-day state of emergency was declared. Company officials fled the city, and on 9 April the government canceled the contract (Finnegan 2002; Marvin and Laurie 1999; Perreault 2006). Water supply was brought back under public management, yet the poor have seen little improvement in service levels. In addition, they continue to pay as much as 10 times the fee paid by those who are connected to the municipal system (K. Bakker 2008).

For many concerned with water sector reform, privatization is the flagship neoliberal strategy, making Cochabamba an emblematic city and its 'Guerra del Agua' a watershed moment. The city's story highlights key issues: privatization, profit-driven exclusion, repression, violence, and resistance. Neoliberalization, however, presents a breadth of (often subtler) initiatives for

Environment and Society: Advances in Research 1 (2010): 46–75 © Berghahn Books
doi:10.3167/ares.2010.010103

water sector reform (K. Bakker 2003a). As such, the focus on privatization, while understandable, can obscure a range of issues that are key to redressing the continued exclusion of one-third of the world's population from safe and reliable water and sanitation services (Budds and McGranahan 2003; Wolff and Hallstein 2005).[1] Moreover, given its urban focus, privatization presents no solutions to the most immediate and intractable problems facing water and sanitation, which are often concentrated in rural and peri-urban areas.[2]

Briefly, since the 1970s (and escalating in the 1990s), water management has undergone significant philosophical, institutional, and organizational changes in many parts of the world. Examples include the introduction of private sector principles into public sector management; the involvement of new actors in governance; the extension of private property rights; liberalization, deregulation, and reregulation; new tariff structures; the creation of water markets; a greater focus on the environment and conservation; private sector partnerships (PSP); and decentralization. These initiatives have had impacts on piped water supply, raw water, watershed management, and transnational water politics. In this article, I focus on municipal supply, my area of expertise, delving only briefly into other aspects of water management.

In the next section of the article, I analyze the arguments and strategies for neoliberalization in the water sector. In the following section, I examine the trends that emerged with respect to the deployment of the central strategies. I then assess the degree to which the reforms yielded their forecasted benefits. Research conducted by both proponents and opponents indicates that many of the predicted gains were not realized. As a result, new proposals have emerged to reform the reforms, four of which are examined in the penultimate section. The article concludes with a discussion about the contradictions that pervade the neoliberal program and their implications for its analysis.

Why Neoliberalize the Water Supply? Why Not?

What Is Meant by Neoliberalization?

In the 1970s, a potent neoliberal discourse gained momentum in response to the fiscal crises facing many Western states. This discourse attributed the 'fiscal crisis of the state' to excessive spending, bloated government, and the disproportionate power of labor (see O'Connor 1973; Workman 1996). The prescription was to promote fiscal austerity and to reduce the public sector bureaucracy, regulatory 'red tape', barriers for business, government involvement in service delivery, and the power of unions (see, e.g., Harris 1994; World Bank 1981).[3]

Since then, neoliberalism has come to represent both a diverse set of policy reforms and an array of distinct experiences and applications. It "has many authors, and many birthplaces" (Peck 2007: 4) and is universalized neither in its origins nor its applications (Larner 2000). That said, the Thatcher and Reagan governments of the 1980s most strongly exemplify the ideology that underpins neoliberalism (Harvey 2005; Peck 2007). They represent an assertive period of neoliberalism, which has since been reconstituted in a variety of ways across space (Brenner and Theodore 2002; Larner 2003; Peck 2001a).

In general, neoliberalization constitutes a series of related institutional and organizational processes that are aimed at creating new—and expanding existing—spaces for the accumulation of capital (Block 1977; Duménil and Lévy 2004). Harvey (2003: 45) calls this "accumulation by dispossession," whereby wealth is concentrated in the hands of the few through the disenfranchisement of the many. Common techniques include deregulation, privatization, liberalization, the rise of market-based regulatory mechanisms, and the creation of markets for what were recently conceived as public goods (Harvey 2005; Kolko 1988).

In response to the fiscal crises, governments in the global North moved (to various degrees) to cut social spending, to shift from state-based to market-based regulatory mechanisms, and to open their markets to liberalized trade (Harvey 2005). In the global South, reforms were pushed through by international lender organizations such as the World Bank, the International Monetary Fund, the Inter-American Development Bank, and the Organisation for Economic Co-operation and Development (OECD) (Peet 2003; Stiglitz 2002). Neoliberal restructuring of state institutions became a condition for lending, first through structural adjustment programs (SAPs) and subsequently through 'good governance' approaches (Gore 2000; Hout 2007). SAPs called for significant cuts in social spending, reduced regulation, and economic liberalization to attract foreign capital (Mohan et al. 2000). Good governance would focus on creating and strengthening the necessary institutions (e.g., predictability and accountability) for "the conduct of public and private business" (World Bank discussion paper, cited in Hout 2007: 25).

Rather than the 'hollowing out of the state' through reduced taxation and regulatory power, the state is said to become 'differently powerful' through a variety of mechanisms to ensure the enrollment of labor and citizens (Gamble 1988; Peck 2001b). Peck and Tickell (2002) refer to this process as 'roll-out' neoliberalization (following 'roll-back'). It involves the deployment of new regulations to enhance the power of the state in terms of surveillance and control, often of vulnerable groups (Aguiar 2006; I. Bakker 2003; Peck 2001b), and to solidify and enhance the neoliberal project (Lemke 2001).

In the water sector, the roll-out or reregulatory phase of neoliberalization speaks to another related process. In order to preserve the neoliberal order, the state must also restrain it. The processes of 'reregulation', following those of 'deregulation', have included measures to reassert the state social welfare function to redress some of the deleterious effects of earlier neoliberal restructuring (K. Bakker 2003b). Such trends in other sectors have been interpreted as "slow reversals" of neoliberalism (Harvey 2005: 87), a reconsideration of tactics (Jessop 2002: 470), and a new approach that combines neoliberalism with social investment (Larner and Butler 2005). Such reregulation reflects Polanyi's (1944) 'double movement' thesis, whereby the state must regulate anew to redress the deleterious effects of earlier rounds of liberalization.

Applying the Ideas to Water

Just as neoliberal responses to the fiscal crisis of the state were transposed to the global South, so too were they transposed across many sectors. In environmental governance, theories advancing neoliberal reforms have secured substantial influence (Bernstein 2001). The approach variously termed 'market environmentalism', 'liberal environmentalism', and 'ecological modernization' (see Bernstein 2001; Hajer 1995) has been described by Dryzek (1997: 102) as notable for its "hostility to environmental management on the part of government administrators—except … in establishing the basic parameters of designed markets."

It is through market environmentalism that neoliberal ideas have gained prominence in water management. Many highlight the inclusion of water as an economic good in the Dublin Principles—adopted by the United Nations at the 1992 International Conference on Water and the Environment—as a starting point for the reformulation of water along neoliberal lines (Figueres, Rockstrom, and Tortajada 2003; Rogers et al. 2002; World Bank 2004). This and other recent water reforms, however, cannot be severed from the longer historical evolutions of both water management and neoliberalism. Rather than being wholly original, they form part of a broader set of ideas and trends. As Budds and McGranahan (2003: 112) assert, instead of being "grounded in experiences from the water and sanitation sector," water reforms are driven by "international political changes and policy shifts … from the 1970s onwards." More generally, for global water governance, Conca (2006) shows how certain dominant neoliberal ideas came to be articulated in the formulation of transnational water management dictums.

Neoliberal ideas were responding to real challenges. Widespread problems in water management had gained attention by the mid-1970s. In many industrialized countries, water infrastructure suffered from underinvestment and disrepair (Gandy 1997; Kaïka 2005), water quality from insufficient treatment of municipal waste and industrial pollution (Hassan 1998), and utility capacity from rising water demand (Winpenny 1994). Simultaneously, the social and environmental effects of large water projects—including staples of the 'hydraulic age'[4] such as dams and diversions—were coming under question (Conca 2006). In the European Union and Canada, combinations of these issues led to new legislation for water protection and allocation (de Loë and Kreutzwiser 2007; Kaïka 2003).

If, in the global North, the eroding myth was that of the 'technical sublime', that is, the triumph of urban-networked infrastructure over nature (Kaïka and Swyngedouw 2000), in the global South the eroding myth was that of 'technical universalism' (Coutard 2008; McGranahan et al. 2001). In what Bond et al. (2002: 271) would come to call "droughts for the poor and floods for the rich," vast tracts of the urban population in the global South were without access to public water and sewerage networks. In rural and peri-urban areas, the numbers were even more alarming (WHO and UNICEF 2008). Research highlighted the fact that governments failed to extend services to the poor, who often paid higher fees for lower-quality water obtained through a mix of sources that included private vendors and raw water (K. Bakker 2003a; Swyngedouw 2004). In many places, the situation was found to be particularly burdensome for women and girls, who were often responsible for collecting and purifying water (Bennett et al. 2005).

In the case of water management, neoliberal ideas again offered seemingly straightforward solutions to complex, protracted, and geographically diverse problems. The proposed solutions followed the discourses of market environmentalism, in which the development of markets and market-simulating mechanisms were presented as the most effective means of realizing environmental protection (K. Bakker 2003b). Examples included the market allocation of scarce resources, the establishment of private property rights, and the inclusion of new actors in resource governance (Anderson 1982; Weidner 2002).

Responsibility for the challenges described above was attributed to 'state failure' (Parker and Kirkpatrick 2005). From the neoliberal perspective, water is an increasingly scarce resource that cannot meet demand projections under current management practices, a problem driven by its historic and widespread undervaluation (Winpenny 1994). Market-based, full cost pricing would engender greater efficiency, appropriate allocation to the highest-value uses, and enable utilities to recover their costs and reinvest in infrastructure (Anderson and Leal 1988; Lee 1999).

Coincident with this logic is the neoliberal argument that the regulation of water would be better orchestrated through the development of markets and the injection of private sector discipline. Private sector management of water supply would reduce costs, opportunistic management, and 'regulatory capture' (i.e., situations in which state regulatory agencies advance commercial or special interests instead of the public interest), while increasing investment, transparency, and efficiency (Lee and Jouravlev 1997). This is because public sector management and ownership are said to suffer from diffuse property rights, politicization of resource and infrastructure management, and regulatory capture (ibid.). The issues as defined by proponents of neoliberal reforms, along with the counter-arguments of those critical of such programs, are outlined in table 1.

Those favoring public sector management argue that (1) the profit motive of private corporations leads to higher, not lower, prices, further marginalizing the poor; (2) privatization is insufficient in the absence of strong regulatory institutions; (3) the environment will suffer, due to increased incentives for externalities; (4) competition is limited to the contract, as

Table 1: Neoliberal policy prescriptions for water reform—reasonings and critiques

Issue	Policy Presscription	Reasoning	Critique
Water scarcity/ allocation	Private property rights	Create the proper incentives for management and cost control while developing the necessary conditions for markets to evolve	Enclosure of the 'commons' (communal resources), resulting in marginalized access and the discounting of environmental uses
	Creation of water markets	This scarce resource will be allocated to its most valuable uses, avoiding the politicization of water resource development	Vital resources, which are essential for life, cannot be left to the vagaries of the market
Water scarcity/ underinvestment in infrastructure	Full cost pricing	Will lead to improved allocation and reduced wastage, as well as improved infrastructure and service delivery	Will reduce access by low-income users and may be unrealistic, given the investments required
	Reduction of subsidies	Subsidies generally benefit industrial and wealthy customers at the expense of the poor; reducing subsidies will send proper price signals and improve resource allocation	Detrimental to low-income users, who (in less developed countries) have heretofore subsidized the services of those with service connections
Poorly performing water utilities	Public sector: from provider to regulator	Avoidance of 'regulatory capture', whereby the provider regulates itself, leading to better regulation and utility performance	Better regulation is necessary, but only governments can ensure that the social and environmental goals of water provision are met
	Private sector management	Reduction in government interference and the politicization of water management; introduction of competition, leading to greater efficiency, infrastructure investment, and cost reduction	Profit motive will lead to higher prices and the exclusion of the most vulnerable users; sufficient institutions are lacking to regulate the private sector; water is a 'natural monopoly' for which competition is limited

water constitutes a 'natural monopoly' and only a few international firms control the market;[5] (5) the state can access cheaper financing than the private sector; and (6) water is a human right as opposed to a commodity (Barlow and Clarke 2002; Hall and Lobina 2007; Lobina and Hall 2008; Shiva 2002). Others simply highlight the fact that, in many cases, too little was known to embark on such "wholesale national restructuring of a sector's institutions" without prior experimentation (Cairncross 1987: 182).

General Trends: Key Targets of Reform

Overview

Commonly identified trends in water supply reform include the introduction of economic principles into various types of management, an increased focus on the environmental implications of water resource development, and a realignment of how governments, utilities, and a variety of non-state actors engage in the production and consumption of the resource (Conca 2006; Kallis and de Groot 2003a; Sauri and del Moral 2001). Research has also found that these currents are not always compatible and can manifest themselves differently across contexts (Jones 1998; Kallis and de Groot 2003b; Saleth and Dinar 2000). Trends underscored in the literature are summarized in table 2. While the analyses identify many common trends, they also show overlap and conflation of key processes. For example, marketization is defined as including deregulation, and decentralization and privatization are grouped as a single trend. Such conflation and overlap are common and inhibit effective comparison among instances of market environmentalism (K. Bakker 2007, 2009).

Table 2: Trends identified as a result of neoliberalization in the water sector

Study	Kallis and de Groot (2003a)	Jones (1998)	Saleth and Dinar (2000)
Region covered	European Union	Organisation for Economic Co-operation and Development (OECD)	Mexico, Chile, Brazil, Spain, Morocco, Israel, South Africa, Sri Lanka, Australia, China, India
Trends identified	• Naturalization: growth in the environmental focus of management • Marketization: array of processes involving, for example, deregulation and PSP • Rationalization: integration of environmental, social, and economic goals into planning	• Consolidation of water utilities into larger units • Decentralized management • Government shift from provider to regulator with an increase in PSP • Greater reliance on tariffs to achieve a broad range of social, environmental, and economic goals	• Importance of market-based allocation • Decentralization combined with privatization • Integrated water resources management: growth in national water plans and coordination of water files across government departments • Economic viability and physical sustainability: cost recovery, water quality standards, and pollution regulation

But such conflation is also revealing. First, it points to ambiguity within the processes themselves, whose deployment has not been homogeneous and has involved a range of experiences, levels of implementation, and reformulations (K. Bakker 2005). Second, it suggests a lack of distinction within neoliberalism when identifying policies to deal with the environment, social welfare, and economic stability. Instead, it ventures to address all issues with similar policy mechanisms. Finally, the conflation implies an internal avoidance of specificity, with many programs (such as deregulation and privatization) intended to reinforce one another. As such, rather than simply seeking to achieve particular goals external to the program (e.g., improved service provision), neoliberal reforms serve to reinforce neoliberalism itself (Lemke 2001: 203; Peck 2001a).

Specificity, however, is essential in order to provide an account of the relationship between neoliberalization and water management. Karen Bakker's (2007) organization of the processes involved in the neoliberalization of resource management provides the most explicit framing available in the literature. Arguing that neoliberal reforms should be understood in terms of targets and types of reform that can be articulated distinctly or as part of the same process, Bakker (ibid.: 435, table 1) divides these targets and types of reform into three groups—those pertaining to resource management institutions, resource management organizations, and resource governance. Table 3 presents a modified version of Bakker's table, adding in particular a summary of unfolding trends. The table gives an indication of the possible range of experiences that have

Table 3: Specifying neoliberal reforms with respect to water resources

Resource Area	Target	Type of Reform	Trend
Institutions (rules, norms, laws)	Property rights	Privatization (exclusive water rights, asset sale)	Peaked in the 1990s, increasingly less common
	Trade rules	Liberalization (GATS, NAFTA)	Controversial; some reregulation to avoid bulk water exports
	Legislation and regulations	Deregulation (removal of legislative 'barriers' to private sector entry, including 'costly' environmental legislation)	Trend seen in the 1980s and early 1990s; identified with SAPs in the global South
	Legislation and regulations	Reregulation (new regulations in response to neoliberal reforms)	Evident especially in the late 1990s in Western countries
Organizations (management entities)	Asset management	Private sector partnerships	Decreasing after peak in the late 1990s
	Organizational structure	Corporatization and alternative service delivery	Increasing in some countries (e.g., Canada)
Governance (decision-making practices and actors)	Resource allocation	Introduction of markets and commercialization (e.g., distance-based or peak-use pricing reforms)	Development of new and different types of markets. Pricing reforms are widespread but subject to a variety of challenges.
	Performance incentives/ sanctions	New public management (e.g., cost-benefit analysis, benchmarking). Commercialization in terms of full cost recovery.	Increasing importance in the 1990s
	Participation (users, intermediaries)	Devolution or decentralization	Increasing with different types of actors in new roles

Source: Table modified from Bakker (2007: 435, table 1).

evolved across different contexts. Following Bakker's categorization, trends with respect to the neoliberal reform of water resource institutions, organizations, and governance are discussed in turn in the following sections.

Reforming Institutions—Deregulation, Reregulation, Liberalization

In terms of institutions, water sector reform was largely directed at opening up new spaces for markets and the private sector by reworking regulations and the function of the state. In general, the neoliberal discourse vis-à-vis the state was that its appropriate role lay in regulation, whereas the private sector ought to deliver services (Lee 1999; Osborne and Gaebler 1992; World Bank 1993). Interestingly, this would frequently involve rolling back environmental regulations and market protections as a precursor to private sector engagement and investment (McCarthy and Prudham 2004). In the global South, international financial institutions (IFIs) demanded certain reforms as conditions for loans (Koeberle et al. 2005). In Africa, reforms required liberalization (the opening of domestic markets) and the privatization of state assets (Ferguson 2006). In many Western countries, reforms often involved a weakening of environmental and labor protections to render the public sector more attractive to PSP.[6]

Despite the emphasis on the regulatory role of governments, rarely did the development of the new institutions and mechanisms precede PSP or the rolling back of existing regulations. In England and Wales, for example, the reregulation of the water industry arrived several years after privatization in response to extreme price hikes, ensuing water poverty, and the return of cholera to the island (K. Bakker 2003b). The rapidity with which the regulatory regime was established has been found to have engendered several shortcomings that have proven difficult to correct (Mumssen and Williamson 2002: 39, 60). In Ontario, Canada, reregulation followed a water quality disaster in which 7 people died and more than 1,000 others became seriously ill—some permanently (Prudham 2004; Snider 2004). In the global South, researchers argue that the policies themselves presupposed certain institutions and social relations that were not strongly present, would take decades to develop, and were in fact weakened by the decline in wealth and state authority following structural adjustment (Castro 2006, 2007; Dávila-Poblete and Rico 2005; Prasad 2006: 684).

Perhaps, given the lack of coordination between these two processes, deregulation and reregulation can prove contradictory, with the former inhibiting the implementation of the latter. In British Columbia, Canada, for instance, Patrick (2009) has found that earlier rounds of deregulation have made it difficult for municipalities to implement the province's recent legislation that protects source water. On another level, international trade agreements such as NAFTA and GATS have allowed foreign companies to challenge the domestic strengthening of environmental laws to protect ground water quality when it has involved banning a polluting substance (see McCarthy 2004; Warner and Gerbasi 2004: 864–866).

Regulatory amendments have also been used to impose changes at the level of organizations and governance. In order to make entry into the water sector more attractive to private corporations, utilities were compelled to cut costs, to reduce expenditures (Hassan 1998; Mumssen and Williamson 2002: 45–46), and to operate (and be evaluated) according to business principles, such as cost-benefit analysis (Schwartz 2008). These reforms can create difficulties in meeting the costs of improved environmental regulation (Jones 1998) and can face problems where existing institutional capacity is weak (Schwartz 2008). In Latin America, Castro (2008) expects such reforms to persist (irrespective of the introduction of PSP) and to have long-term social consequences.

Reforming Organizations—Finding a Niche for the Private Sector

The private sector has participated in public service delivery (including water supply) in a myriad of forms throughout the post-war period—that is, the participation of the private sector has not been simply a feature of neoliberalism (Salamon 2002).[7] What is new is the idea that PSP presents a universal solution to all challenges (Gleick et al. 2002; Prasad 2006). There are many means by which the private sector becomes involved in public service delivery (e.g., contracting, grants). With neoliberalization, the privatization of services (full divestiture, including asset sale) and various types of private sector contracting to operate, maintain, and/or build infrastructure were increasingly encouraged by IFIs in low-income countries and by both government and private initiatives in the OECD (Lorrain 1995).[8]

Privatization and PSP contracts have followed certain geographical trends. Notable currents include a concentration first in Latin America in the 1990s and then in East Asia and the Pacific after 2002, as well as a paucity of projects in Africa (see fig. 1) and non-urban areas (Jimenez and Perez-Foguet 2009). According to Budds and McGranahan (2003: 108), the initial concentration in Latin America can be credited to the region's large cities and sizable middle class, the indebtedness of the utilities, and the depth of existing neoliberal reforms. Contract disputes (and cancellations) would also exhibit clustering, notably with respect to large contracts in Latin America and East Asia (Jimenez and Perez-Foguet 2009: 11). As of 2008, close to 60 projects in these regions had been canceled or were under distress, representing 34 percent of investment.[9]

In terms of the type of private sector involvement, there are few examples of full or even partial divesture; rather, concessions were most common until the early 2000s. Since then, 'greenfield projects' (involving the building, operation, and transfer of facilities where none had existed before) have grown in prominence (see fig. 2). However, despite the increasing number of greenfield contracts, the overall level of investment has declined since the late 1990s. In 2008, investment was less than one-third of its 1997 level (fig. 2). The type of private sector partner is also shifting—from international corporations to local and regional companies, especially in China (Marin 2009: 29).

Notwithstanding the growth in the number of PSP projects, PSP remains a relatively minor approach worldwide. Approximately 90 percent of water utilities in the world's 400 largest cities remain publicly owned and operated, while PSP participation in small cities and rural areas is highly uncommon (Hall et al. 2010: 2). Moreover, recent growth in PSP has been driven by an increase of incidence in China (involving Chinese companies), with some new projects also found in North Africa after 2005 (see fig. 1). This uneven geography was likewise a feature of the Latin American experience, with heavy concentration in Argentina (Castro 2008).

Privatization and PSP have not materialized to the promised extent, a conclusion shared by both proponents and opponents of the reform (Budds and McGranahan 2003; Estache 2005; Jimenez and Perez-Foguet 2009; Wolff and Hallstein 2005). The issue continues to present challenges, however, given that (1) the preparatory reforms have initiated "far-reaching transformations with potentially long-term consequences" (Castro 2008: 66), including legal and institutional reforms; (2) even where contracts have been rescinded, difficult challenges remain, with Argentina facing 30 separate lawsuits; and (3) the will to pursue PSP has not lessened (Castro 2008: 66; Prasad 2006). On this last point, various authors, while conceding the shortcomings of PSP, look for other avenues to improve the conditions for investing companies (see, e.g., Estache 2005; Franceys 2008; Lee and Jouravlev 1997).

Figure 1: PSP projects in developing countries, 1991–2008

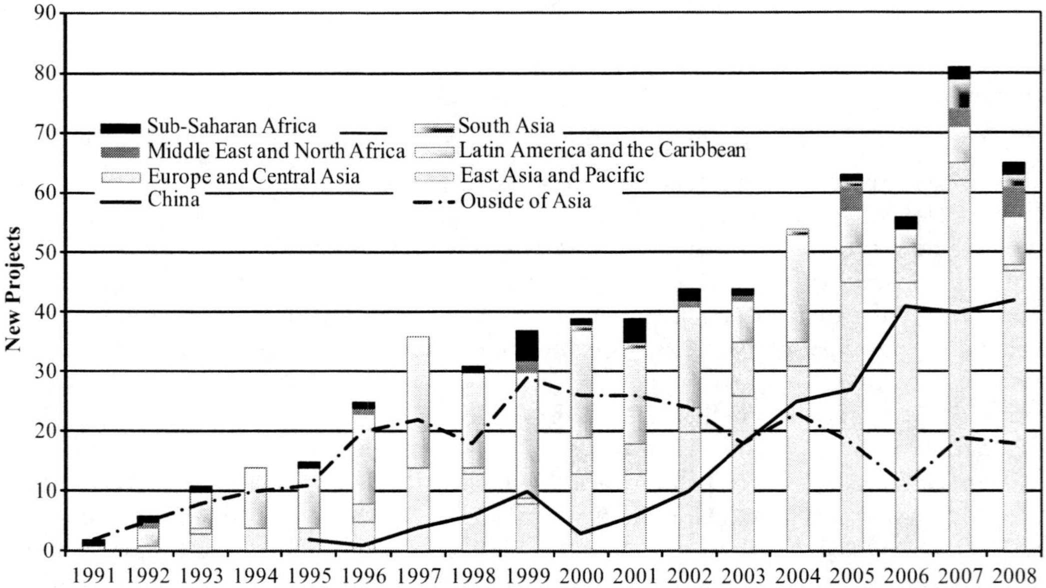

Source: Data taken in 2009 from the Private Participation in Infrastructure Database of the Public-Private Infrastructure Advisory Facility (PPIAF). See http://www.ppiaf.org/ppiaf/page/private-participation-infrastructure-database.

Figure 2: Type of PSP contract, 1991–2008

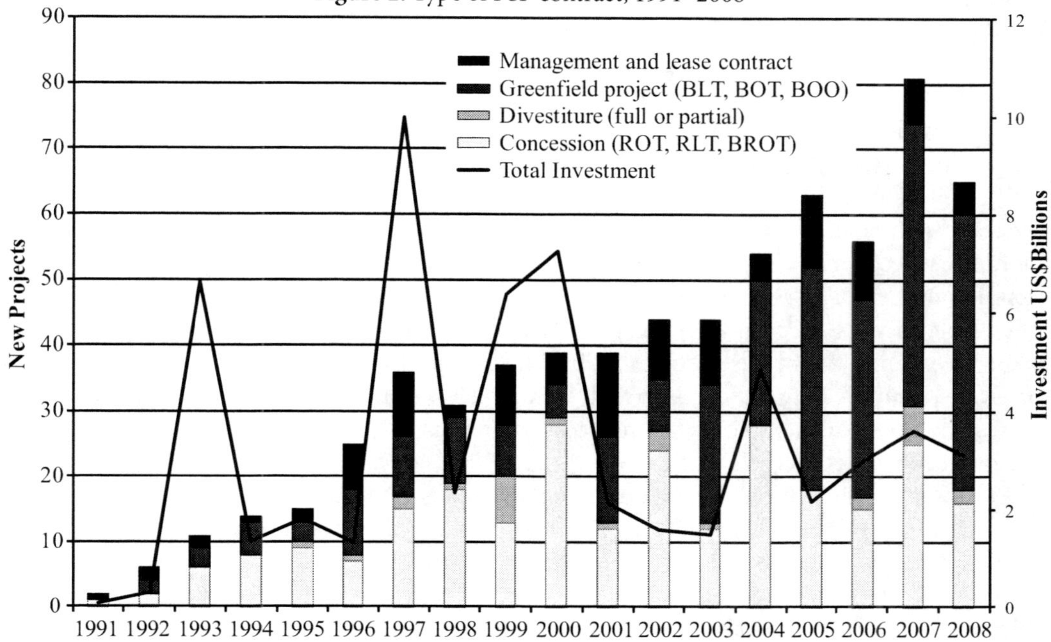

Source: Data taken in 2009 from the Private Participation in Infrastructure Database of the PPIAF. See http://www.ppiaf.org/ppiaf/page/private-participation-infrastructure-database.

Reforming Governance 1—Commercialization, Commodification, Marketization

Market allocation and market-based regulation of resources are among the key tenets of market environmentalism in the water sector. Such practices are often predicated on the introduction of private property rights, new pricing structures, and new technologies (e.g., smart meters) that enable further commercialization. Like PSP, market environmentalism is not a new phenomenon, nor is it uniformly applied. Its prominence has increased under neoliberalization (Gleick et al. 2002), but the program must be adjusted according to the context, and 'results' cannot be guaranteed (Boelens 2009).

The application of commercial principles to water management has a lengthy pedigree. The long-term influence of economic interests in water resource development in California is made patent in Worster's (1985) *Rivers of Empire*. Similarly, the interaction of commodification and decommodification of water supply in Cameroon over the last century demonstrates, according to Page (2005: 293), that "the commodification of public water supplies is not new, permanent or inevitable." With regard to water tariffs in Latin America, Gilbert (2007) argues that although the balance of pricing and subsidization may have been challenged, the idea of charging for water services was never seriously questioned.

Nonetheless, the application of economic tools has not remained static. Several trends in water pricing, subsidization, and allocation have emerged over the neoliberal period. Across the OECD, prices have increased—in some countries by an average of as much as 8 percent per year (Jones 1998). Tariffs are being used to achieve "a wider range of economic, environmental, and social objectives" (ibid.: 640), utilities have moved from fixed to volumetric charging with more complex tariff structures and metering, the principle of full cost recovery is now widespread, costs have increased to meet new regulations, and both the level and transparency of subsidies have been challenged (ibid.). In the case of the global South, subsidies have been questioned on the grounds that they have not benefited—but rather have disadvantaged—the most needy

(Komives et al. 2005) and that the poor are willing to pay for improved services that can be achieved only through better-funded utilities (Rogers et al. 2002).

Despite the widespread drive to treat water as an economic good, the dictum remains controversial and is only partially implemented in practice. The EU Water Framework Directive (WFD) provides an example of the contradictory compromises that can emerge to accommodate varying perspectives (Kaïka 2003: 304–306). The WFD incorporates the user pays principle (UPP), prescribes full cost recovery (including environmental impacts), and advances tools for economic analysis as a basis for water management (DG Environment 2008). Yet it requires only that full cost recovery be 'taken into account', leaves the inclusion of environmental costs to the discretion of member states, and defines water as "not a commercial product like any other but, rather, a heritage" (European Parliament 2000; see also Kaïka 2003: 304–306). From a practical perspective, Jones (1998) has argued that in OECD countries full cost recovery for water and wastewater (and the elimination of subsidy) is unlikely, given the degree of investment needed in infrastructure. For example, following privatization in Catalonia, cross-subsidies that favored residential consumers actually increased (Miralles 2008: 159).

In terms of marketization, several prominent examples of the creation of water markets exist, for example, the Southern United States, Australia, and Spain. However, Chile is often promoted as the model to be replicated (Dinar and Saleth 2004; Lee 1999). A key lesson that can be derived from experiences with water markets is that the extent and nature of the markets developed are subject to mixtures of local conditions and external influences. On the one hand, markets are politically produced. Haughton (2002) finds that foreign governments and IFIs have crucial influence over the creation of new and modified water markets in the global South. Instead of a technocratic depoliticization of water management, the new markets reflect the political outcome of "complex multiagency multiscalar negotiations" (ibid.: 798).

On the other hand, local conditions can constrain the ability to produce the desired market arrangements. In western India, water markets did not emerge in an idealized form; rather, they had to be re-formed through interaction with already established local norms and rules (Dubash 2004). In the Andean countries, the profusion of existing water rights regimes that are long established, numerous, and locally specific has posed a significant challenge to the institution of a universal market system (Boelens 2009). Finally, in the English case, Karen Bakker (2003b) has shown that despite significant effort, the result of water sector reform was full privatization, limited commercialization and unsuccessful commodification.

Reforming Governance 2—New Actors: From Government to Governance

Of significant interest to many researchers of neoliberalism has been the apparent shift from government to governance over the last 25 years. This change entails the decentralization of state authority to lower levels of government and supra-national authorities, as well as the introduction of new actors (including private corporations, NGOs, and the voluntary sector) into the delivery, oversight, and decision-making procedures surrounding the management of public services.[10]

With respect to neoliberal water sector reform, the promotion of participation is among the items that Karen Bakker (2007: 442) finds in common among both proponents and opponents of PSP. The World Bank, for example, calls for improved mechanisms to promote the 'voice' and participation of citizens, including "direct involvement by users, nongovernmental organizations and other groups of citizens" as a necessary element of public sector reform (World Bank 2000: 23). Citizen participation is intended to improve the transparency and responsiveness of organizations, resulting in more effective and locally appropriate services that, in some cases, lead to community ownership and operation (cf. Subramanian et al. 1997).

In practice, while it is widely agreed that a transition from government to governance is occurring, the evenness and effectiveness of the introduction of new actors is variable. The WFD, for example, arguably encodes a shift from government to governance, in particular, through the introduction of water basin councils and a general rise in the influence of unelected bodies, such as private firms, lobby groups, and NGOs (Page and Kaïka 2003). Yet their ability to participate is limited by their size and financial capacity (ibid.). Timothy Moss and colleagues, moreover, find a range of 'intermediaries' working among users, service providers, and regulators, significantly affecting relationships between the physical networks and localities (T. Moss 2009; T. Moss et al. 2009). New actors, however, can have conflicting goals. For example, the private sector can be resistant to regulation and limit incentives for user participation (Kaïka 2003; Page and Bakker 2005).

The rise in prominence of new actors is likewise important in the global South. There, a range of public and private groups have long been involved in water supply (Gulyani et al. 2005; Kooy and Bakker 2008). Since the early 1980s, organizations that include development associations, NGOs, and civil society agencies have increasingly been promoted to strengthen governance and to supplement (or even to replace) the state in service provision (Ferguson 2006). In water management, this has led to the growing involvement of a range of non-governmental, community, and civil society groups in rural and peri-urban areas, as well as the private sector in cities (Cairncross 1987; McFarlane 2008).

Have the Reforms Yielded the Anticipated Results?

Lower Prices and Improved Allocation

A key tenet of market environmentalism is that water is underpriced and thus poorly allocated. Nevertheless, the theory also argues that the introduction of private sector discipline will result in lower prices through increased efficiency and reduced costs—for example, through staffing reductions and competition for contract renewals (Lee and Jouravlev 1997: 52; Shirley 2002). Generally, this claim has not been born out in practice. In the case of England and Wales, Hassan (1998: 194) describes a "spectacular" increase in prices by an average of 70 percent in the initial years following privatization (1989–1994), fueled in part by controversially high salaries within the privatized companies. The price increases led to problems of affordability and user disconnections, which would later be prohibited by law (K. Bakker 2003b; Graham 1997). In Bolivia, similar increases in tariffs following privatization meant that many users were denied their customary water service (Assies 2003; Marvin and Laurie 1999).

Price increases were often among the preparatory reforms enacted within public utilities in order to encourage PSP in the sector. The UK, Greece, and South Africa are examples (Bond 2002; Hassan 1998; Kaïka 2005). In Athens, these preparatory price increases were not rolled back after the anticipated reforms were abandoned, suggesting that privatization was used as a means of increasing prices rather than as a route to their reduction (Kaïka 2005: 160). In South Africa, the focus on cost recovery through the UPP (as opposed to cross-subsidization) has resulted in significant water debt among low-income users (McInnes 2005), as well as cutoffs and reduced water access for the poor (Loftus 2006; Ruiters 2007; Smith and Hanson 2003).

Reduced access due to high prices raises questions with respect to the predicted improvements in water allocation under full cost pricing and the introduction of water markets. Thus, beyond the enhanced agricultural activity that was taken to reflect improved allocation in the Chilean case (Rosegrant and Binswanger 1994), the assessment should also include issues such as improved environmental protection and user access. Such an approach has been taken in Amsterdam. There, practitioners find market-based valuation techniques useful in selecting policy options that may or may not promote the further commercialization of the resource. For example, they may promote nature protection strategies and limit the extension of metering (Dalhuisen et al. 2003).

Effective and Efficient Utilities

Neoliberal reforms were also predicted to yield more effective and efficient water utilities by reducing costs and wastage through improved business practice. Below I examine the results in terms of improved service quality, reduced wastage, and improvements in conservation. With respect to service quality, some PSP projects in West Africa have been found to have generated improvements in (1) service quality and reliability, (2) effectiveness and productivity, and (3) increased revenue and balanced books (Jaglin 2002: 233).[11] Experience in Kenya, on the other hand, has demonstrated that raising prices does not necessarily translate into service improvements; there, even the 'non-poor' are faced with insufficient water access despite increased prices (Gulyani et al. 2005). Similarly, private sector operations in Tanzania are perceived to have a variety of detrimental impacts on users (Kjellén 2006).

In terms of greater efficiency with respect to water produced by utilities and consumed by users, the literature demonstrates that while 'environmental crises' like droughts and other forms of water shortage may encourage neoliberal reforms in the water sector, such reforms will not necessarily improve water efficiency and conservation within the utility. In the case of Athens, the 1989–1990 drought partially enabled a rescripting of water as an economic good, including the restructuring of the rate system to encourage demand management and eventually the partial privatization of the utility (Kaïka 2005: 153–160). The price increases led to an initial drop in demand of 20 percent, yet the drive for conservation later came into conflict with incentives to expand the system under partial privatization (Kallis and Coccossis 2003). Drought likewise yielded unexpected consequences for conservation under privatization in England and Wales. In 1995, Yorkshire experienced a severe drought, but users resisted compliance with conservation demands, reasoning that they were buying water from a private company. Yorkshire Water, beholden to water quantity and price-cap regulations, was forced to tanker water into the region (K. Bakker 2000; Haughton 1998; Osborn and Marvin 2001).

Research in a variety of contexts has shown that public utilities are more likely to engage in water conservation. Gleick et al. (2002: iv) find that conservation programs are generally scaled back or canceled following contracting to the private sector due to the link between revenue and sales. This concurs with findings in California that show public utilities to be more pro-active with regard to conservation than private utilities (Kallis et al. 2010). In the Canadian context, the scope of conservation programming was shown to be reduced under decreasing government leadership (through arm's-length business models) of water utilities (Furlong and Bakker 2010). However, insufficient regulation is the key issue. A public regulator can legitimate and drive conservation whether the utility is public or private (Kallis et al. 2010).

Environmental Benefits

The relatively poor performance of neoliberal reforms on conservation can be contrasted with better performance on the environment more generally. Both environmental and neoliberal initiatives have the potential to drive one another. In the case of England and Wales, privatization of water utilities was both reinforced by and facilitative of new environmental directives at the European level that are credited with the improvement of water quality in the UK. The new European regulations made the transfer of costs to the private sector more attractive, while this transfer of costs made acquiescing to the extension of European law politically feasible (Hassan 1998: 163–166).

Nevertheless, conflicts can arise between the different goals of market environmentalism. In England, improvements to the environment were accompanied by questions of user access. In Karen Bakker's (2003b) terms, environmental externalities prior to privatization were traded for social externalities after privatization. In Buenos Aires, the water quality situation also improved following privatization as a result of upgrades in wastewater treatment. Yet significant price increases elevated tensions between the operator and the regulator (see Idelovitch and Ringskog 1995: 45–46).

International trade agreements, on the other hand, may present issues in terms of bulk water transfer and water quality. The potential for governments to lose authority over bulk water exports under NAFTA, for example, has raised alarm among activists and academics (Barlow and Clarke 2002; Boyd 2003). According to Gleick et al. (2002: 15–16), it is difficult to assess the potential impacts of agreements like GATT, NAFTA, and GATS on bulk water trade. However, should raw water be classified as a product, it would become subject to World Trade Organization rules that would restrict government action once transfers are initiated.[12] Concern over such an eventuality under NAFTA became the catalyst for a ban on bulk water transfers by all Canadian provinces and territories—a rare element of convergence in Canadian water legislation (Hill et al. 2008).

Investment in Infrastructure and Delivery to the Poor

Improved water delivery for the poor was another forecasted outcome of neoliberal reforms. In particular, PSP was expected to enable service extension to non-serviced areas through new efficiencies and cost savings. The private sector was to provide much-needed investment for the improvement and extension of water and sewerage infrastructure (GWP 2000). Since the 1990s, however, the levels of investment in water and sewerage have not increased appreciably. Instead, governments—not the private sector or development assistance—have financed infrastructure improvements, while PSP has concentrated on the more lucrative, as opposed to the most needed, projects (Budds and McGranahan 2003; Estache 2005; Jimenez and Perez-Foguet 2009). Jaglin (2002: 233) makes the observation that issues of social justice and improved services for the poor were presented as indirect consequences of neoliberalization, that is, as would-be by-products of other reforms. Consequently, there was little proper planning for their realization, and the results have been unimpressive.

The experience of private sector involvement in water supply has been geographically segmented, avoiding the poorest areas globally, regionally, and locally. In order to ensure a high probability of return on investment, private corporations must seek service areas where the demand and the ability to pay for water are both high and secure, usually valued at over $100 million (Jimenez and Perez-Foguet 2009: 12). This mandate implies a focus on urban areas. According to Karen Bakker (2003a: 329), the private sector is unlikely to take an interest in conurbations with a population of less than 500,000. Even in urban settings, private operators exhibit reluctance to extend services to poor neighborhoods, generally by defining the least lucrative areas as beyond the city limits (e.g., Cartegena, Colombia, and La Paz, Bolivia) or by excluding 'informal' settlements within the contracted area due to their lack of legal land tenure (e.g., Cordobá and Buenos Aires, Argentina) (Budds and McGranahan 2003: 110).[13]

In terms of PSI in infrastructure, a similar pattern is visible. At its peak, the private sector's contribution was only 5 percent of total investment. This level has declined since the 1990s, with the poorest countries least able to attract investment (see fig. 3). In fact, Africa received only 0.95 percent of private investment, with sub-Saharan Africa (other than South Africa) being particularly neglected (Jimenez and Perez-Foguet 2009: 2, 12). Moreover, the length and controversial nature of PSP contracts have a tendency to reduce actual levels of investment (ibid.: 5). Finally, given the lack of viability of investing in underserviced areas, Prasad (2006: 688) finds that the private sector either 'cherry-picks' the most lucrative service areas or looks to development funds or the renegotiation of contracts in order to extend services to the poor.

Another postulate in support of PSP in the global South is that it would indirectly benefit the poor by freeing up scarce government resources to fund other essential services (de Azevedo 1998, cited in Haughton 2002: 793). The data, however, indicate that the majority of investment in water supply has come from the public sector, even where PSP contracts are found. OECD and World Bank data show that, since its peak in 1997, private sector investment (PSI) in water and sanitation has decreased by more than two-thirds, even though the number of PSP contracts has increased (OECD 2009: 211) (fig. 3). Indeed, Estache (2005) finds that increased PSP is highly correlated with increases in public sector investment and fiscal deficit but not with increases in private investment or GDP. Thus, while he notes some correlation between privatization and increased access and regulation, he concludes that it is the public sector that invests (ibid.: 8–9).

This means a reliance on official development assistance (ODA) and government funds to support infrastructure development. ODA, however, provides barely a fraction of the investment that is needed in developing and middle-income countries (Budds and McGranahan 2003: 100; Jimenez and Perez-Foguet 2009). Moreover, ODA has a tendency to bypass the poorest countries, instead favoring middle-income countries (Prasad 2006: 678). Concurrently, local government contributions (although the most significant) remained essentially static throughout the 1990s, due to donor requirements for reductions in public spending, compounded by a general lack of resources (Jimenez and Perez-Foguet 2009: 1–2). The result is that, in some locations,

Figure 3: Private sector investment compared with new projects in water, 1991–2008

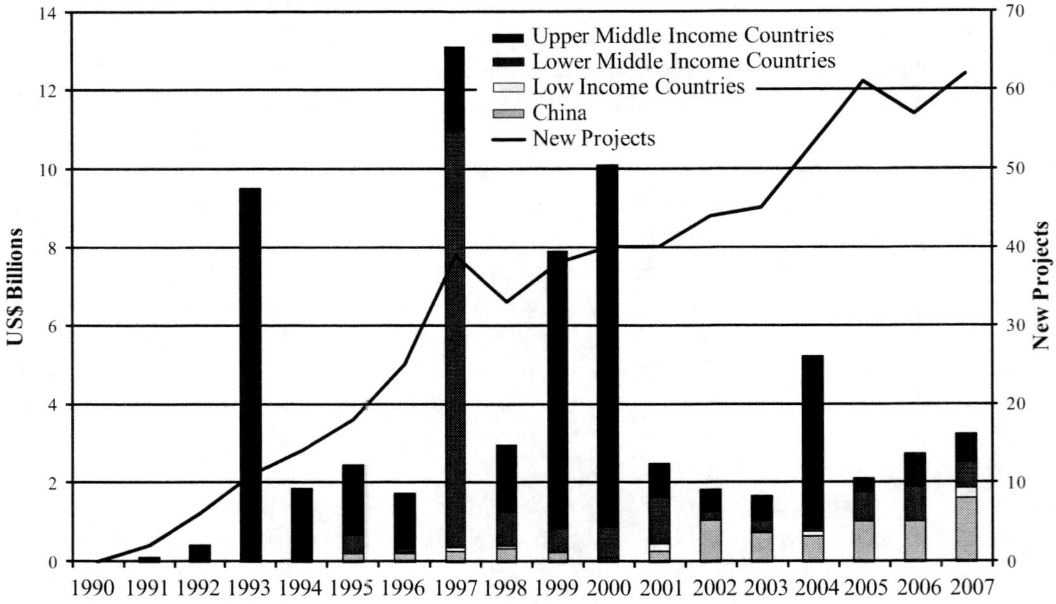

Source: Data taken in 2009 from the Private Participation in Infrastructure Database of the PPIAF. See http://www.ppiaf.org/ppiaf/page/private-participation-infrastructure-database. Note: Low income: GNI per capita < $875; lower middle income: GNI $876–$3,465; upper middle income: GNI $3,466–$10,725.

progress toward the UN's Millennium Development Goals (MDGs) is actually regressing (WHO and UNICEF 2006).[14]

Prominent private sector representatives have themselves argued that the expectations for private investment were unrealistic. The Business and Industry CEO Panel, for example, questioned whether the private sector represented a "realistic solution to underinvestment in water systems" (J. Moss et al. 2003: 12; see also Wolff and Palaniappan 2004). The chief executive of Saur Group, which manages close to 6,000 water contracts, has similarly articulated the limits of PSI, given a premature emphasis on concession contracts, increased risk in developing countries, unrealistic standards, and the inability of users to pay for the required investments without subsidy (Talbot 2002). The World Bank has likewise acknowledged that the risks associated with service to the poor are too large a burden for the private sector (World Bank 2005a, 2005b, cited in Prasad 2006: 676).

Although there is broad consensus that PSI has not worked for poor or low-income countries, there is some debate as to whether OECD countries have themselves seen the expected levels of investment. According to the World Bank (Estache 2005), PSI has worked well for high-income countries. For Castro (2008: 69), however, in instances of high-profile privatization contracts in Europe, the promised investment has not materialized. In the case of Thames Water in England, funding came from revenues (96.4 percent) and debt (3.6 percent), while in Athens it came from revenues (80.6 percent) and subsidies.[15]

New and Reformed Strategies

Adjusting the Program

Given the variety of challenges to producing the desired reforms and achieving the promised results, proponents of neoliberalization have advanced adjustments to the reforms. Examples include the development of 'pro-poor' contracting and regulatory mechanisms (Komives 2002; Trémolet and Hunt 2006), differentiated models for private sector involvement (Crean 2002; Hardoy and Schusterman 2000), the promotion of GATS to stimulate private investment in water and sanitation (Franceys 2008), and the facilitation of water markets through the Internet (Meyers 2002).

Once again these changes in water management reflect broader shifts in neoliberal thinking rather than policies developed specifically for the water sector. One example is the 'new minimalist approach', which focuses on pro-poor programming and the expansion of the informal sector while retaining key neoliberal tenets, such as private property rights, administrative streamlining, the removal of subsidies, and contracting out (Altenburg and Drachenfels 2006). Similarly, the post-Washington Consensus (PWC) retains neoliberal principles while seeking a strong focus on participation so as to avoid negative impacts on the poor (Ruckert 2006). For the water sector, PWC means continued support for privatization with a new sensitivity for effective regulatory institutions, competition, and the mitigation of impacts on the poor through various methods of subsidization (ibid.: 57, 59). This transfer of costs to the public sector is described by Ruckert (2007: 101) as "accumulation by subsidization."

Pro-poor PSP Projects

Pro-poor water and sewerage contracts are an example of the implementation of adjusted neoliberal prescriptions. Experience is widespread, with projects in Latin America, Africa, and Asia.

Three key themes in the pro-poor literature are (1) the promotion of differentiated (cheaper) services for poor households, (2) participation to determine users' willingness to pay (and thus the appropriate level of service), and (3) government subsidy to meet service needs for the poor (Debomy et al. 2005; Komives 2001; Trémolet and Hunt 2006). Where a service area has mixed income and service levels, for example, some prescribe an open PSP model that involves other actors in service delivery, as well as a government fund to support services for the poor (Hardoy and Schusterman 2000). While these moves appear to establish a form of cherry-picking, the diversity of PSP approaches that have emerged (including pro-poor) also demonstrates the limitations of neoliberal water sector reform, with the English model as "the exception rather than the rule" (Haughton 2002: 804).

Among the early examples of pro-poor PSP projects is that of La Paz-El Alto, Bolivia. There, the government designed a contract specifically aimed at expanding an equal level of service to unconnected residents (Komives 2002). The contract stipulated the tariffs and connection fees, the type of connection (in-house), and the minimum schedule for network expansion. The contract was awarded to Aguas del Illimani in 1997. Despite positive results, including timely network extension in the first two years (Komives 2001), in 2003 residents engaged in a mass protest against the utility (Laurie and Crespo 2007).

There are differing perspectives on what went wrong. According to the World Bank, the concessionaire was meeting agreed targets, but it required greater flexibility in determining the level and type of service (as neighborhood consultation could reveal the desire for a cheaper level of service), the removal of barriers to servicing the poor (e.g., issues with land tenure), and financial incentives consistent with extending services (Komives 2001: 62). Laurie and Crespo (2007), on the other hand, reject differentiated service types (in this case, condominial water and sewerage connections),[16] highlight the fact that the contract received only one bid, and explore the hidden costs of the PSP contract, such as the elimination of subsidies for public institutions, which led to high municipal water debt and water cutoffs to hospitals and schools in 2003. Others point out that the project did not make any money in its first three years, thus raising questions about its repeatability (Budds and McGranahan 2003: 109).

Community Management

The advancement of community-led or participatory management of local water supplies is a proposal that appeals to both proponents and opponents of neoliberalization. The advancement of participatory, community-led management for certain segments of society has existed in World Bank literature since the early 1990s as part of its advocacy of privatization. In its report *Water Resources Management*, the World Bank (1993) highlights the use of community associations for service delivery, as well as participation to understand the level of service desired by the poor. In its report of the following year, the World Bank (1994: 8, 10) details the role of community sectors in providing services to rural areas and low-income settlements where "commercial and competitive activity is constrained." For similar reasons but different purposes, community management is also advocated as a response to the neglect of rural and low-income areas by traditional state utilities and the private sector. Those who promote community management often see it as an alternative that is capable of addressing the specificity of local needs, that is more readily achievable in many neglected areas than waiting for the state, and that gives a sense of empowerment and ownership to local residents (see K. Bakker 2008).

Community management has also met with reservations. A key concern is its potential to facilitate cherry-picking by the private sector, leaving already disadvantaged groups with inferior services from which the state and the private sector have both divested (K. Bakker 2008). In

Jaglin's (2002: 232) terms, community-run supplies reflect "a compromise between the ambition to provide universal access to water and the principle of cost recovery." Such programs thus contain elements of differentiated services, but the costs are transferred to user organizations rather than the government, which might otherwise provide subsidies (ibid.). According to Jaglin's research, participatory approaches can help to extend services but can also result in unstable systems, given the difficulty of raising capital. They thus risk producing a lower level of service provision that can be "very difficult to upgrade" (ibid.).

Other researchers express unease with the romanticization of 'community', just as they might with the idealization of the private sector (McGranahan and Mulenga 2009). Rather than uniform and uniquely just results, proponents should expect to confront a range of outcomes, not all of which are in opposition to neoliberalization. As Page (2003) shows, community-led management is not antithetical to commodification and can take place as an active disengagement from state services, having repercussions for the solvency of the wider state-run system. Moreover, participatory processes can be complex and time-consuming (Sabatier et al. 2005), may not exhibit fairness (Cooke and Kothari 2001), and can be hindered by a shortage of political capital or funds (Blair 2000). Karen Bakker (2007: 445–446) lists various other alternatives to neoliberalization, finding the most progressive to be those that involve reforming state governance while strengthening "alternative local models of resource management," which, like public utility partnerships, may be replicable across contexts.

Public Utility Partnerships

Public utility partnerships (PUPs) involve the development of capacity-building relationships between publicly owned utilities. PUPs may be local, regional, or international and generally involve one public utility providing support to one or more others that have not yet achieved the same level of sophistication (technical, managerial, etc.) or service provision (water quality and quantity, coverage, etc.). Activities have included training, consulting, management and administration services, financial remodeling, and joint investment initiatives (Hall 2000: 3). Such relationships have been growing over the past 20 years, and, with 130 PUPs in 70 countries, "far more countries host PUPs than host PPPs [i.e., PSPs] in water" (Hall et al. 2010: 10).

PUPs have been supported through EU initiatives—such as the Baltic Sea Joint Comprehensive Environmental Action Programme and the TACIS (Technical Assistance to the Commonwealth of Independent States) program—with many of them operating between Western European countries and those of the former Eastern bloc (Hall 2000). The United Nations Secretary-General's Advisory Board on Water and Sanitation (UNSGAB) operates the Water Operators Partnerships (WOPs) program directed at "providing support for capacity building of public water operators" (UNSGAB 2007). Funding has come from a range of donors including the World Bank, the EU (which promised 40 million euros for 2010), and the International Bank for Reconstruction and Development (Hall 2000).

Researchers at the Public Services International Research Unit (PSIRU) of the University of Greenwich find that PUPs are successful in capacity building and knowledge transfer. They attribute this to "underlying objectives and motivations, the basis of the partnership and the configuration of accountability networks," as well as to mutual trust and a public sector ethos (Lobina and Hall 2006: 3, 6). The researchers do, however, identify some potential issues. These include instances when PUPs involve the restructuring of the host utility (usually corporatization), joint ventures (when the visiting partner obtains some level of ownership), and the subsequent introduction of PSP (Hall 2000: 8). Like PSP, PUPs can take on many forms. Beyond the straight capacity-building relationship between two public sector utilities, PUPs have included relationships whereby publicly owned water corporations have engaged in contracted services both within their own countries and internationally, for profit and as development projects.

Public Corporations

Publicly owned corporations for water supply are not new, but interest in them is growing, given their embodiment of certain private sector principles—such as cost recovery, business-like management, and independence from government (Schwartz and Blokland 2002: 221)—and the opportunities they present for amalgamating small utilities (Rouse 2009). In the South African context, Smith (2004) refers to such organizational models as the 'second wave' of neoliberalism. Others go further, classifying the model as a form of privatization as it "is often seen as a first step towards more explicit forms of private sector involvement" (McDonald and Ruiters 2005: 3). However, experiences are not easily generalized. Supported by federal regulation, Bogotá's municipal water corporation EAAB has achieved coverage and water quality levels that place it among the top-performing utilities in Latin America (Gilbert 2007; Murillo and Nieto 1997).

In Colombia, the municipal corporation model has been in place for water supply in many cities for more than half a century: in Bogotá since 1914, Cali since 1931, and Medellín since 1955. In the Netherlands, the model's dominance has grown rapidly since 1975 (Blokland and Schwartz 1999). Both cases present interesting takes on PUPs. The Empresas Públicas de Medellín (EPM)[17] is a public corporation that is wholly owned by the city of Medellín, to which it pays a dividend ($274 million in 2008).[18] It has contracts for water and energy supply across Colombia, both lucrative ones (e.g., Bogotá, now concluded) and others directed at development (e.g., Quibdo). The company has made extensive social investments, including service extension, the building of public libraries, youth programs, and community lighting, among others (EMP 2008b).

In the Netherlands, while public limited corporations (PLCs) are responsible for water supply, public ownership of these PLCs is guaranteed by amendments made to the Water Supply Act in 2004. In 2006, Holland's two largest PLCs formed a subsidiary, Vitens-Evides International (VEI), with the goal of supporting progress toward the MDGs through WOPs.[19] VEI's funding is raised through donor agencies, among them the Netherlands Development Corporation, the World Bank, the European Bank for Reconstruction and Development, and the Asian Development Bank.[20] VEI currently has nine international projects in seven countries, including two management contracts (Ghana and Malawi) with a total budget of 36.35 million euros (VEI 2009).[21] The management contracts are established on a cost recovery basis following the same philosophy as WOPs. Nevertheless, such contracts can be controversial. The Ghanaian contract was the subject of such extensive criticism on the part of the National Coalition against Privatization (NCAP 2009) that the Ghanaian government decided to look into canceling the contract.[22]

Conclusion

This article presents a review of the literature on water supply neoliberalization, assessing its conceptual roots, trends, outcomes, and reformulations. First, I detailed the particular prescriptions for water sector reform under neoliberalization, contextualizing them within the neoliberal ideas from which they derive. I then presented the actual trends in water sector reform over the neoliberal period, focusing on reforms to the institutions, organizations, and governance of water management. Building on this discussion of trends, I next analyzed the degree to which the program has brought about the results predicted by its proponents. On most indicators, the program falls short of meeting its claims. In cases where improvements are seen, they may come at the expense of regression in other areas.

This is among the key findings of the article—that contradictions exist in the program between the prescriptions and the predicted outcomes, among the reforms themselves, and between the ideology and the ability to impose it. First, in terms of programs and goals, obvious examples include the incompatibility between user-pay full cost recovery and declining prices; between PSP and increased investment in infrastructure; and between business-driven management and improved water conservation. The limits of the program in meeting the range of outcomes envisioned is perhaps unsurprising, given the inherent assumption that vastly different policy goals (e.g., increased access for the poor and environmental improvements) can be met through like organizational reforms (e.g., PSP). Second, the policy prescriptions themselves can come into conflict with one another. For instance, it was shown how earlier rounds of deregulation can inhibit the effectiveness of future reregulation. Moreover, PSP and privatization led to the need for further regulation, as opposed to deregulation, and imposed limits to participation that would include the empowerment of non-state actors beyond the private sector, such as community groups. Finally, there is the contradiction between the ideology and the ability to impose it. Most prominent is the stated doctrine of reduced government involvement in terms of both regulation and service delivery. In practice, this results in the need for reregulation in order to preserve the neoliberal order (e.g., by mitigating its most deleterious consequences, such as water cutoffs), as well as a dependence on public sector subsidies to ensure the system's maintenance (Ruckert's "accumulation by subsidization").

Given these challenges, as opposed to direct implementation, a constant revision of programs and of their focus and goals is necessary. In these reformulations we see the attempt of proponents of neoliberalism to narrow the scope of their ambition, acknowledging, for instance, their inability to extend service to low-income areas or to secure reinvestment in infrastructure.

The evidence of contradictions and of the limited achievement of stated goals does not complete the need for analysis. As Ferguson (1990: 254) states, the most important aspect of a reform is "not so much what it fails to do but what it does do; it may be that its real importance in the end lies in the 'side effects.'" Thus, for example, beyond the fact that privatization has been limited, it is important to recognize that ultimately PSP has proved to be a regional and urban force, as opposed to a global one; that it has moved from a multinational to a national corporate program; that it has driven and been driven by reregulation; and that it has created possibilities for a host of new actors beyond the state. PSP hints at a resource neoliberalization that seeks opportunity rather than uniformity. It does not need to pit the state against the private sector, nor is it always obliged to seek out new and radical methods. Instead, it may draw on existing practices (such as state subsidization and even service delivery) to sustain its programs while reformulating its goals.

ACKNOWLEDGMENTS

This work was conducted under the auspices of a Social Sciences and Humanities Research Council (SSHRC) Postdoctoral Fellowship at the Stockholm Environment Institute (SEI). I am grateful to the SEI for providing the necessary facilities, as well as a stimulating intellectual environment in which to research and write, and to the SSHRC of Canada for providing financial support. I would also like to thank Peter Rudberg and two anonymous reviewers for their thoughtful comments.

KATHRYN FURLONG is an Assistant Professor in Geography at the Université de Montréal. Her research examines governance issues related to the delivery of municipal services, particularly water supply. Her PhD dissertation (University of British Columbia, 2007), examining water utility restructuring under neoliberalization in Ontario, was co-selected as best dissertation by the Economic Geography Specialty Group of the Association of American Geographers in 2008. Her recent and forthcoming published work analyzes the assumed stability of large technical networks, the contradictions in alternative service delivery, and the hidden theories embedded in the water wars thesis. Her current research focuses on the municipally owned corporation for utility services in the Republic of Colombia.

NOTES

1. Recent assessments show that 2.5 billion people lack access to improved sanitation facilities, including 1.2 billion who have none at all (WHO and UNICEF 2008: 2), and that 884 million people are still reliant on unimproved drinking water (ibid.: 23).
2. Private sector partnerships (PSP) are rarely, if ever, found in rural areas of the global South because these regions "often present a commercially unattractive combination of low demand and high cost of service provision" (Andrés et al. 2008: 46). Yet 86 percent of those lacking access to improved drinking water and 74 percent of those lacking access to improved sanitation live in rural areas (WHO and UNICEF 2008: 53).
3. For a critique of these neoliberal reforms, see Harvey (2005) and Stiglitz (2002).
4. The 'hydraulic age' refers to the period in water management defined by supply-driven management, large infrastructure projects, and universal water provision at low or no cost to the user (see Goubert 1986; Melosi 2000).
5. In 2001, five companies represented 80 percent of the population served through PSP in the global South: Suez (36 percent), Saur (15 percent), Veolia (12 percent), Agbar (11 percent), and Thames Water (6 percent). Between 2001 and 2007, their share of the market decreased to less than 60 percent, as new private operators grew in prominence (Marin 2009: 29–30).
6. For the circumstances and outcomes of these reforms in Ontario, Canada, see Prudham (2004).
7. Already in 1982, the private sector (non-profit and for-profit) delivered 60 percent of "government-financed human services" at the local level in the US (Salamon 2002: 3). The phenomenon is also widespread among European welfare states (except Scandinavia) (ibid.: 6).
8. There are several contracting options. See Kessides (1993: 19) for a comparison of the various organizational arrangements, including contracting, build-operate-transfer (BOT), leasing, etc.
9. See the World Bank Group's Web page, "Private Participation in Infrastructure Database: Sector Data Snapshots Water and Sewerage," http://ppi.worldbank.org/explore/ppi_exploreSector.aspx?sectorID=4 (accessed 1 July 2010).
10. A number of authors have written about this shift. See, for example, Brenner (2004), Jessop (1999), Jouve (2005), and Rhodes (1997).
11. Jaglin (2002) draws on the work of Lavigne (1995), Plane (1998), and Rivera (1996).
12. For a full debate on issues surrounding water and GATS, see the special issue of *Progress in Development Studies*, edited by Brown et al. (2008).
13. Here Budds and McGranahan (2003) draw on the work of Carbonel (2000), Nickson (2001a, 2001b), and Trémolet (2002).
14. Between 1990 and 2004, the number of people lacking access to safe drinking water and sanitation in sub-Saharan Africa increased by 23 percent and 30 percent, respectively (WHO and UNICEF 2006: 2, 10).
15. The data, which come from the PRINWASS project coordinated by J. E. Castro (see Azpiazu and Schorr 2004; Kallis and Coccossis 2004), are available at http://www.prinwass.org.

16. Condominial connections reduce costs by using smaller diameter pipes and shallower burying methods. In the case of La Paz-El Alto, community labor was also used to lay the pipes and was compensated through reduced connection fees (Laurie and Crespo 2007).
17. The EPM is responsible for water, sewerage, and energy services.
18. The level of the dividend paid to the municipality by EPM is set at 30 percent of profits. In 2008, the dividend constituted 19.6 percent of the municipal budget (EPM 2008a, 2008b).
19. Vitens-Evides International, "Position Paper," http://www.vitensevidesinternational.com/downloads.html (accessed 1 July 2010).
20. Vitens-Evides International, "Corporate Brochure," http://www.vitensevidesinternational.com/downloads.html (accessed 1 July 2010)
21. Information on the Malawi contract is not included in the 2008 report as it began in 2009. The information can be found on the VEI Web site at http://www.vitensevidesinternational.com/projects/malawi.html.
22. "I'll Look into AVRL Operations—Minister-Designat," Ghana Business News, 24 February 2009, http://ghanabusinessnews.com/2009/02/24/i%E2%80%99ll-look-into-avrl-operations-%E2%80%93-minister-designate/ (accessed 1 July 2010).

REFERENCES

Aguiar, Luis L. M. 2006. "Janitors and Sweatshop Citizenship in Canada." *Antipode* 38 (3): 440–461.

Altenburg, Tilman, and Christian von Drachenfels. 2006. "The 'New Minimalist Approach' to Private-Sector Development: A Critical Assessment." *Development Policy Review* 24 (4): 387–411.

Anderson, Terry L. 1982. "The New Resource Economics: Old Ideas and New Applications." *American Journal of Agricultural Economics* 64 (5): 928–934.

Anderson, Terry L., and Donald R. Leal. 1988. "Going with the Flow: Expanding the Water Markets." *Cato Policy Analysis* 104. http://www.cato.org/pubs/pas/pa104.html.

Andrés, Luis A., J. Luis Guasch, Thomas Haven, and Vivien Foster. 2008. *Impact of Private Sector Participation in Infrastructure: Lights, Shadows, and the Road Ahead.* Washington, DC: World Bank.

Assies, Willem. 2003. "David versus Goliath in Cochabamba: Water Rights, Neoliberalism, and the Revival of Social Protest in Bolivia." *Latin American Perspectives* 30 (3): 14–36.

Azpiazu, Daniel, and Martín Schorr. 2004. *Cross-Comparative Report on the Economic-Financial Dimension.* Oxford: PRINWASS Project, University of Oxford.

Bakker, Isabella. 2003. "Neo-Liberal Governance and the Reprivatization of Social Reproduction: Social Provisioning and Shifting Gender Orders." Pp. 66–82 in *Power, Production and Social Reproduction*, ed. I. Bakker and S. Gill. New York: Palgrave MacMillan.

Bakker, Karen. 2000. "Privatizing Water, Producing Scarcity: The Yorkshire Drought of 1995." *Economic Geography* 76 (1): 4–27.

______. 2003a. "Archipelagos and Networks: Urbanization and Water Privatization in the South." *Geographical Journal* 169 (4): 328–341.

______. 2003b. *An Uncooperative Commodity: Privatizing Water in England and Wales.* Oxford: Oxford University Press.

______. 2005. "Neoliberalizing Nature? Market Environmentalism in Water Supply in England and Wales." *Annals of the Association of American Geographers* 95 (3): 542–565.

______. 2007. "The 'Commons' versus the 'Commodity': Alter-Globalization, Anti-Privatization and the Human Right to Water in the Global South." *Antipode* 39 (3): 430–455.

______. 2008. "The Ambiguity of Community: Debating Alternatives to Private-Sector Provision of Urban Water Supply." *Water Alternatives* 1 (2): 236–252.

______. 2009. "Neoliberal Nature, Ecological Fixes and the Pitfalls of Comparative Research." *Environment and Planning A* 41 (8): 1781–1787.

Barlow, Maude, and Tony Clarke. 2002. *Blue Gold: The Battle against the Corporate Theft of the World's Water*. London: Earthscan.

Bennett, Vivienne, Sonia Dávila-Poblete, and María Nieves Rico, eds. 2005. *Opposing Currents: The Politics of Water and Gender in Latin America*. Pittsburgh, PA: University of Pittsburgh Press.

Bernstein, Steven. 2001. *The Compromise of Liberal Environmentalism*. New York: Columbia University Press.

Blair, Harry. 2000. "Participation and Accountability at the Periphery: Democratic Local Governance in Six Countries." *World Development* 28 (1): 21–39.

Block, Fred L. 1977. *The Origins of the International Economic Disorder: A Study of United States International Monetary Policy from World War II to the Present*. Berkeley: University of California Press.

Blokland, Maarten, and Klaas Schwartz. 1999. "The Dutch Public Water PLC." Pp. 63–80 in *Private Business, Public Owners: Government Shareholdings in Water Enterprises*, ed. M. Blokland, O. Braadbaart, and K. Schwartz. The Hague: Dutch Ministry of Housing, Spatial Planning and the Environment in collaboration with the Water Supply and Sanitation Council.

Boelens, Rutgerd. 2009. "The Politics of Disciplining Water Rights." *Development and Change* 40 (2): 307–331.

Bond, Patrick, ed. 2002. *Unsustainable South Africa: Environment, Development and Social Protest*. Scottsville, South Africa: University of Natal Press.

Bond, Patrick, Greg Ruiters, and Robyn Stein. 2002. "Droughts and Floods: Water Prices and Values in the Time of Cholera." Pp. 255–299 in *Unsustainable South Africa: Environment, Development and Social Protest*, ed. P. Bond. Scottsville, South Africa: University of Natal Press.

Boyd, David R. 2003. *Unnatural Law: Rethinking Canadian Environmental Law and Policy*. Vancouver: University of British Columbia Press.

Brenner, Neil. 2004. "Urban Governance and the Production of New State Spaces in Western Europe, 1960–2000." *Review of International Political Economy* 11 (3): 447–488.

Brenner, Neil, and Nik Theodore. 2002. "Cities and the Geographies of 'Actually Existing Neoliberalism.'" *Antipode* 34 (3): 349–379.

Brown, Ed, Jonathan Cloke, and Mansoor Ali, eds. 2008. "Special Issue: GATS and Development: The Case of the Water Sector." *Progress in Development Studies* 8 (1): 1–128.

Budds, Jessica, and Gordan McGranahan. 2003. "Are the Debates on Water Privatization Missing the Point? Experiences from Africa, Asia and Latin America." *Environment and Urbanization* 15 (2): 87–113.

Cairncross, Sandy. 1987. "The Private Sector and Water Supply in Developing Countries: Partnerships or Profiteering?" *Health Policy and Planning* 2 (2): 180–182.

Carbonel, Alain. 2000. "Supply of Water and Sanitation to Low-Income Households in La Paz under the Aguas Del Illimani Concession." Paper presented at "Infrastructure for Development: Private Solutions and the Poor," Private Provision of Infrastructure Advisory Facility, Department for International Development, and World Bank, London.

Castro, José Esteban. 2006. *Water, Power and Citizenship: Social Struggle in the Basin of Mexico*. St Antony's Series, ed. J. Zielonka. Houndmills, UK, and New York: Palgrave Macmillan in association with St Antony's College, Oxford University.

______. 2007. "Systemic Conditions Affecting the Universalisation of Water and Sanitation Services: A Sociological Exploration." *Journal of Comparative Social Welfare* 23 (2): 105–119.

______. 2008. "Neoliberal Water and Sanitation Policies as a Failed Development Strategy: Lessons from Developing Countries." *Progress in Development Studies* 8 (1): 63–83.

Conca, Ken. 2006. *Governing Water: Contentious Transnational Politics and Global Institution Building*. Cambridge, MA: MIT Press.

Cooke, Bill, and Uma Kothari, eds. 2001. *Participation: The New Tyranny*. London: Zed Books.

Coutard, Oliver. 2008. "Placing Splintering Urbanism: Introduction." *Geoforum* 39 (6): 1815–1820.

Crean, Michael J. 2002. "A Public Sector Alternative for the New Millennium." Pp. 439–458 in *Reinventing Water and Wastewater Systems: Global Lessons for Improving Water Management*, ed. P. Seidenstat, D. Haarmeyer, and S. Hakim. New York: John Wiley and Sons.

Dalhuisen, Jasper M., Caroline A. Rodenburg, Henri L. F. de Groot, and Peter Nijkamp. 2003. "Sustainable Water Management Policy: Lessons from Amsterdam." *European Planning Studies* 11 (3): 263–281.

Dávila-Poblete, Sonia, and María Nieves Rico. 2005. "Global Water and Gender Policies: Latin American Challenges." Pp. 30–49 in *Opposing Currents: The Politics of Water and Gender in Latin America*, ed. V. Bennett, S. Dávila-Poblete, and M. N. Rico. Pittsburgh, PA: University of Pittsburgh Press.

de Azevedo, Sérgio. 1998. "Law and the Future of Urban Management in the Third World Metropolis." Pp. 258–273 in *Illegal Cities: Law and Urban Change in Developing Countries*, ed. E. Fernandes and A. Varley. London: Zed Press.

de Loë, Rob, and Reid Kreutzwiser. 2007. "Challenging the Status Quo: The Evolution of Water Governance in Canada." Pp. 85–103 in *Eau Canada: The Future of Canada's Water*, ed. K. Bakker. Vancouver: University of British Columbia Press.

Debomy, Sylvie, Donald T. Lauria, and Omar S. Hopkins. 2005. "Pro-Poor Subsidies for Water Connections in West Africa: A Preliminary Study." Water Supply and Sanitation Working Notes 4, World Bank Group.

DG Environment. 2008. "Water Note 5: Economics in Water Policy: The Value of Europe's Waters." European Commission Water Information System for Europe (WISE), Brussels.

Dinar, Ariel, and R. Maria Saleth. 2004. *Institutional Economics of Water: A Cross-Country Analysis of Institutions and Performance*. Washington, DC: World Bank.

Dryzek, John S. 1997. *The Politics of the Earth: Environmental Discourses*. Oxford: Oxford University Press.

Dubash, Navroz. 2004. "Water, Markets, and Embedded Institutions in Western India." Pp. 218–243 in *Liberation Ecologies: Environment, Development, Social Movements*, ed. R. Peet and M. Watts. London: Taylor and Francis.

Duménil, Gérard, and Dominique Lévy. 2004. *Capital Resurgent: Roots of the Neoliberal Revolution*. Trans. D. Jeffers. Cambridge, MA: Harvard University Press.

EPM (Empresas Públicas de Medellín). 2008a. *Informe Financiero 2008*. Medellín, Colombia: Empresas Públicas de Medellín.

______. 2008b. *Informe Social 2008*. Medellín, Colombia: Empresas Públicas de Medellín.

Estache, Antonio. 2005. "PPI Partnerships versus PPI Divorces in LDCs." Policy Research Working Paper Series No. 3470. World Bank, Washington, DC.

European Parliament. 2000. "Directive 2000/60/EC of the European Parliament and of the Council of 23 October 2000 Establishing a Framework for Community Action in the Field of Water Policy." Pp. 1–73, vol. L: Official Journal L 327.

Ferguson, James. 1990. *The Anti-Politics Machine: "Development," Depoliticization, and Bureaucratic Power in Lesotho*. Cambridge: Cambridge University Press.

______. 2006. *Global Shadows: Africa in the Neoliberal World Order*. Durham, NC, and London: Duke University Press.

Figueres, Caroline, Johan Rockstrom, and Cecilia Tortajada, eds. 2003. *Rethinking Water Management: Innovative Approaches to Contemporary Issues*. London: Earthscan.

Finnegan, William. 2002. "Leasing the Rain." *New Yorker*, 8 April, 45–53.

Franceys, Richard. 2008. "GATS, 'Privatization' and Institutional Development for Urban Water Provision: Future Postponed?" *Progress in Development Studies* 8 (1): 45–58.

Furlong, Kathryn, and Karen Bakker. 2010. "The Contradictions of 'Alternative' Service Delivery: Governance, Business Models, and Sustainability in Municipal Water Supply." *Environment and Planning C: Government and Policy* 28 (2): 349–368.

Gamble, Andrew. 1988. *The Free Economy and the Strong State: The Politics of Thatcherism*. Basingstoke: Macmillan Education.

Gandy, Matthew. 1997. "The Making of a Regulatory Crisis: Restructuring New York City's Water Supply." *Transactions of the Institute of British Geographers* 22 (3): 338–358.

Gilbert, Alan. 2007. "Water for All: How to Combine Public Management with Commercial Practice for the Benefit of the Poor?" *Urban Studies* 44 (8): 1559–1579.

Gleick, Peter H., Gary Wolff, Elizabeth L. Chalecki, and Rachel Reyes. 2002. *The New Economy of Water: The Risks and Benefits of Globalization and Privatization of Fresh Water*. Oakland, CA: Pacific Institute.

Gore, Charles. 2000. "The Rise and Fall of the Washington Consensus as a Paradigm for Developing Countries." *World Development* 28 (5): 789–804.

Goubert, Jean-Pierre. 1986. *The Conquest of Water: The Advent of Health in the Industrial Age*. Oxford: Blackwell.

Graham, Stephen. 1997. "Liberalized Utilities, New Technologies and Urban Social Polarization: The UK Experience." *European Urban and Regional Studies* 4 (2): 135–150.

Gulyani, Sumila, Debabrata Talukdar, and R. Mukami Kariuki. 2005. "Universal (Non)Service? Water Markets, Household Demand and the Poor in Urban Kenya." *Urban Studies* 42 (8): 1247–1474.

GWP (Global Water Partnership). 2000. *Towards Water Security: Framework for Action 2000*. Stockholm and London: Global Water Partnership.

Hajer, Maarten A. 1995. *The Politics of Environmental Discourse: Ecological Modernization and the Policy Process*. Oxford: Clarendon Press.

Hall, David. 2000. *Water Partnerships: Public-Public Partnerships and 'Twinning' in Water and Sanitation*. London: Public Services International Research Unit (PSIRU), University of Greenwich.

Hall, David, and Emanuele Lobina. 2007. "Profitability and the Poor: Corporate Strategies, Innovation and Sustainability." *Geoforum* 38 (5): 772–785.

Hall, David, Emanuele Lobina, and Violeta Corral. 2010. *Replacing Failed Private Water Contracts*. London: Public Services International Research Unit (PSIRU), University of Greenwich.

Hardoy, Ana, and Ricardo Schusterman. 2000. "New Models for the Privatization of Water and Sanitation for the Urban Poor." *Environment and Urbanization* 12 (2): 63–75.

Harris, Mike. 1994. "The Common Sense Revolution." Policy document of the Progressive Conservative Party of Ontario, Toronto.

Harvey, David. 2003. *The New Imperialism*. Oxford: Oxford University Press.

______. 2005. *A Brief History of Neoliberalism*. Oxford: Oxford University Press.

Hassan, John. 1998. *A History of Water in Modern England and Wales*. Manchester: Manchester University Press.

Haughton, Graham. 1998. "Private Profits—Public Drought: The Creation of a Crisis in Water Management for West Yorkshire." *Transactions of the Institute of British Geographers* 23 (4): 419–433.

______. 2002. "Market Making: Internationalisation and Global Water Markets." *Environment and Planning A* 34 (5): 791–807.

Hill, Carey, Kathryn Furlong, Karen Bakker, and Alice Cohen. 2008. "Harmonization versus Subsidiarity in Water Governance: A Review of Water Governance and Legislation in the Canadian Provinces and Territories." *Canadian Water Resources Journal* 33 (4): 315–332.

Hout, Wil. 2007. *The Politics of Aid Selectivity: Good Governance Criteria in World Bank, US, and Dutch Development Assistance*. Abingdon and New York: Routledge.

Idelovitch, Emanuel, and Klas Ringskog. 1995. "Private Sector Participation in Water Supply and Sanitation in Latin America." Paper provided by World Bank Latin America and the Caribbean Region Department, Washington, DC.

Jaglin, Sylvy. 2002. "The Right to Water versus Cost Recovery: Participation, Urban Water Supply and the Poor in Sub-Saharan Africa." *Environment and Urbanization* 14 (1): 231–245.

Jessop, Bob. 1999. "The Changing Governance of Welfare: Recent Trends in Its Primary Functions, Scale, and Modes of Coordination." *Social Policy and Administration* 33 (4): 348–359.

______. 2002. "Liberalism, Neo-Liberalism, and Urban Governance: A State Theoretical Perspective." *Antipode* 34 (3): 452–472.

Jimenez, Alejandro, and Agusti Perez-Foguet. 2009. "International Investments in the Water Sector." *International Journal of Water Resources Development* 24 (1): 1–14.

Jones, Tom. 1998. "Recent Developments in the Pricing of Water Services in OECD Countries." *Water Policy* 1: 637–651.

Jouve, Bernard. 2005. "From Government to Urban Governance in Western Europe: A Critical Analysis." *Public Administration and Development* 25 (4): 285–294.

Kaïka, Maria. 2003. "The Water Framework Directive: A New Directive for a Changing Social, Political and Economic European Framework." *European Planning Studies* 11 (3): 299–316.

______. 2005. *City of Flows: Modernity, Nature, and the City*. New York and London: Routledge.

Kaïka, Maria, and Erik Swyngedouw. 2000. "Fetishizing the Modern City: The Phatasmagoria of Urban Technological Networks." *International Journal of Urban and Regional Research* 24 (1): 120–139.

Kallis, Giorgos, and Harry Coccossis. 2003. "Managing Water for Athens: From the Hydraulic to the Rational Growth Paradigm." *European Planning Studies* 11 (3): 245–261.

______. 2004. *Greece Country Strategic Report*. Oxford: University of Oxford.

Kallis, Giorgos, and Henri L. F. de Groot. 2003a. "Shifting Perspectives on Urban Water Policy in Europe." *European Planning Studies* 11 (3): 223–228.

______, eds. 2003b. "Special Issue: Urban Water Policy in Europe." *European Planning Studies* 11 (3): 223–299.

Kallis, Giorgos, Isha Ray, Julian Fulton, and James E. McMahon. 2010. "Public versus Private: Does It Matter for Water Conservation? Insights from California." *Environmental Management* 45 (1): 177–191.

Kessides, Christine. 1993. "Institutional Options for the Provision of Infrastructure." World Bank Discussion Paper No. 212. World Bank, Washington, DC.

Kjellén, Marianne. 2006. *From Public Pipes to Private Hands: Water Access and Distribution in Dar es Salaam, Tanzania*. Vol. 15 of Stockholm Studies in Human Geography. Stockholm: Stockholm University Press.

Koeberle, Stefan, Harold Bedoya, Peter Silarsky, and Gero Verheyen, eds. 2005. *Conditionality Revisited: Concepts, Experiences, and Lessons Learned*. Washington, DC: World Bank.

Kolko, Joyce. 1988. *Restructuring the World Economy*. New York: Pantheon.

Komives, Kristin. 2001. "Designing Pro-Poor Water and Sewer Concessions: Early Lessons from Bolivia." *Water Policy* 3: 61–79.

______. 2002. "Water Concession Design and the Poor." Pp. 149–162 in *Reinventing Water and Wastewater Systems: Global Lessons for Improving Water Management*, ed. P. Seidenstat, D. Haarmeyer, and S. Hakim. New York: John Wiley and Sons.

Komives, Kristin, Vivien Foster, and Jonathan Halpern. 2005. *Water, Electricity, and the Poor: Who Benefits from Utility Subsidies?* Washington, DC: World Bank.

Kooy, Michelle, and Karen Bakker. 2008. "Splintered Networks: The Colonial and Contemporary Waters of Jakarta." *Geoforum* 39 (6): 1843–1858.

Larner, Wendy. 2000. "Theorizing Neoliberalism: Policy, Ideology, and Governmentality." *Studies in Political Economy* 63: 5–26.

______. 2003. "Neoliberalism?" *Environment and Planning D: Society and Space* 21: 509–512.

Larner, Wendy, and Maria Butler. 2005. "Governmentalities of Local Partnerships: The Rise of a 'Partnering State' in New Zealand." *Studies in Political Economy* 75 (Spring): 85–108.

Laurie, Nina, and Carlos Crespo. 2007. "Deconstructing the Best Case Scenario: Lessons from Water Politics in La Paz–El Alto, Bolivia." *Geoforum* 38 (5): 841–854.

Lavigne, Jean-Claude. 1995. "Deux expériences africaines: Côte d'Ivoire et Guinée." Pp. 135–158 in *Gestions urbaines de l'eau*, ed. D. Lorrain. Paris: Economica.

Lee, Terence R. 1999. *Water Management in the 21st Century: The Allocation Imperative*. Cheltenham: Edward Elgar.

Lee, Terence R., and Andrei Jouravlev. 1997. *Private Participation in the Provision of Water Services: Alternative Means for Private Participation in the Provision of Water Services*. Santiago: Economic Commission for Latin America and the Caribbean (ECLAC).

Lemke, Thomas. 2001. "'The Birth of Biopolitics': Michel Foucault's Lecture at the Collège de France on Neo-Liberal Governmentality." *Economy and Society* 30 (2): 190–207.

Lobina, Emanuele, and David Hall. 2006. "Public-Public Partnerships as a Catalyst for Capacity Building and Institutional Development: Lessons from Stockholm Vatten's Experience in the Baltic Region." Paper presented at IRC and UNESCO-IHE Symposium on "Sustainable Water Supply and Sanitation: Strengthening Capacity for Local Governance," Delft, the Netherlands. Public Services International Research Unit (PSIRU), University of Greenwich, London.

______. 2008. "The Comparative Advantage of the Public Sector in the Development of Urban Water Supply." *Progress in Development Studies* 8 (1): 85–101.

Loftus, Alex. 2006. "Reification and the Dictatorship of the Water Meter." *Antipode* 38 (5): 1023–1045.

Lorrain, Dominique, ed. 1995. *Gestions urbaines de l'eau*. Paris: Economica.

Marin, Philippe. 2009. "Public-Private Partnerships for Urban Water Utilities: A Review of Experiences in Developing Countries." Trends and Policy Options No. 8. The International Bank for Reconstruction and Development and the World Bank, Washington, DC.

Marvin, Simon, and Nina Laurie. 1999. "An Emerging Logic of Urban Water Management, Cochabamba, Bolivia." *Urban Studies* 36 (2): 341–357.

McCarthy, James. 2004. "Privatizing Conditions of Production: Trade Agreements as Neoliberal Environmental Governance." *Geoforum* 35 (3): 327–341.

McCarthy, James, and Scott Prudham. 2004. "Neoliberal Nature and the Nature of Neoliberalism." *Geoforum* 35 (3): 275–283.

McDonald, David A., and Greg Ruiters, eds. 2005. *The Age of Commodity: Water Privatization in Southern Africa*. London and Sterling, VA: Earthscan.

McFarlane, Colin. 2008. "Governing the Contaminated City: Infrastructure and Sanitation in Colonial and Post-Colonial Bombay." *International Journal of Urban and Regional Research* 32 (2): 415–435.

McGranahan, Gordon, Pedro Joacobi, Jacob Songsore, Charles Surjadi, and Marianne Kjellén. 2001. *The Citizens at Risk: From Urban Sanitation to Sustainable Cities*. London: Earthscan.

McGranahan, Gordon, and Martin Mulenga. 2009. "Community Organization and Alternative Paradigms for Improving Water and Sanitation in Deprived Settlements." Pp. 173–187 in *Water and Sanitation Services: Public Policy and Management*, ed. J. E. Castro and L. Heller. London: Earthscan.

McInnes, Peter. 2005. "Entrenching Inequalities: The Impact of Corporatization on Water Injustices in Pretoria." Pp. 111–132 in *The Age of Commodity: Water Privatization in Southern Africa*, ed. D. A. McDonald and G. Ruiters. London: Earthscan.

Melosi, Martin. 2000. *The Sanitary City: Urban Infrastructure in America from Colonial Times*. Baltimore: Johns Hopkins University Press.

Meyers, Patrick J. 2002. "Creating Liquidity in an Illiquid Market: How the Internet Can Facilitate the Development of Efficient Water Markets." Pp. 409–426 in *Reinventing Water and Wastewater Systems: Global Lessons for Improving Water Management*, ed. P. Seidenstat, D. Haarmeyer, and S. Hakim. New York: John Wiley and Sons.

Miralles, Antonio. 2008. "The Link between Service Privatization and Price Distribution among Consumer Types: Municipal Water Services in the Spanish Region of Catalonia." *Environment and Planning C: Government and Policy* 26 (1): 159–172.

Mohan, Giles, Ed Brown, Bob Milward, and Alfred B. Zack-Williams. 2000. *Structural Adjustment: Theory, Practice and Impacts*. London: Routledge.

Moss, Jack, Gary Wolff, Graham Gladden, and Eric Guttieriez. 2003. "Valuing Water for Better Governance: How to Promote Dialogue to Balance Social, Environmental, and Economic Values?" White paper, CEO Panel for Business and Industry, Pacific Institute, Paris.

Moss, Timothy. 2009. "Intermediaries and the Governance of Sociotechnical Networks in Transition." *Environment and Planning A* 41 (6): 1480–1495.

Moss, Timothy, Will Medd, Simon Guy, and Simon Marvin. 2009. "Organising Water: The Hidden Role of Intermediary Work." *Water Alternatives* 2 (1): 16–33.

Mumssen, Yogita U., and Brian Williamson. 2002. "Comparative Analysis of Regulation: Water and Other Network Utilities." Pp. 31–62 in *Reinventing Water and Wastewater Systems: Global Lessons for Improving Water Management*, ed. P. Seidenstat, D. Haarmeyer, and S. Hakim. New York: John Wiley and Sons.

Murillo, Antonio Estrada and Enrique Giraldo Nieto. 1997. *El Agua en la Historia de una Ciudad*. Vol. 2. Bogotá: Empresa de Acueducto e Alcantarillado de Bogotá (EAAB-ESP).

NCAP (National Coalition against Privatization of Water). 2009. "For Water Services You Deserve … AVRL Must Go Now! Support GWCL Workers!" National Coalition against Privatization, Accra, Ghana.

Nickson, Andrew. 2001a. "The Cordoba Water Concession in Argentina." Building Municipality Capacity for Private Sector Participation Series, Working Paper No. 442 05, GHK International, London.

______. 2001b. "Establishing and Implementing a Joint Venture: Water and Sanitation Services in Cartagena, Colombia." Building Municipality Capacity for Private Sector Participation Series, Working Paper No. 442 03, GHK International, London.

O'Connor, James. 1973. *The Fiscal Crisis of the State*. New York: St. Martin's Press.
OECD (Organisation for Economic Co-operation and Development). 2009. *Managing Water for All: An OECD Perspective on Pricing and Financing*. Paris: OECD.
Osborn, Suzie, and Simon Marvin. 2001. "Restabilizing a Heterogeneous Network: The Yorkshire Drought 1995–96." Pp. 68–77 in *Urban Infrastructure in Transition: Networks, Buildings, Plans*, ed. S. Guy, S. Marvin, and T. Moss. London: Earthscan.
Osborne, David E., and Ted Gaebler. 1992. *Reinventing Government: How the Entrepreneurial Spirit Is Transforming the Public Sector*. Reading, MA: Addison-Wesley.
Page, Ben. 2003. "Communities as Agents of Commodification: The Kumbo Water Authority in Northwest Cameroon." *Geoforum* 34 (4): 483–498.
______. 2005. "Paying for Water and the Geography of Commodities." *Transactions of the Institute of British Geographers* 30 (3): 293–306.
Page, Ben, and Karen Bakker. 2005. "Water Governance and Water Users in a Privatised Water Industry: Participation in Policy-Making and in Water Services Provision: A Case Study of England and Wales." *International Journal of Water* 3 (1): 38–60.
Page, Ben, and Maria Kaïka. 2003. "The EU Water Framework Directive: Part 2. Policy Innovation and the Shifting Choreography of Governance." *European Environment* 13 (6): 328–343.
Parker, David, and Colin Kirkpatrick. 2005. "Privatisation in Developing Countries: A Review of the Evidence and the Policy Lessons." *Journal of Development Studies* 41 (4): 513–541.
Patrick, Robert J. 2009. "Source Water Protection in a Landscape of 'New Era' Deregulation." *Canadian Geographer/Le Géographe canadien* 53 (2): 208–221.
Peck, Jamie. 2001a. "Neoliberalizing States: Thin Policies/Hard Outcomes." *Progress in Human Geography* 25 (3): 445–455.
______. 2001b. *Workfare States*. New York and London: Guilford Press.
______. 2007. "Remaking Laissez-Faire." *Progress in Human Geography* 32 (1): 3–43.
Peck, Jamie, and Adam Tickell. 2002. "Neoliberalizing Space." *Antipode* 34 (3): 380–404.
Peet, Richard. 2003. *Unholy Trinity: The IMF, World Bank and WTO*. London: Zed Books.
Perreault, Thomas. 2006. "From the *Guerra del Agua* to the *Guerra del Gas*: Resource Governance, Neoliberalism and Popular Protest in Bolivia." *Antipode* 38 (1): 150–172.
Plane, Patrick. 1998. "Les services publics africains à l'heure du désengagement de l'État." *Problèmes économiques* 2587: 21–27.
Polanyi, Karl. 1944. *The Great Transformation*. New York: Octagon Books.
Prasad, Naren. 2006. "Privatisation Results: Private Sector Participation in Water Services after 15 Years." *Development Policy Review* 24 (6): 669–692.
Prudham, Scott. 2004. "Poisoning the Well: Neoliberalism and the Contamination of Municipal Water in Walkerton, Ontario." *Geoforum* 35 (3): 343–359.
Rhodes, R. A. W. 1997. *Understanding Governance: Policy Networks, Governance, Reflexivity and Accountability*. Berkshire: Open University Press.
Rivera, Daniel. 1996. *Private Sector Participation in the Water Supply and Wastewater Sector*. Washington, DC: World Bank.
Rogers, Peter, Radhika de Silva, and Ramesh Bhatia. 2002. "Water Is an Economic Good: How to Use Prices to Promote Equity, Efficiency, and Sustainability." *Water Policy* 4: 1–17.
Rosegrant, Mark W., and Hans P. Binswanger. 1994. "Markets in Tradable Water Rights: Potential for Efficiency Gains in Developing Country Water Resource Allocation." *World Development* 22 (11): 1613–1625.
Rouse, Michael. 2009. "The Market-Centred Paradigm." Pp. 139–152 in *Water and Sanitation Services: Public Policy and Management*, ed. J. E. Castro and L. Heller. London: Earthscan.
Ruckert, Arne. 2006. "Towards an Inclusive-Neoliberal Regime of Development: From the Washington to the Post-Washington Consensus." *Labour, Capital and Society* 39 (1): 34–67.
______. 2007. "Producing Neoliberal Hegemony? A Neo-Gramscian Analysis of the Poverty Reduction Strategy Paper (PRSP) in Nicaragua." *Studies in Political Economy* 79 (Spring): 91–118.
Ruiters, Greg. 2007. "Contradictions in Municipal Services in Contemporary South Africa: Disciplinary Commodification and Self-Disconnections." *Critical Social Policy* 27 (4): 487–508.

Sabatier, Paul A., Will Focht, Mark Lubell, Zev Trachtenberg, Arnold Vedlitz, and Marty Matlock, eds. 2005. *Swimming Upstream: Collaborative Approaches to Watershed Management*. Cambridge, MA: MIT Press.

Salamon, Lester M. 2002. "The New Governance and the Tools of Public Action: An Introduction." Pp. 1–47 in *The Tools of Government: A Guide to the New Governance*, ed. L. M. Salamon and O. V. Elliott. Oxford: Oxford University Press.

Saleth, R. Maria, and Ariel Dinar. 2000. "Institutional Change in Global Water Sector: Trends, Patterns, and Implications." *Water Policy* 2: 175–199.

Sauri, David, and Leandro del Moral. 2001. "Recent Developments in Spanish Water Policy: Alternatives and Conflicts at the End of the Hydraulic Age." *Geoforum* 32 (3): 351–362.

Schwartz, Klass. 2008. "Mimicking the Private Sector: New Public Management in the Water Supply and Sanitation Sector." *International Journal of Water* 4 (3/4): 159–179.

Schwartz, Klass, and Maarten Blokland. 2002. "Government-Owned Public Limited Companies in the Dutch Water Sector." Pp. 221–246 in *Reinventing Water and Wastewater Systems: Global Lessons for Improving Water Management*, ed. P. Seidenstat, D. Haarmeyer, and S. Hakim. New York: John Wiley and Sons.

Shirley, Mary M., ed. 2002. *Thirsting for Efficiency: The Economics and Politics of Urban Water System Reform*. Washington and Amsterdam: World Bank and Pergamon.

Shiva, Vandana. 2002. *Water Wars: Privatization, Pollution, and Profit*. Cambridge, MA: South End Press.

Smith, Laila. 2004. "The Murky Waters of the Second Wave of Neoliberalism: Corporatization as a Service Delivery Model in Cape Town." *Geoforum* 35 (3): 375–393.

Smith, Laila, and Susan Hanson. 2003. "Access to Water for the Urban Poor in Cape Town: Where Equity Meets Cost Recovery." *Urban Studies* 40 (8): 1517–1548.

Snider, Laureen. 2004. "Resisting Neo-Liberalism: The Poisoned Water Disaster in Walkerton, Ontario." *Social and Legal Studies* 13 (2): 265–289.

Stiglitz, Joseph. 2002. *Globalization and Its Discontents*. New York: Norton.

Subramanian, Ashok, N. Vijay Jagannathan, and Ruth Meinzen-Dick, eds. 1997. "User Organizations for Sustainable Water Services." World Bank Technical Paper No. 354. World Bank, Washington, DC.

Swyngedouw, Erik. 2004. *Social Power and the Urbanization of Water*. Oxford: Oxford University Press.

Talbot, Jean-François. 2002. "Is the Water Business Really a Business?" World Bank Water and Sanitation Lecture Series. World Bank, Washington, DC.

Trémolet, Sophie 2002. "Rural Water Service: Is a Private National Operator a Viable Business Model?" Public Policy for the Private Sector, Viewpoint Note 249. World Bank, Washington, DC.

Trémolet, Sophie, and Catherine Hunt. 2006. "Taking Account of the Poor in Water Sector Regulation." Water Supply and Sanitation Working Notes, Note No. 11 World Bank, Washington, DC.

UNSGAB (United Nations Secretary-General's Advisory Board on Water and Sanitation). 2007. "Water Operators Partnerships (WOPs)." http://www.unsgab.org/hapi/wops/index.htm (accessed 1 July 2010).

VEI (Vitens-Evides International). 2009. "Annual Report 2008." http://www.vitens.nl/English/pressand-media/Documents/VitensJaarverslag2008ENG.pdf.

Warner, Mildred, and Jennifer Gerbasi. 2004. "Rescaling and Reforming the State under NAFTA: Implications for Subnational Authority." *International Journal of Urban and Regional Research* 28 (4): 858–873.

Weidner, Helmut. 2002. "Capacity Building for Ecological Modernization: Lessons from Cross-National Research." *American Behavioral Scientist* 45 (9): 1340–1368.

WHO (World Health Organization) and UNICEF (United Nations Children's Fund). 2006. *Meeting the MDG Drinking Water and Sanitation Target: The Urban and Rural Challenge of the Decade*. Geneva: World Health Organization and United Nations Children's Fund.

______. 2008. *Progress on Drinking Water and Sanitation: Special Focus on Sanitation*. New York: WHO/UNICEF Joint Monitoring Programme for Water Supply and Sanitation.

Winpenny, James. 1994. *Water as an Economic Resource*. London: Routledge.

Wolff, Gary, and Eric Hallstein. 2005. *Beyond Privatization: Restructuring Water Systems to Improve Performance*. Oakland, CA: Pacific Institute.

Wolff, Gary H., and Meena Palaniappan. 2004. "Public or Private Water Management? Cutting the Gordian Knot." *Journal of Water Resources Planning and Management* 130 (1): 1–3.

Workman, Thom. 1996. *Banking on Deception: The Discourse of Fiscal Crisis*. Halifax: Fernwood.

World Bank. 1981. *Accelerated Development in Sub-Saharan Africa: An Agenda for Action*. Washington, DC: World Bank.

______. 1993. *Water Resources Management: A World Bank Policy Paper*. Washington, DC: World Bank.

______. 1994. *World Development Report 1994: Investing in Infrastructure*. New York: Oxford University Press.

______. 2000. *Reforming Public Institutions and Strengthening Governance*. Washington, DC: World Bank.

______. 2004. *Water Resources Sector Strategy: Strategic Directions for World Bank Engagement*. Washington, DC: World Bank.

______. 2005a. *Economic Growth in the 1990s: Learning from a Decade of Reform*. Washington, DC: Oxford University Press for the World Bank.

______. 2005b. *World Development Report 2006: Equity and Development*. New York: Oxford University Press for the World Bank.

Worster, Donald. 1985. *Rivers of Empire: Water, Aridity, and the Growth of the American West*. New York: Pantheon Books.

Controversies in Climate Change Economics

Robert Eastwood

ABSTRACT: This article is a non-technical review of the economics of global policy on reducing greenhouse gas emissions. Quite a lot is known about the likely physical consequences of anthropogenic climate change, but much uncertainty remains. In particular, account needs to be taken of possible catastrophes such as ice sheet melting. How are we to balance the known costs of taking action to reduce greenhouse gas emissions in the present against the uncertain benefits of such action for future generations? How convincing is the case for substantial measures to be undertaken now? If the case for such action is accepted, should emissions be controlled via Kyoto-style national emissions targets or by the imposition of carbon taxes? How can the challenges of burden sharing between developed and developing countries be addressed?

KEYWORDS: cap-and-trade, carbon tax, climate change, emissions permit, greenhouse gas, Kyoto Protocol

The judgments in the Fourth Assessment Report (FAR) of the Intergovernmental Panel on Climate Change (IPCC) regarding anthropogenic global warming can be summarized as follows:

1. The warming trend in the last half of the twentieth century amounted to 0.13 °C per decade, or 0.65 °C for the 50-year period, and it is 'very likely' that most of this increase was due to anthropogenic greenhouse gas (GHG) emissions.[1] If atmospheric GHG concentrations were to stay at the level of the year 2000, estimated warming in the twenty-first century would be a further 0.6 °C (IPCC 2007: 30–39, table 3.1).
2. In the absence of policies to reduce emissions (i.e., under 'business-as-usual'), estimated warming in the twenty-first century would be between 1.8 °C and 4.0 °C, depending on which of six scenarios is chosen. However, each of these estimates is subject to substantial uncertainty, so that the overall range of warmings that are judged 'likely' from one scenario to another is between 1.1 °C and 6.4 °C (IPCC 2007: table 3.1).

The science behind these judgments involves three main steps. First, the climate-relevant impacts of human activity (principally GHG emissions) must be estimated. Second, the so-called radiative forcings (RF) generated by these impacts must then be estimated. This terminology arises from the fact that each impact can be expressed numerically in terms of the change in the intensity of incoming solar radiation that would have the same effect. For instance, because of the greenhouse effect, the impact of doubling atmospheric CO_2 is equal to 3.8 watts per square

Environment and Society: Advances in Research 1 (2010): 76–95 © Berghahn Books
doi:10.3167/ares.2010.010104

meter, equivalent to a rise of about 1.7 percent in incoming solar radiation (IPCC 2007: fig. 2.4; see also Shaviv 2006). Third, climate sensitivity, defined as the effect on equilibrium global temperature of a given RF, must be estimated. Since the RF of doubling atmospheric CO_2 is known, climate sensitivity can be neatly re-expressed as the rise in equilibrium global temperature that such a doubling, from the pre-industrial level of about 280 parts per million (ppm), would produce. The FAR gives the 'likely' range for climate sensitivity so defined to be 2 °C to 4.5 °C and considers it 'very unlikely' that it is below 1.5 °C (IPCC 2007: 38).

For the economist seeking to analyze climate change policy, the key lesson from the above is that the amount of anthropogenic warming under business-as-usual is almost certainly positive, but it is also subject to a great deal of uncertainty. This uncertainty is both economic and scientific. Economic uncertainty is reflected in the six scenarios considered in the FAR, which vary according to what is assumed about economic development, population growth, and the speed and direction of technological change (IPCC 2007: 44). Corresponding to these scenarios are atmospheric concentrations of GHGs equivalent to CO_2 concentrations in 2100 ranging from 600 ppm to 1550 ppm (ibid.: table 3.1, note c). Scientific uncertainty, indicated by the wide range of 'likely' values for climate sensitivity, arises from various sources, of which by far the most important is uncertainty about the link between warming and the formation of clouds of different types and at different altitudes (ibid.: fig. 2.4).

One further source of uncertainty should be mentioned. Any case for action on climate change depends not on the anthropogenic component of warming discussed above but on *total* warming. If anthropogenic warming happened to be against a background of natural cooling at a comparable rate, then—far from being a threat—it would be viewed as delivering us from a new ice age.[2] That such a notion may not be entirely fanciful follows from the established fact that for the past million years, the earth has experienced a series of ice ages lasting around 100,000 years each, interspersed with interglacial periods, warmer by some 5 °C, lasting around 10,000–15,000 years (Broecker and Kunzig 2008; Jansen et al. 2007; King 2006). It is from this perspective that one feature of current warming becomes vital—its rapidity. As noted above, the upper end of the 'likely' range for anthropogenic warming in the twenty-first century is 4.5 °C, corresponding to about 5 °C over the 150-year period starting in 1950. Although our understanding of natural temperature changes over such short time scales is imperfect, we have no scientific basis for expecting any significant compensation from natural processes over the next one or two centuries, so a rough equation of warming with anthropogenic warming is reasonable.

Given the scientific background, arriving at an assessment of climate policy requires a number of steps. But the case in principle for global policy action is straightforward: it is that GHG emissions create a 'global externality'. A private action that releases a gram of CO_2 into the atmosphere imposes future costs on other persons across the globe that are not taken account of by the emitter. These costs are the same wherever and however the emissions take place. Externalities mean, in general, that uncoordinated private actions produce bad outcomes.

To illustrate this principle in the simplest possible way, consider the case of a single emitter and a single victim. The emitter will rationally increase its level of emissions until the gain derived from the last unit of emissions falls to zero. The victim of emissions, however, will (usually) be willing to pay something—say $1—to avoid that last unit, so a negotiated agreement whereby the victim bribes the emitter with between $0 and $1 to cut emissions by one unit benefits both parties. Extension of this reasoning leads to the idea of an 'efficient' level of emissions, which negotiation should achieve, where the benefit to the emitter and the cost to the victim, for small changes in emissions, are equal, and the scope for further benefits to both parties from emissions reduction has been exhausted.[3]

The real situation regarding GHG emissions represents the opposite extreme to this idealized example. Rather than a single victim, there are billions, and rather than the damage being contemporaneous, its burden will fall mostly on people who are not yet born. Therefore, instead of private negotiation, public interventions such as taxes or quotas are required to reduce emissions to an efficient level. Since the externality spills across national boundaries, any coordinated action at only a national level will not lead to sufficient emissions reduction: to accomplish this, binding international agreements are essential. Moreover, in contrast to the idealized case of the preceding paragraph, it is impossible for the current losers from mitigation effort to be compensated by the future gainers. As a result, social value judgments regarding the relative worth of gains and losses to different individuals and at different times are inescapable.[4] The global and intertemporal nature of the GHG externality, together with its magnitude, underlies Nicholas Stern's (2007: xviii) comment that this constitutes "the greatest market failure the world has ever seen."

The Economic Theory of Climate Policy

To understand the economic theory of climate policy, one must start with the objectives of the policy. The approach that is typical in economics is both 'utilitarian' and 'consequentialist'. The utilitarian approach means that, ultimately, the costs and benefits of action are assessed by adding up estimated gains and losses to individual persons, now and in the future. These gains and losses can be expressed in monetary terms (the dollar amount that would compensate a person for damage related to climate change) or in 'real' terms (the required compensation in terms of some commodity or 'basket' of commodities). The consequentialist approach means that policy actions are judged solely in terms of their consequences, without taking into account, for instance, the political process that leads to them.

The utilitarian-consequentialist framework dodges some issues and raises others. Among issues dodged, the most important is any notion that the environment should be valued as such, without reference to the sum of human utilities. According to the IPCC, 30–40 percent of all known species could become extinct by 2100 as a result of climate change. Utilitarianism implies that this is of concern only to the extent that it is possible to trace consequences of this for individual well-being (Heal 2009). Among issues raised, two are paramount: income distribution and uncertainty.

As regards income distribution, both inter- and intra-generational, how should gains and losses to different individuals at the same or different times be valued relative to one another? The inter-generational question has two aspects to it. First, is there any justification for discriminating against persons simply because they belong to future generations? Such 'pure time preference'[5] is hard to justify ethically, other than as a response to the possibility that some event will wipe out humanity before the future arrives. This consideration leads to positive, but very low, 'discounting'[6] of future gains, as discussed below (see Stern 2007: 31). Second, if we assume that the general pattern of rising prosperity during the past 200 years will be replicated in the next 200, how should this be taken into account? In that case, mitigation efforts today will, on average, benefit individuals in the future who are better off than those who bear the mitigation costs today. Therefore, optimal mitigation depends inescapably on how society values extra units of consumption to richer and poorer people, respectively—in other words, it depends on society's degree of 'inequality aversion' . If higher future prosperity is assumed and inequality aversion is positive, then gains in future consumption should be valued less than gains today, providing another reason for discounting, on top of any pure time preference. When the degree of inequality aversion is higher, future gains should be discounted more and mitigation in the present should be lower.

This line of reasoning, however, must be modified to take into account intra-generational inequality, as it is likely for several reasons that mitigation costs will fall on the relatively rich today, while the future benefits will accrue to the relatively poor. It is inevitable that the bulk of any mitigation now will occur in the richer countries of the world, imposing costs that will fall mainly on individuals in these nations. The two most important categories of future beneficiaries will be (1) those whose livelihoods depend directly on agriculture in areas where agricultural production is particularly vulnerable to climate change, notably because of increased water stresses that are associated either with changes in rainfall patterns or glacier-fed river flow, and (2) those living in areas vulnerable to sea level rise and for whom relocation would be costly. Clearly, these categories are largely made up of relatively poor people in developing countries. In sum, it is far from certain that future gainers from present mitigation will be any richer than current losers, in which case the argument for discounting, via inequality aversion, would evaporate.

Turning to uncertainty, an analysis of its consequences for climate policy within the utilitarian framework begins from a presumption that individuals are 'risk-averse' and that policymakers' decisions should respect this. At the individual level, risk aversion is indicated by (indeed, can be defined by) an unwillingness to take fair gambles. It seems a reasonably secure generalization that, with respect to gambles that are large enough so that the pleasure of gambling per se can be neglected, individuals do display considerable risk aversion. Few of us would stake the value of our house on the toss of a coin.

We have seen that forecasts of climate change are subject to a high degree of uncertainty. This uncertainty is greatly compounded by an additional lack of certainty over the human consequences of any given temperature rise. In general, and assuming risk aversion, taking account of these uncertainties tends to shift optimal climate policy in the direction of higher mitigation now. Why is this so?

Most analysts have answered this question by dividing it into two parts, making a distinction between what, for want of a better word, I will call 'ordinary' (bell curve) uncertainty and catastrophe (e.g., see Stern 2007). For the first, suppose we are considering a choice between business-as-usual and a given program of mitigation, which would stabilize atmospheric GHGs at some given level. To keep matters simple, assume, as in figure 1, that under mitigation future consumption (per head, say) is certain at C_{MIT}. Under business-as-usual (BAU), suppose there is uncertainty regarding C_{BAU}, indicated by the bell curve probability distribution in the figure. Expected consumption

Figure 1: Bell curve uncertainty and the case for mitigation

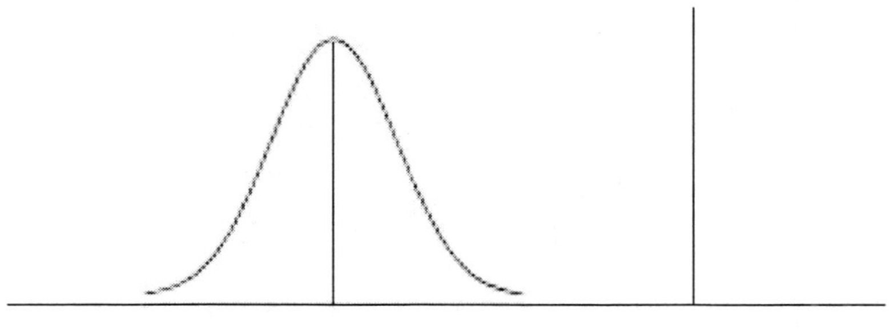

Note: The curve in the diagram indicates that consumption under business-as-usual (BAU) is uncertain, with mean C_{BAU} and probability distribution indicated by the bell curve. Future consumption under mitigation is certain at C_{MIT}.

is C^E_{BAU}, but risk aversion implies that the bell curve outcome is worse than if BAU had entailed C^E_{BAU} for certain. Taking uncertainty into account therefore tilts the argument in favor of any given mitigation program and thus will clearly raise the optimal amount of mitigation.

However, this example does not make provision for possible catastrophes—irreversible events with large-scale consequences for humanity. Global warming seems to expose us to several of these, of which the two most important are ice sheet melting, leading to sea level rise,[7] and permafrost melting, which would release large quantities of methane (a greenhouse gas that is 23 times more potent than CO_2) into the atmosphere.

In terms of figure 1, taking account of catastrophes would imply amending the bell curve model to move some of the probability mass, for the case of BAU, well to the left of where the bell curve is in the diagram. And this would strengthen the case for mitigation, not merely by lowering the expected (or average) BAU outcome, but because risk aversion would lead us to give special weight to the catastrophe outcomes in our assessment. The difficulty, however, is that we do not—and possibly cannot—know enough to pin this down in a quantitative way with any confidence. We know too little about the consequences of catastrophes and how their probabilities depend on mitigation. As Dasgupta (2008: 167) puts it:

> Moreover the Earth system is driven by interlocking non-linear processes running at differing speeds and operating at different spatial scales. Doing little about climate change would involve Earth crossing an unknown number of tipping points … in the global climate system. We have no data on the consequences if Earth were to cross those tipping points. They could be good in some places, disastrous in others. And even if we did have data, they would probably do us little good, because Nature's processes are irreversible. One implication of Earth system's deep non-linearities is that estimates of climatic parameters based on observations from the recent past are unreliable for making forecasts about the state of the world at concentration levels of 560 p.p.m. or more. The uncertainties are therefore enormous.

Allowance for catastrophe risk is made in many of the attempts to quantify the costs and benefits of mitigation, but the modeling is inevitably crude and is one reason why modelers are usually at pains to stress the 'indicative' nature of their estimates (e.g., see Stern 2007: chap. 6).

As will be illustrated below, such estimates are critically dependent on the extent to which future expected gains resulting from mitigation are discounted relative to current gains as a result of abstaining from mitigation. We have seen how the extent of discounting will depend on both inequality aversion and risk aversion, so it appears that a parameter for each of these must be chosen. However, utilitarian analysis in practice (e.g., as applied in Stern 2007) uses a single parameter to represent both risk and inequality aversion. This can be justified using the 'veil of ignorance' argument that originated in Harsanyi (1955; see also Dasgupta 2008: 147).[8]

The Case for Stabilizing Atmospheric GHG Concentrations: The Stern Review

In 2005, the UK government asked Nicholas Stern (now Lord Stern) to lead the world's first major review of the economics of climate change. The work was carried out by a team of economists in the UK Treasury, with input from independent academics on a consultancy basis, and was published as *The Economics of Climate Change: The Stern Review* (Stern 2007). The report itself (hereafter the Stern Review) was not peer-reviewed, but it has spawned a vast academic literature, both critical and supportive,[9] and has inspired comparable work elsewhere, notably Garnaut (2008).

The central claim of the Stern Review is that "the costs of stabilizing the stock of GHGs in the range 450–550 ppm CO_2e are considerably less than the costs of delayed action" (Dietz et al. 2007: 248). As percentages of world income, the costs are put at about 1 percent and the benefits, in the Review's 'base case', at 10.9 percent. If taken at face value, these estimates imply an overwhelmingly strong case for global action. How are the numbers arrived at, and what degree of confidence should be attached to them?

To begin with, the Review distinguishes among market impacts, 'catastrophic' impacts, and non-market impacts (Stern 2007: chap. 6). Non-market impacts include health, the environment, and consequences associated with large-scale migrations of people. Market impacts are tied to projections of outputs in different sectors, and catastrophic impacts are modeled—inevitably, rather arbitrarily—via projected risks of large output losses that are assumed to rise with global temperature.

In order to generate sectoral output projections for a variety of scenarios, the Stern Review uses an integrated assessment model known as PAGE2002 (Policy Analysis of the Greenhouse Effect 2002). The two scenarios needed to arrive at the 10.9 percent figure above are BAU (with 'baseline' climate assumptions) and GHG stabilization. Only the BAU scenario is simulated: the GHG stabilization scenario is a benchmark that simply assumes that world income (and consumption) will grow, for certain, at a given constant rate of 1.3 percent per annum in per capita terms.

In order to take account of uncertainty, BAU is simulated 1,000 times with different values of key parameters, such as climate sensitivity, inserted in the different simulations. At this point, the utilitarian calculus is employed, to add up social utility for each of the (assumed equiprobable) 1,000 paths, allowing average (or 'expected') social utility to be calculated. This inherently meaningless number is converted into something meaningful using the following device.

First, social utility for the benchmark path, along which per capita consumption rises at an assumed constant rate (1.3 percent per annum) from its current value—C_0, say—is calculated. In just the same way, the social utility for paths on which consumption grows at just the same rate through time, but from a starting value of consumption lower than C_0 by various percentages, is calculated. One of these paths will have social utility equal to the average expected utility under BAU. The percentage by which C_0 has to be shrunk to get this result is called the 'balanced-growth-equivalent' (BGE) cost of BAU.

To assert that the BGE cost of BAU is 10.9 percent therefore means that BAU would be as bad as reducing average consumption in the world by 10.9 percent now and at all dates in the future. This number is roughly commensurate with the estimate of the 1 percent cost of GHG stabilization.[10] If the two numbers are accepted, then the case for mitigation is overwhelmingly strong.

The estimate of 10.9 percent however depends on, and is very sensitive to, two key parameters. One is Stern's 'eta' (η), which represents inequality aversion. The second is Stern's 'delta' (δ), which represents pure time preference. Stern applies the 'risk of planetary extinction' argument to choose δ equal to 0.1 percent per annum. For η, Stern chooses a value of +1. What η=1 means is that the relative social valuation of (small) gains to richer and poorer persons is in inverse proportion to their relative levels of consumption: if your consumption is one-third of mine, then a little extra consumption for you is judged three times as valuable as it would be for me. Put another way, a utilitarian-consequentialist government would be willing to transfer some consumption from me to you even if up to two-thirds of it was wasted in the process. Higher values of η reflect higher inequality aversion: η=2, for example, implies in the above example that the relative valuation is nine times rather than three times. Accordingly, the transfer from me to you would be implemented even if 89 percent was lost or wasted en route, so that only one-ninth (11 percent) got through. Stern justifies the lower value for inequality aversion mainly on the grounds that anything higher would be grossly inconsistent with social preferences as revealed, for instance, in both domestic and international (foreign aid) policies toward poverty and inequality that are currently pursued in First World countries.

The sensitivity of the estimated cost of BAU to η and δ and whether or not uncertainty is taken into account are well illustrated in figure 2 below. The Stern Review's base case is shown at the bottom. If all allowance for uncertainty is eliminated (the case at the left of the figure), so that central projections for both temperature and the material consequences of any given temperature rise are employed, then BAU would be equivalent to taking only 3.5 percent off everyone's consumption, both now and in the future, rather than 10.9 percent. If, instead, uncertainty is left in the analysis (the case at the right of the figure), but δ is raised to 1.5 percent per annum and η to +2, the BAU cost falls to a paltry 1.1 percent, eliminating the argument for action. Making both changes (the case at the top of the figure) cuts the cost to 0.6 percent.

The sensitivity of the BAU cost to η and δ can be understood in terms of the power of compound interest, even with an apparently small interest rate, over a long period of time. Suppose, for instance, that we imagine a world in which consumption is growing at 1 percent per annum, and we take η=2 and δ =1.5. In order for it to be worth sacrificing one unit of consumption today to raise the consumption of a person living 100 years from now, the increase would have to be 33 units. However, with η=1 and δ =0, the required increase is only 3 units. So with the higher parameter values, future losses are heavily discounted, and the cost of BAU is correspondingly low.

The case against Stern's assumptions on η and δ is based on the view that what is assumed about these parameters, rather than being derived from "the lofty vantage point of the world social planner" (Nordhaus 2007: 691),[11] should reflect what is observed in both market interest rates and estimated rates of return on investment in plant and machinery and human capital

Figure 2: The roles of equity and risk in the Stern Review's formal modeling: Balanced growth equivalent costs of BAU under alternative assumptions

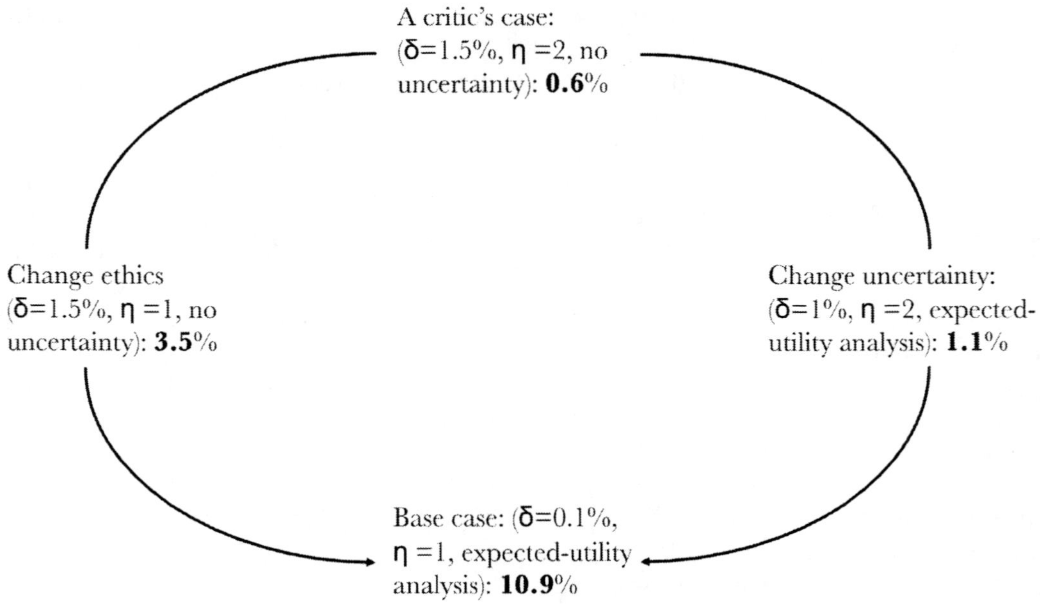

Source: Dietz et al. (2007: 245).

(see also Deaton 2007; Weitzman 2007). Stern's assumptions imply a real rate of discount of 1.4 percent per annum, which means that any investment yielding as little as 1.4 percent (or above) should be made.[12] Nordhaus sets this against estimated pre-tax rates of return on US corporate investment of 6.6 percent and on human capital investment of 6–20 percent, as well a real return on 20-year US Treasury securities of 2.4 percent (ibid.: 689–690). *Prima facie*, there are two implications, one from the standpoint of how a given quantum of investment should be allocated between investment in mitigation and other types of investment and the other related to what total investment should be, that is, how the balance between consumption in the present and consumption in the future should be struck.

On the allocation of investment, it might appear foolish to forgo non-mitigation investments yielding 6.6 percent per year in favor of mitigation investments that, even if they yielded as little as 1.4 percent per year, would show a social profit in the Stern calculus. On the balance between consumption and investment, it might seem wrong to forcibly squeeze consumption in the present to finance investments yielding (in the limit) as little as 1.4 percent per annum, when we can observe that private persons would require an average return of 2.4 percent over a 20-year horizon to make voluntary reductions in their consumption.

However, there are three serious difficulties with these arguments from market observation. The first difficulty is that no capital markets exist for the time horizons relevant to investments in mitigation (i.e., 100 years or more), so high current rates of return on non-mitigation investments tell us nothing about whether lower rates of return on mitigation investments at far longer horizons are worth having or not. The second difficulty, also related to the long horizons intrinsic to this issue, is that future generations are not represented in capital markets today. These markets can coordinate the preferences only of those currently alive in respect of savings versus consumption with what investments can deliver over relatively short time horizons, hardly extending beyond 20 years. A 'lofty vantage point' is indeed needed if the interests of future generations are to be represented. The third difficulty is that it is hard to see how available data on interest rates and returns to capital can be of any help in deciding how to react to unknown catastrophe risks at a horizon of 100 years or more.

These risks not only are essential to Stern's double-digit estimates of the cost of BAU, as figure 1 demonstrates, but also are recognized by some prominent critics of Stern's cost-benefit methods. For example, Weitzman (2007: 721) argues that "the focus of the economics of climate change [should shift] from the middle range of the distribution of what might happen with ΔT at a IPCC-4 mean of ≈ 2.8°C a hundred years hence to thinking more about what might be in the tails with $\Delta T > 6°C$, which is just two IPCC-4 standard deviations out for a century from now, meaning a probability ≈ 3 percent (and presumably a yet higher probability *after* 2105). This thick tail is where most of the cost–benefit action may well be even if—or perhaps precisely because—our estimates of the probabilities involved are themselves so highly uncertain."

For the remainder of the article, I will assume a goal of stabilizing atmospheric stocks of greenhouse gases in the range 450–550 ppm of CO_2 equivalent (the current level is about 430 ppm and is rising at over 2 ppm per year). Depending on whether the higher or lower figure is the target, this will require emissions to peak within 20 years and to fall 30–70 percent by 2050 (Dietz et al. 2007: 230).

Emissions Reductions: When, How Much, in What Sectors, and Where?

Before discussing policy, I will present a quantitative overview that involves three questions regarding stabilization. First, how do the time paths of total GHG emissions under BAU and stabilization at various levels compare? Second, from the technological viewpoint, what are the cheapest ways of achieving any given emissions reductions? Stern's estimate of about 1 percent as the cost of stabilization in the range 500–550 ppm CO_2 equivalent assumes that mitigation will be efficient, that is, achieved in the least costly way.[13] Third, what is implied for the pattern of emissions reductions across sectors and countries?

Figure 3, borrowed from the Stern Review, shows the overall scale of the challenge posed by stabilization. Global GHG emissions in the year 2000 were about 42 $GtCO_2$ equivalent ($GtCO_2eq$), which corresponds globally to about 6.5 tonnes per head. Energy-related emissions account for 65 percent (24 percent being from power generation), and most non-energy-related emissions come from land use changes, such as deforestation, and from agriculture. Under BAU, emissions are projected to double by 2050, driven by rises in per capita income and population.[14] To stabilize the atmospheric concentration at 450 ppm by the year 2050, without overshooting, requires a dramatic departure from BAU, with annual emissions cut by 70 $GtCO_2eq$—that is, a reduction of over 80 percent. Stabilization at 550 ppm requires a cut of 50 $GtCO_2eq$, or 60 percent.

How can emissions reductions on such scales be most efficiently achieved? Enkvist et al. (2010) have tried to answer this question, assembling data from a wide range of sources and basing their analysis only on technologies that either exist or are judged to be available soon (notably, those that involve carbon capture and storage).[15] Figure 4 summarizes their results. On the horizontal axis are global reductions in emissions in 2030, amounting to 38 gigatonnes of CO_2eq per year, estimated to be potentially achievable at a cost of less than 80 euros per tonne of CO_2eq, and which, if initiated now, would lead to a path for atmospheric GHGs peaking at 510 ppm CO_2eq before eventually stabilizing at 450 ppm CO_2eq. This estimate is broadly in line with the corresponding projection in the Stern Review, as indicated in figure 3.[16]

In figure 4, the sources of emissions reductions are listed, left to right, in order of increasing cost. For the first 13 $GtCO_2eq$, the costs are negative, meaning that these changes would pay for

Figure 3: Global emissions trajectories under BAU and alternative stabilization scenarios

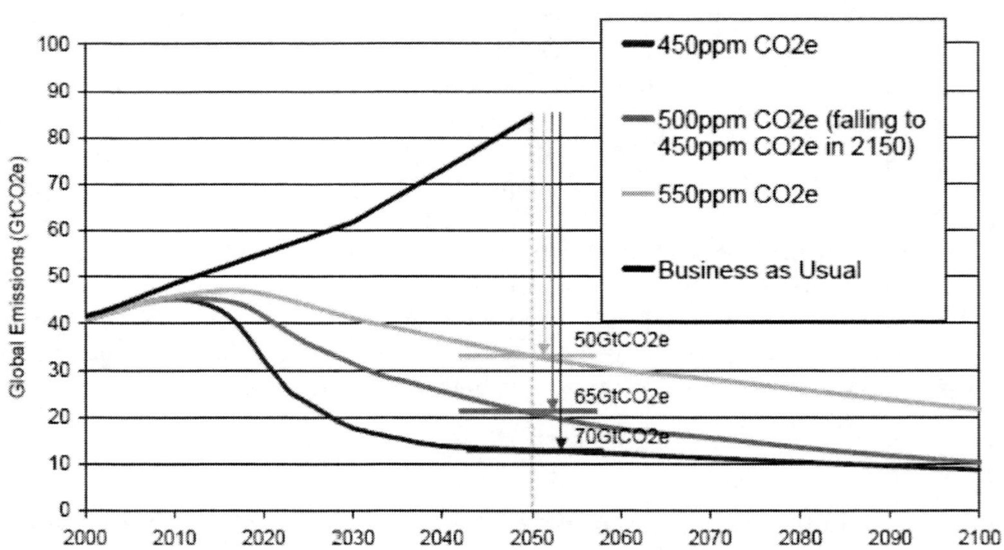

Source: Stern (2007: 206, fig. 8.4).

Figure 4: Efficient emissions reductions by 2030

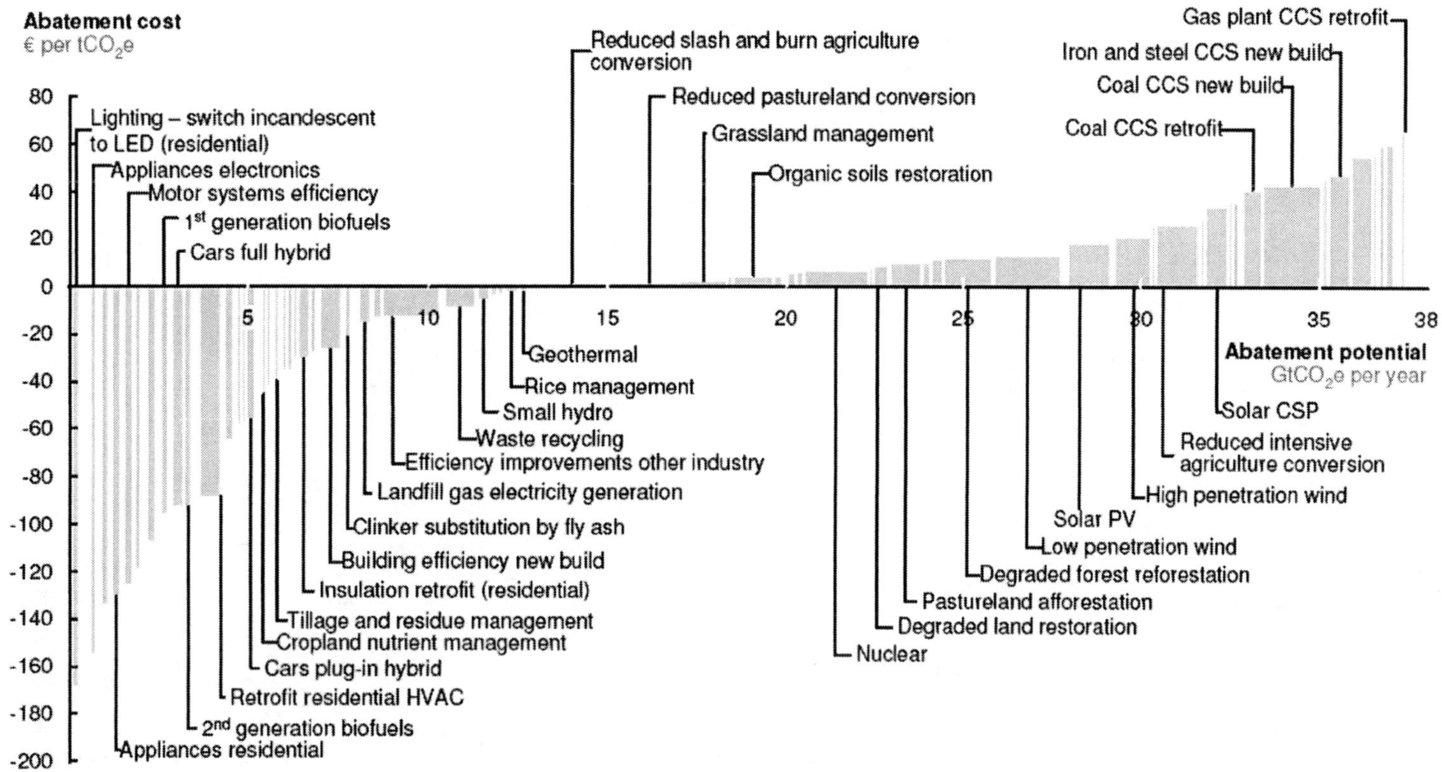

Source: Enkvist et. al (2010).

themselves. Efficiency improvements, both residential and industrial (e.g., heating, ventilation, air conditioning), together with improved fuel efficiency in vehicles, are important negative cost categories. The fact that these opportunities remain unexploited has important lessons for policy, to be discussed later. Increasing emissions reduction from 13 up to 38 $GtCO_2eq$ entails the employment of technologies that are costly, raising the marginal cost of reduction to 80 euros per tonne of CO_2eq, as noted above. Nuclear and wind power account for about one-fourth of the positive cost component, with the rest more or less evenly split between land use changes and, at the high cost end, carbon capture and storage.[17] Taking the negative and positive cost components together, efficient mitigation is found to be "highly fragmented across sectors and regions" (Enkvist et al. 2007). Industry accounts for only 47 percent, of which power generation contributes 26 percent, and, of the rest, forestry accounts for 21 percent. No less than 72 percent of efficient mitigation is located in developing regions (Enkvist et al. 2010: 9 and Exhibit 7). These facts raise a number of policy challenges.

First, it is far easier to present large-scale immovable emitters, such as power stations, with an effective price for CO_2 emissions than is the case for small, scattered, and/or mobile producers and consumers. These include small-scale farmers in developing countries who are deciding how much fertilizer to use or how often to till the soil.

Second, afforestation and avoided deforestation, much of it in Africa and Latin America, account for over one-fifth of efficient mitigation. Creating appropriate financial incentives in this case raises special difficulties, since it is necessary to set a forestation baseline with respect to which afforestation and avoided deforestation can be defined for the purposes of paying for the 'negative emissions' involved. Also difficult may be defining the associated property rights (i.e., who is to be paid) and establishing mechanisms, both to connect buyer and seller (the buyer might be a developed country power station, which would acquire permits to emit) and to verify that the forestation change has actually occurred. Not for nothing was deforestation excluded from the Clean Development Mechanism (CDM) in the Kyoto Protocol.[18]

Third, even when it is possible to bring the price mechanism to bear—indirectly, for instance, via the energy prices faced by consumers—the evidence in figure 4 makes it clear that reductions of negative cost emissions are far from being fully exploited. While bounded rationality may perhaps be partly responsible for a house owner failing to install house insulation that would pay for itself over time, for example, there are other factors that come into play. Credit markets are imperfect, and the house owner may be unable to borrow the capital sum required, or may be able to do so but only at a cost that would make the investment uneconomic. Moreover, for the incentives to work in full, the house owner must be confident that any savings in heating costs will be capitalized into the price of the house in the event that he or she decides to move. It is not enough for the individual house owner to be rational: he or she must assume that other market participants are rational as well.[19]

Fourth, the high proportion of efficient mitigation that is located in developing countries raises difficult questions about the international sharing of the costs of mitigation. Even if the distribution of responsibility for the present level of atmospheric GHGs were to be left to one side, it is incontestable that the future benefits from mitigation actions in one place will confer benefits across the globe.[20] If efficient mitigation opportunities happen to be geographically distributed in a similar way to future benefits, then perhaps an international agreement to reduce emissions can be achieved without international financial transfers being required. In this case, each country may calculate that the mitigation costs that it has to bear will be compensated by gains to its own citizens in the future. The facts are hard to judge here, since the benefits—as well as the costs—of (efficient) mitigation are probably skewed toward poor countries in general.

Policies to Stabilize Atmospheric GHGs

The design of climate policy can be approached at several levels. From a Marxist standpoint, runaway CO_2 emissions have been viewed as a symptom of 'excess capitalism', raising the prospect of "a widespread reversal of many of the systems that constitute capitalism as it turns into its own gravedigger" (Urry 2010: 97). Indeed, the worst scenarios envisaged in the Stern Review, which was written from a quite different perspective, are hardly more encouraging than this. Where Marxist and neoliberal approaches may differ, however, concerns the validity of climate change policies that do not question the existing political and economic order. Here I assume that this is worthwhile—in other words, that the implementation of such policies is not inconceivable within the existing order—and I restrict myself to such approaches.

From this perspective, the first and most important challenge is to establish a system of market incentives that will motivate emissions reductions by making market participants pay a price for emitting. The basic choice is between a carbon tax and a cap-and-trade system, as discussed below.

Second, policy must provide appropriate investment incentives. For example, the design life of a coal-fired power station is around 50 years, so the investment decision in this case will depend not so much on the emissions price today, but on expectations of what the time path of future emissions prices will be over the next 50 years. The lesson for policy is that long-term credibility of the policy regime is vital if lock-in of high-emission technologies is not to occur and if, via a higher time path for energy prices, appropriate investment in low-emission technologies such as nuclear power and wind turbines is to be stimulated.

Third, appropriate incentives for the development of new technologies are needed. Part of this is achievable via credible and enduring emissions pricing, as above, but there is also a strong case for public subsidy, due to technological spillovers—gains that cannot be captured by the originator of the advance. Research and development do matter, but demonstration is equally important. For example, the first producer to demonstrate carbon capture and storage (CCS) on a commercial scale will give an important stimulus to its development globally (while at the same time finding that further, induced advances render the original innovation obsolete).[21] Lock-in and the absence of a global deal on global emissions pricing create a particularly strong case for the development and international transfer of CCS technology, especially to China, where the expansion of power generation based on cheap and plentiful coal reserves is rapid and where total emissions now exceed those of the United States.[22]

Fourth, since the first-best solution of direct emissions pricing is practicable only where emissions can be directly monitored at a reasonable cost—thus limiting the application of this strategy to fixed industrial establishments—indirect pricing will often be needed, for example, taxing fuel rather than fuel-related emissions. This can be combined with regulatory or tax measures to stimulate procedures such as carbon capture that can lower emissions per unit of fuel used.

Fifth, bounded rationality and imperfections in credit and insurance markets create a situation that calls for regulation and/or subsidy. This is the case when dealing with the insulation of buildings and restrictions on house building in areas potentially more at risk of flooding as sea levels rise.

Sixth, 17.4 percent of global CO_2 emissions—more than the US total—derive from forestry, and about 80 percent of these are negative emissions associated with deforestation (IPCC 2007: fig. 2.1).[23] Creating appropriate incentives in this area is especially difficult, as noted earlier. Even if a baseline can be defined so that net gains of forest from that baseline can earn emissions credits, verifying such gains on a continuous basis is expensive. In addition, since property rights over forest are often poorly defined, it may not be clear to whom the associated payments should be made. One response to this might be to advocate titling and the privatization of forests, but such 'clarification' of property rights might easily lead to more—not less—forest clearing to make way for farming (Stern 2008b: 15).

Seventh, policy must not ignore adaptation to climate change. It is true that adaptation effort, in contrast to mitigation, mostly generates benefits that are reaped by the individual adapting, so that market failure is not inevitable. Nevertheless, there are some areas in which market responses will not suffice. For example, defenses against rising sea levels are a quintessential public good that will be set in motion only as a result of collective action. Global warming is also forecast to intensify drought in many areas; however, technological spillovers, as well as credit market deficiencies and imperfect information, mean that research and development efforts to generate drought-resistant crop varieties will be insufficient if left to the market.

Cap-and-Trade versus a Carbon Tax: Kyoto, Copenhagen, and Beyond

At a simple level, there is a fundamental equivalence between price and quantity methods for limiting CO_2 emissions, as illustrated in figure 5. In the figure, the line labeled 'marginal benefit' shows the value, at the margin, of extra emissions to those who emit. This can be most simply thought of as the extra profit that would be earned as a result of emitting one more unit of CO_2 (some of the gain would accrue to consumers as well, via lower prices of emissions intensive products), and this clearly must decline as total emissions increase. In the absence of any policy, total emissions will be E_0. The line labeled 'marginal social cost' represents the social cost to future generations, at the margin, of extra emissions. This line slopes upward if additions to the global stock of GHGs progressively create more damage. However, the slope might be quite gentle if the diagram is interpreted as referring to emissions over the next five years, for instance, because of the possibility of compensating for extra emissions now by spreading out a subsequent reduction over several decades.

Point A in the figure is optimal: raising emissions from zero up to E* yields benefits at the margin that exceed costs, but the limit is reached at A, where marginal cost and marginal benefit are equal. Point A can be achieved with a carbon tax equal to OP*. With this tax, firms will limit their emissions in aggregate to E*, and revenue equal to rectangle OP*AE* will accrue to the government. Equivalently, A can be achieved if the government creates amount E* of emissions permits and distributes these by auction or 'grandfathering' (i.e., assignment to firms on the basis of history). Providing that the permits are freely traded, the market price will settle at P*, and both aggregate emissions and their distribution across firms will be the same as in the case of the tax.[24] A hybrid system that yields the same result is also possible, whereby there are some grandfathered tradable 'perpetual' permits (fewer than E*), together with an elastic supply of short-term permits made available by the government at price P* (McKibbin and Wilcoxen 2002).

Two issues are central to choice among these mechanisms—uncertainty and income distribution. On uncertainty, note that the marginal benefit curve in figure 5 also represents the inherently unpredictable demand for emissions. Suppose there is an unforeseen fall in emissions demand. In the figure, line CAB shifts to the left, as indicated by the dotted line. Under a carbon tax, emissions fall (point d), while under a permit system, emissions stay the same but their price falls (point f). This is exactly what happened in the EU's emissions trading scheme in April 2006, when the price of permits fell almost to zero. Determining which outcome is preferable depends on the time horizon.[25] Over the short term, with the marginal social cost of

Figure 5: Carbon taxation versus emission permits

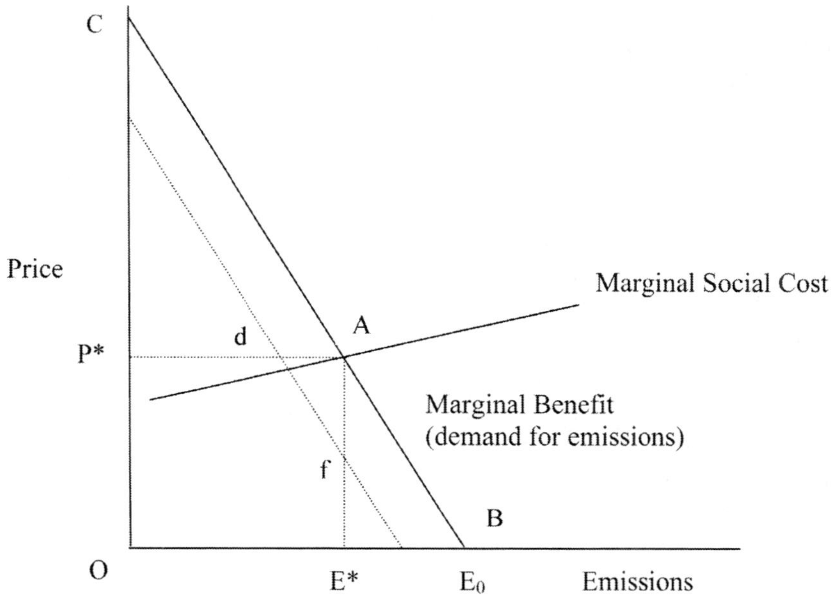

emissions relatively constant, it makes more sense in a recession to cut emissions substantially. Saving emissions 'cheaply' now is worthwhile in relation to the future benefits, whether taken in the form of higher emissions or a lower eventual GHG stock. Therefore, a tax system is superior to a permit system. Over the long term, the argument is reversed. Climate 'tipping points' may mean that the costs associated with extra emissions become very high, so it is better to control quantity, allowing price to vary. The way out of this apparent dilemma is to control emissions via a carbon tax and to revise the tax rate at, say, five-year intervals in light of trends in emissions and atmospheric GHG concentrations. A merit of McKibbin and Wilcoxen's (2002) hybrid system is that it handles uncertainty in just the same way as a carbon tax system.

Issues of income distribution, both nationally and internationally, create very serious obstacles to the introduction of policies to limit emissions. Consider the national level first. The essential difficulty is that both a tax system and a pure auction-based permit system imply substantial revenue transfers from firms and consumers to the government. In figure 5, these revenues—either in the form of tax receipts or as proceeds from the auctioning of permits—equal the area of rectangle OP*AE*, and they relate, of course, to the total volume of emissions, not to the much smaller reduction in emissions. Firms and consumers lose, between them, area OP*AB, equal to the revenue transfer to the government plus a 'deadweight' loss ABE*, sacrificed for the benefit of future generations. The government must return the revenue to the citizenry (or spend it), but such large-scale recycling of revenue raises uncomfortable political questions over who is to gain or lose and how the process can be made free of corruption. The avoidance of such large financial transfers is the second claimed advantage of McKibbin and Wilcoxen's (2002) hybrid scheme. Nevertheless, to the extent that this system cushions producers by giving them free perpetual permits, the revenues available to relieve consumers of some of the cost of energy price increases are necessarily reduced.

At the global level, the same issues arise, albeit in a different form. Suppose, for concreteness, that a global scheme were to begin by covering only emissions related to fossil fuels. Suppose also that choke points in the distribution of fossil fuels could be identified so as to allow for the credible and verifiable measurement of fossil fuel use and therefore the application either of a global carbon tax or an emissions permit system.[26] With a global carbon tax, revenues would accrue to national governments for them to spend as they saw fit, subject only to the restriction that this spending did not undo the emissions-reducing incentives of the tax.

To reproduce exactly this outcome with a cap-and-trade system, it would be necessary to forecast, country by country, the emissions that the carbon tax would have generated and then simply assign the equivalent number of permits to each country, with countries being free to assign permits internally.[27] If the forecasts were accurate, the result would be no international trading of permits—'cap-without-trade', in effect.

As shown above, in the absence of uncertainty and leaving aside issues about how countries would assign emissions permits internally under cap-and-trade, these two systems would come to the same thing. However, global equity issues arise at several levels. One possible ethical standpoint is that foreign aid and emissions control should not be mixed up, in which case one would ask whether a global carbon tax would be more or less neutral as regards the international distribution of income. Otherwise put, would the deadweight costs of emissions control in this case, country by country, be in rough proportion to the future benefits, also calculated country by country? It was noted earlier that efficient emissions reduction is probably skewed toward poor countries, which might seem to suggest that they would be net losers. However, against this is the fact that a substantial proportion of these reductions is forestry related and would therefore lie outside the scope of a carbon tax.[28] So perhaps a global carbon tax, supplemented by a mechanism that would allow rich countries to pay for forest preservation and expansion in poor countries, would be distributionally neutral in the sense of this paragraph. The permit-based equivalent system would then have little to be said for it, given that, by design, international trade in such permits would be negligible.

In practice, international climate policy has not been based on the principle of distribution neutrality outlined above. The Kyoto Protocol, which led to the setting up of the first national emissions trading scheme in the UK in 2004 (which was merged into the EU Emissions Trading Scheme in 2005), was based on the principle that national emissions caps would be set 5 percent below a baseline of actual emissions in 1990. This might not have departed too far from distributional neutrality as far as the 'Annex 1' industrialized countries were concerned. But also included in the Protocol was the CDM, the baseline-and-credit scheme that allowed developed countries to finance emissions reductions in developing countries.

It was hoped that a global successor to Kyoto, based also on cap-and-trade rather than a carbon tax, would be achieved at the 2009 Copenhagen summit, but this hope was not realized.[29] Central to why it was possible to arrive at agreement on cap-and-trade among the developed countries at Kyoto but impossible to extend this agreement to the world as a whole at Copenhagen is the question of distribution. An alternative ethical principle to 'forward-looking' distribution neutrality, as outlined above, is one that holds that countries' rights are to their share of stocks of GHGs in the atmosphere, rather than additions to those stocks—an approach that is known as the Brazilian Proposal (see Walker and King 2008: 171). As almost all of the rise in atmospheric GHGs since the Industrial Revolution is attributable to economic activity in rich countries, the application of this principle would mean that most, if not all, emissions permits under cap-and-trade would be allocated to poor countries in the first instance, with subsequent sales to rich countries, generating substantial international financial transfers.

Any agreement along these lines seems remote, for three main reasons. First, there is no consensus on the Brazilian Proposal itself. As Buiter (2009) contends, going back to 1800 in order to argue for distribution neutrality forward from that date would involve, in principle, the impossible task of trying to assess all of the consequences of actions in rich countries for the welfare of poor countries from 1800 on. There could be no basis in principle for considering the GHG consequences in isolation. Second, even were there to be such a consensus, agreeing on a methodology for translating this into an allocation of emission permits by country would seem an impossible task. Third, the financial transfers from rich to poor countries that would flow from any such agreement would be government to government, with no guarantee that flows to poor countries would mean flows to poor people. According to Cooper (2006: 2), such international trading in emissions rights "entails potentially large transfers between law-abiding citizens in rich countries such as the USA, Canada, Europe, and Japan to corrupt officials and their favored oligarchs in countries less meticulous about the rule of law."

Conclusions

The argument of this article may be summarized in the following propositions.

1. Human action since 1800 has led to atmospheric GHG concentrations rising far above the range in which they have fluctuated in the past 600,000 years, and under BAU, given the plentiful remaining stocks of fossil fuels in the world, this rise will continue at a significant rate for the foreseeable future.
2. Global warming in the past 150 years has been very rapid by historical standards. The evidence is consistent with it being largely or wholly of anthropogenic origin, and under BAU, 'likely' warming by 2100 is in the range 1.1 °C–6.4 °C.
3. GHG emissions create a global externality and, accordingly, a *prima facie* case for global action to reduce them. The size of the externality is subject to a great deal of uncertainty, both scientific and economic, and stabilizing GHG concentrations at 550 ppm CO_2eq or below is therefore best seen as taking out insurance against catastrophic outcomes, most importantly serious sea level rises and significant parts of the globe being rendered uninhabitable.
4. Creating a credible and enduring price for emissions must be the central plank of policy, so that the direction of technological advance and investment choices, as well as current consumption choices, can be skewed toward low-emission choices. But supporting policies are needed to create appropriate incentives for dispersed hard-to-tax activities and activities with negative emissions, and to allow for spillovers associated with technological advance.
5. The limbo in which international cooperation in this area finds itself after the Copenhagen summit is due largely to the misguided conflation of concerns over global poverty and/or inequality with the question of global GHG emissions.
6. This reasoning also helps to explain why an agreement on cap-and-trade among the rich countries did prove possible at Kyoto, while, thus far, global cap-and-trade has not been achieved. The implication is that a global carbon tax, together with international transfers focused on technology transfer and forest-related negative emissions, may be the best avenue for future international cooperation.

ROBERT EASTWOOD is Senior Lecturer in Economics at Sussex University, specializing in Development Economics, in particular the links from demography to economic growth and poverty in developing countries. His recent and forthcoming published works include "Farm Size" (2009, with Michael Lipton and Andrew Newell) and "Demographic Transition in Sub-Saharan Africa: How Big Will the Economic Dividend Be?" (2011, with Michael Lipton). He was an expert witness to the recent All Party Parliamentary Group on Population, Development and Reproductive Health and is a member of the Steering Group of the current African Economic Research Consortium project on reproductive health, economic growth, and poverty reduction in Africa.

NOTES

1. In IPCC terminology, 'very likely' means with a probability of above 90 percent. For 'likely', the corresponding figure is 66 percent (IPCC 2007: 27).
2. Interest in the possibility of global cooling was widespread in the 1970s (see numerous references cited at http://en.wikipedia.org/wiki/Global_cooling). The entire development of modern civilization, from the beginning of settled agriculture in the Fertile Crescent some 11,000 years ago, has coincided with—and may even be viewed as having been courtesy of—the current interglacial warm period.
3. The idea that negotiation can resolve the externality problem is known in economics as the 'Coase theorem', named after Nobel Laureate Ronald Coase (1960).
4. It has, however, been noted by Foley (2008) that if mitigation investments were entirely financed by reducing non-mitigation investments, the burden of mitigation could be shifted to the future beneficiaries, so that no one in the current generation would need to bear a cost. Implicit in the mainstream discourse, including both Stern (2007) and his critics, is that full intertemporal compensation of this kind is not possible in practice.
5. 'Zero pure time preference' means that if there are two persons, one living today and one in the future, both of whom are enjoying the same level of consumption, then society values equally an extra unit of consumption to each of them. 'Positive pure time preference' means that the extra unit to the person living today is more highly valued.
6. This idea can be expressed in terms of an annual 'discount rate'. For example, a discount rate of 5 percent means that an extra unit of consumption next year is of equal social value to $1/(1.05) \approx 0.95$ extra units this year. Even if pure time preference is zero, positive discounting is justified if the potential gainer next year is (already) better off than the potential gainer this year. The effect of compounding is that even a small, constant positive discount rate can make consumption gains in 50 or 100 years from now almost valueless in comparison with consumption gains today.
7. It is estimated that a rise of atmospheric CO_2 to 550 ppm would cause irreversible melting of the Greenland ice sheet, raising global sea levels by 6.5 meters. Simple physical models suggest that this process would take 3,000 years. Against this, for reasons that are beginning to be understood, loss of ice from Greenland has recently been much faster than the models predict, suggesting that a catastrophe involving this ice sheet cannot be regarded as an impossibility. Similar uncertainties apply to the West Antarctic ice sheet (King 2006; Walker and King 2008: chap. 5).
8. The essential idea behind the 'veil of ignorance' argument can be conveyed by considering an idealized situation in which there will be two persons in the future, A and B, with different consumption levels, and the planner today needs a measure of their total welfare, taking into account the preferences of each person. Imagine that A and B are somehow around to be consulted and that neither of them knows now whether they will be occupying their own or the other's shoes in the future—the two possibilities being assumed equiprobable. This construction has the effect of converting inequality into risk.

9. See, for instance, "Yale Symposium on the Stern Review," http://www.ycsg.yale.edu/climate/forms/FullText.pdf (accessed 12 October 2010), and other references cited below.
10. In principle, the time path of stabilization costs could be similarly converted into BGE equivalent terms, but in fact Stern's (2007: 239) figure of 1 percent is an estimate of the cost as a share of GDP in 2050.
11. Nordhaus (2007: 691) refers to this as "Government House utilitarianism."
12. This assumes, as in Stern's no-climate-change baseline, growth in consumption per head of 1.3 percent per annum (see, e.g., Stern 2007: chap. 2, technical annex, equation 8).
13. Stern (2007: 239) estimates that the cost for the range 450–500 is about three times as great.
14. To illustrate approximately where this projected doubling during 2010–2050 comes from, consider that 40 years of growth of global GDP per head at 1.7 percent per annum, together with the expected rise of global population from 6.5 to 9 billion, would cause global GDP to double. If emissions per unit of GDP did not change under BAU, then emissions would also double.
15. For another analysis of technology and its application to climate problems, see Pacala and Socolow (2004).
16. The reduction in emissions is somewhat larger than the equivalent amount in figure 3, mostly because Enkvist et al.'s (2010) estimate for BAU emissions in 2030 is 10 percent higher than that in the Stern Review.
17. In contrast to the Stern Review and an estimate by Enkvist et al. (2007), Enkvist et al. (2010) add up the overall cost of this reduction and estimate it to be *negative*. This change from the 2007 estimate of Enkvist et al., which was approximately in line with the Stern Review's figure of 1 percent of world GDP, results from an upward revision in the assumed future path of world energy prices.
18. The CDM is a 'baseline-and-credit scheme', inasmuch as emissions credits are created, on a project-by-project basis, in respect of emissions cuts below a defined baseline. Transaction costs are high, often exceeding $500,000 per project (Ellis and Kamel 2007, cited in Stern 2008b: 15)
19. House insulation is an example of an investment where the return is reasonably certain. Where an investment would reduce risk, market failure is also common (Kleindorfer and Kunreuther 2000). For example, losses from Hurricane Katrina in the US Gulf states were eight times higher for those who had failed to install recommended hurricane proofing than for those who had installed it. Hurricane proofing of $2.5 million saved $500 million in damages (Dankwa 2006).
20. For a trenchant expression of the view that climate negotiations should pay no attention to historical responsibility for the current level of atmospheric GHGs, see Buiter (2009).
21. Stern (2008a) reports a remark made to him by the prime minister of India to the effect that India would introduce CCS technology once it had been demonstrated to work elsewhere.
22. As Cooper (2006: 3) puts it: "Thus here is an arena for practical international cooperation, with Americans and others providing the technical know-how, China providing the experimental ground, and the rich countries together paying the incremental costs—not for power generation, but for potential sequestration."
23. For the 1990s, emissions associated with deforestation are estimated by the IPCC at 5.8 $GtCO_2eq$, which is 13.8 percent of the average of global emissions for 1990 and 2000 (Metz et al. 2007: chap. 9, Executive Summary; see also IPCC 2007: fig. 2.1a).
24. Necessary for this scenario is that the use of permits today confers no right or expectation of receiving permits tomorrow.
25. Technically, the answer depends on the slopes of the marginal cost and benefit curves. For a thorough analysis, see Stern (2007: chap. 14).
26. Cooper (2006: 8) suggests that the International Monetary Fund could be charged with the monitoring role.
27. Cooper (2006: 2) contends that this would make environmental programs the "handmaiden of corruption" in many countries.
28. In figure 5, provided that the marginal benefit curve is a straight line, then the deadweight cost equals $0.5(AE^*)(E^*E_0)$ and so is proportional to the amount of efficient emissions reduction in a country.
29. One potential program for Copenhagen was contained in Stern (2008b).

REFERENCES

Broecker, Wallace S., and Robert Kunzig. 2008. *Fixing Climate: What Past Climate Changes Reveal about the Current Threat—and How to Counter It*. New York: Hill and Wang.

Buiter, Willem M. 2009. "Does Poverty Give a Country the Right to Pollute the Atmosphere?" 24 July. http://blogs.ft.com/maverecon/2009/07/does-poverty-give-a-country-the-right-to-pollute-the-atmosphere/ (accessed 12 October 2010).

Coase, Ronald. 1960. "The Problem of Social Cost." *Journal of Law and Economics* 3: 1–44.

Cooper, Richard N. 2006. "Alternatives to Kyoto: The Case for a Carbon Tax." http://74.125.155.132/scholar?q=cache:Fqt1mLmdUxcJ:scholar.google.com/+Cooper+%22the+case+for+a+carbon+tax%22&hl=en&as_sdt=2000 (accessed 12 October 2010).

Dankwa, David. 2006. "FM Global Touts Underwriting by Engineering as Superior." *Best's Review* 107 (2): 93–93.

Dasgupta, Partha. 2008. "Discounting Climate Change." *Journal of Risk and Uncertainty* 37: 141–169.

Deaton, Angus. 2007. "Letter from America—On Transatlantic Vices, or Stern in America." *Royal Economic Society Newsletter* 139: 3–4.

Dietz, Simon, Dennis Anderson, Nicholas Stern, Dimitri A. Zenghelis, and Chris Taylor. 2007. "Right for the Right Reasons: A Final Rejoinder on the Stern Review." *World Economics* 8 (2): 229–258.

Ellis, Jane, and Sami Kamel. 2007. "Overcoming Barriers to Clean Development Mechanism Projects." OECD/IEA/UNEP Risø Centre. http://www.oecd.org/dataoecd/51/14/38684304.pdf.

Enkvist, Per-Anders, Jen Dinkel, and Charles Lin. 2010. "Impact of the Financial Crisis on Carbon Economics." McKinsey & Company. http://www.mckinsey.com/clientservice/sustainability/pdf/Impact_Financial_Crisis_Carbon_Economics_GHGcostcurveV2.1.pdf.

Enkvist, Per-Anders, Tomas Nauclér, and Jerker Rosander. 2007. "A Cost Curve for Greenhouse Gas Reduction." *McKinsey Quarterly* 1 (February): 35–45.

Foley, Duncan. 2008. "The Economic Fundamentals of Global Warming." Pp. 115–126 in *Twenty-First Century Macroeconomics: Responding to the Climate Challenge*, ed. Jonathan M. Harris and Neva R. Goodwin. Cheltenham: Edward Elgar.

Garnaut. Ross. 2008. *The Garnaut Climate Change Review*. Cambridge: Cambridge University Press.

Harsanyi, John. 1955. "Cardinal Welfare, Individualistic Ethics, and Interpersonal Comparisons of Utility." *Journal of Political Economy* 63 (4): 309–321.

Heal, Geoffrey. 2009. "Controversies in the Economics of Climate Change." London School of Economics and Political Science public lecture. 6 May. http://richmedia.lse.ac.uk/publicLecturesAndEvents/20090506_1830_controversiesInTheEconomicsOfClimateChange.mp3 (accessed 12 October 2010).

IPCC (Intergovernmental Panel on Climate Change). 2007. *Climate Change 2007: Synthesis Report. Contribution of Working Groups I, II and III to the Fourth Assessment Report of the Intergovernmental Panel on Climate Change*. Ed. Core Writing Team, R. K. Pachauri, and A. Reisinger. Geneva: IPCC.

Jansen, E., J. Overpeck, K. R. Briffa, J.-C. Duplessy, F. Joos, V. Masson-Delmotte, D. Olago, B. Otto-Bliesner, W. R. Peltier, S. Rahmstorf, R. Ramesh, D. Raynaud, D. Rind, O. Solomina, R. Villalba, and D. Zhang. 2007. "Palaeoclimate." In *Climate Change 2007: The Physical Science Basis. Contribution of Working Group I to the Fourth Assessment Report of the Intergovernmental Panel on Climate Change*, ed. S. Solomon, D. Qin, M. Manning, Z. Chen, M. Marquis, K. B. Averyt, M. Tignor, and H. L. Miller. Cambridge: Cambridge University Press.

King, David. 2006. "Climate Change: Global Solutions for an International Problem." Ralph Miliband Global Risks and Politics in the 21st Century Lecture Series, London School of Economics and Political Science, 20 November. http://www2.lse.ac.uk/publicEvents/events/2006/20060905t1303z001.aspx.

Kleindorfer, Paul R., and Howard Kunreuther. 2000. "Managing Catastrophe Risk." *Regulation* 23 (4): 26–31. http://www.cato.org/pubs/regulation/regv23n4/kleindorfer.pdf (accessed 12 October 2010).

McKibbin, Warwick, and Peter Wilcoxen. 2002. "The Role of Economics in Climate Change Policy." *Journal of Economic Perspectives* 16 (2): 107–129.

Metz, B., O. R. Davidson, P. R. Bosch, R. Dave, L. A. Meyer, eds. 2007. *Contribution of Working Group III to the Fourth Assessment Report of the Intergovernmental Panel on Climate Change*. Cambridge: Cambridge University Press.

Nordhaus, William. 2007. "A Review of the Stern Review on the Economics of Climate Change." *Journal of Economic Literature* 45 (3): 686–702.

Pacala, Stephen, and Robert Socolow. 2004. "Stabilization Wedges: Solving the Climate Problem for the Next 50 Years with Current Technologies." *Science* 305: 968–972.

Shaviv, Nir. 2006. "Carbon Dioxide or Solar Forcing?" http://www.sciencebits.com/CO2orSolar (accessed 12 October 2010).

Stern, Nicholas. 2007. *The Economics of Climate Change: The Stern Review*. Cambridge: Cambridge University Press.

———. 2008a. "Climate Change, Energy and the Way Ahead." London School of Economics and Political Science public lecture. 27 February. http://richmedia.lse.ac.uk/publicLecturesAndEvents/20080227_1830_climateChangeEnergyAndTheWayAhead.mp3 (accessed 12 October 2010).

———. 2008b. "Key Elements of a Global Deal on Climate Change." London School of Economics and Political Science. http://citeseerx.ist.psu.edu/viewdoc/download?doi=10.1.1.153.3745&rep=rep1&type=pdf (accessed 12 October 2010).

Urry, John. 2010. "Sociology and Climate Change." *Sociological Review* 57 (s2): 84–100.

Walker, Gabrielle, and David King. 2008. *The Hot Topic: What We Can Do about Global Warming*. London: Bloomsbury Publishing.

Weitzman, Martin. 2007. "A Review of the Stern Review on the Economics of Climate Change." *Journal of Economic Literature* 45 (3): 703–724.

Origins, Uses, and Transformation of Extinction Rhetoric

Richard J. Ladle and Paul Jepson

ABSTRACT: The concept of extinction is at the heart of the modern conservation movement, and massive resources have been spent on developing models and frameworks for quantifying and codifying a phenomenon that has been described by American researcher and naturalist Edward O. Wilson as an obscure and local biological process. Scientists, environmentalists, and politicians have repeatedly used extinction rhetoric as a core justification for a global conservation agenda that seeks to influence a wide range of human activities despite the inherent difficulty and uncertainty involved in estimating current and future rates of extinction, or even in verifying the demise of a particular species. In this article we trace the historical origins of the extinction concept and discuss its power to influence policies, agendas, and behaviors. We argue that conservation needs to develop a more culturally meaningful rhetoric of extinction that aligns scientific evidence, cultural frames, institutional frameworks, and organizational interests.

KEYWORDS: climate change, conservation, crisis, extinction, frames, Red Lists, rhetoric

> The death of a species is a more remarkable event than the end of an imperial dynasty. (Orton 1869: 540)

Extinction is typically viewed by contemporary conservation scientists as the logical endpoint of the process of population decline, that is, the point on the graph where the population size curve meets the x-axis and terminates abruptly and finally (Ladle and Jepson 2008). The operational definition given by the International Union for Conservation of Nature (IUCN) is that a species should be considered extinct "when there is no reasonable doubt that the last individual has died" (IUCN 2001: 14). This definition reveals one of the main stumbling blocks for field-based measurements of extinction: the difficulty of ascertaining the continued existence of a species that is certainly exceedingly rare and may also inhabit an isolated area that is difficult to survey effectively. The final death of a species, beyond the gaze of the scientists and conservationists, led Edward O. Wilson (1992: 255) to conclude that extinction is "the most obscure and local of all biological processes."

The IUCN guidelines state that a species can be declared extinct only after exhaustive surveys fail to produce any observations within a time period and geographical range that are appropriate to its life cycle and life form—an unfeasible task for most species (Roberts 2006). Butchart

Environment and Society: Advances in Research 1 (2010): 96–115 © Berghahn Books
doi:10.3167/ares.2010.010105

6

and his colleagues have recently reinstated the defunct category of 'possibly extinct' to apply to those species that are, "on the balance of evidence, likely to be extinct, but for which there is a small chance that they may be extant and thus should not be listed as Extinct until adequate surveys have failed to find the species and local or unconfirmed reports have been discounted" (Butchart, Stattersfield, and Brooks 2006: 9). Interestingly, this necessary 'gray area' has not prevented many species from being prematurely labeled as extinct, allowing them to be 'rediscovered' and, in the process, generating many positive headlines (Ladle et al. 2009).

The above definitions of extinction are concerned with species that are known to science, have been collected and processed, and exist as physical specimens in the world's biological collections. However, there may be anywhere from 1 to 100 million species for which science has no official records at the present time (Lomolino 2004). Extinctions of these 'yet to be discovered' species have been of great importance in the construction of the 'extinction crisis', and they are claimed to outnumber documented extinctions many times over (Whittaker et al. 2005). Such 'unseen' or postulated extinctions are highly dependent on estimates of global species richness and are the source of most of the headline-grabbing figures that periodically appear in the media. Moreover, the majority of such extinctions are in poorly described taxa (groups of organisms) and ecosystems, such as arthropods in tropical forests.

Despite the inherent uncertainty associated with identifying, measuring, and forecasting extinctions (Ladle 2009), the concept of extinction is a prominent element of issues relating to the environment. It has been instrumental in the production of international conservation agreements and associated programs and, for more than a century, has been deployed to attract attention and motivate individual and collective actions that constrain our attitudes and behaviors toward the non-human world.

Extinction is the focal point of a vast scientific literature, concentrating mainly on the development of techniques to measure and forecast extinction rates (reviewed in Lawton and May 1995; see also Ladle 2009). These techniques are based on insights from diverse fields, including paleontology (Jablonski 2001; McKinney 1997), genetics (Keller and Waller 2002), island biogeography (Channell and Lomolino 2000), behavior (Courchamp et al. 1999), and taxonomy (Nee and May 1997). The wide variety of ways in which extinction can be defined, classified, and measured has (perhaps unsurprisingly) led to increasing levels of misrepresentation of this concept in the public domain (Ladle et al. 2005). In an attempt to clarify the diverse meanings of the term 'extinction' and to explain its power to influence institutional and individual behaviors, Ladle and Jepson (2008) presented a new typology of extinction concepts associated with different degrees of scientific certainty concerning the disappearance of a species and the potential for its re-emergence at some point in time. They briefly discussed the social influence and power of these different extinction 'types' in different settings and, in so doing, drew attention to the multiple and changing usages of the term.

In this article we present an extended review and discussion about the origins and agency of the term 'extinction' and examine its historical and contemporary impact in conservation discourse, policy, and management. To provide greater analytical traction, we draw on the concept of the 'frame' (Goffman 1974), which suggests that people make sense of and act within a complex world by gathering together an assemblage of ideas, objects, and practices in frames (i.e., mental models/schemas). The concept has been extended by new social movement theorists (e.g., Benford and Snow 2000) and is gaining popularity in science technology studies (e.g., Callon 1998), policy studies (e.g., Triandafyllidou and Fotiou 1998), including conservation (e.g., Lorimer 2006), and environmental communication (e.g., Davis 1995). Frames are essentially stories constructed from concepts, metaphors, beliefs, and images interacting with the everyday practices and technologies through which we live our lives. While each person constructs his or her own frame and is able to switch between multiple frames, frame construction is largely a social and cultural process. Frames develop over time and can be understood as the sedimented histories of particular ways of understanding and engaging with the world. When frames include collective actions that attract widespread consent, they become institutionalized, guiding policies and scientific, management, and cultural practices (Tarrow 1992).

The premise we explore in this article is that extinction—as a term, concept, fact, and possibility—organizes and assembles frames that incorporate individual and collective anxieties relating to loss, decline, crisis, and so forth. These frames help to explain and give meaning to events in the non-human world and motivate and legitimate ideas for collective action. At the same time, they enable conservation/environment issues to bridge and flow into numerous other frames. Further, we suggest that the rhetorical power and cultural agency of extinction enroll actors not directly concerned with avoiding species extinction, consciously or otherwise, into pronouncements and interventions in the environment/conservation issue frame. These processes serve to transform extinction from a key idea element within wider frames to a distinct issue frame that has come to have an immense influence on social decision making.

In considering these ideas, this article is organized into three sections. The first section presents a brief review of the origins and historical context of the extinction concept. The second section focuses on the power of this concept to transform intentional collective or individual actions (*sensu* Engeström 2006; Eskola 1999) and examines the interplay between the domains of scientific practice, the media, and conservation policy and governance institutions. In the concluding section, we take stock of the influence of extinction on decision-making frames and explore areas of further investigation that would extend our understanding of the cultural, social, and political dimensions of this concept.

The Origins of the Extinction Concept

The history of the concept of species extinction has two separate components. The first involves the realization during the nineteenth century that it is possible for a species to cease to exist. The second entails the recognition that the decline and eventual disappearance of a species can be directly or indirectly attributable to human impacts on the environment (Lowe 1983). Neither of these processes occurred rapidly or completely, and it was probably not until the first decades of the twentieth century that a scientific consensus emerged on the role of humans in extinction, following the high-profile disappearance of several formerly abundant and highly visible species, such as the passenger pigeon.

The slow emergence of a concept of extinction is unsurprising, given the almost complete lack of biogeographical knowledge and the strict biblical interpretations that dominated Western thinking in the natural sciences until the middle of the nineteenth century. Perhaps the best candidate for the first documented extinction is a plant referred to as 'silphium' by Pliny the Elder in AD 77, who commented that its resin was a valuable and effective remedy against illnesses ranging from fevers, coughs, and warts, not to mention its role as a useful contraceptive (Parejko 2003). Pliny goes on to describe a drastic decline in silphium availability and that, within his own lifetime, the plant had not been seen in its native habitat for many years, the last known stalk, valued at its weight in gold, having been sent to the emperor Nero (ibid.). However, it was not until detailed analysis of the fossil remains of species such as the mammoth in the late eighteenth century that extinction emerged as a socially important theme with potential theological implications. The central figure in establishing extinction as an undeniable fact was Georges Cuvier (1769–1832), a professor of animal anatomy at the French National Museum of Natural History in Paris. Cuvier's careful reconstructions of fossil elephants led him to conclude that (1) they were very different from any living species; (2) given their size and dramatic appearance, it was very unlikely that they still existed; and (3) the probable cause of extinction was some form of periodic catastrophe (Rudwick 1997).

By 1859, the year Darwin published his seminal work, extinction was already widely accepted among the academic community, although great doubts still remained about what could have driven enormous animals such as dinosaurs and mastodons out of existence. These doubts extended to more contemporary examples of extinction where, in hindsight, human involvement seems obvious and decisive. A good example is the Great Auk (*Pinguinus impennis*), a large and impressive seabird that was hunted into extinction by European fisherman. According to nineteenth-century accounts, the last specimens were collected by a party of Icelanders in 1844 (Bengtson 1984). Although in hindsight overhunting was clearly to blame for the demise of the Great Auk, contemporary writers had great difficulty accepting the pivotal role of human action. As James Orton (1869: 540) expressed it: "The upheaval or subsidence of strata, the encroachments of other animals, and climatal revolutions—by which of these great causes of extinction now slowly but incessantly at work in the organic world, the Great Auk departed this life, we cannot say."

This reluctance to attribute human causes to the contemporary extinction of species lasted into the twentieth century, with many (but by no means all) commentators being reluctant to point the finger of blame at fellow humans for playing an important role in extinctions. The reasons behind the decline of the passenger pigeon (*Ectopistes migratorius*)—a heavily hunted species that has become iconic to the modern conservation movement—were also actively debated in the scientific literature of the time. In a publication of the American Ornithologists' Union, G. C. Tremaine Ward (1901: 191) stated: "I am not aware of any satisfactory explanation of the phenomena [i.e., the decline of the passenger pigeon]. It is not improbable, some epidemic disease, spreading more rapidly on account of the immense number of individuals, nearly exterminated the species."

As with the acceptance of the fact of extinction, the scientific community slowly coalesced around the idea that this new wave of extinctions was most likely being driven by human action, often through direct persecution. This new understanding mobilized the formation of influential societies, notably the Boone and Crockett Club in New York, founded in 1887 by Theodore Roosevelt, and the Society for the Preservation of the Wild Fauna of the Empire in London, originally founded in 1903 in Africa by British naturalists and American statesmen. These organizations and others like them promoted the social value that the "human conquest of nature carries with it a moral responsibility to ensure the survival of threatened life forms" (Whittaker et al. 2005: 4) and were a driving force behind the creation of wildlife sanctuaries and refuges in North America and colonial territories (Jepson and Whittaker 2002). This social value, along with the reality of extinctions, aligned with and was informed by nineteenth-century Western humanitarian preoccupations relating to cruelty to domestic animals (Thomas 1984) and their extension to practices of "bird slaughter for millinery ornament" (Doughty 1975: 103), market hunting (e.g., for bison tongues), and recreational hunting (Trefethen 1961).

These social trends of the late nineteenth century coincided with the rise of the print media, and newspaper reportage of such events constructed the concepts of human-induced extinction and associated moral responsibilities in the minds of European and East Coast American readerships. In conjunction with the debates surrounding the publication of Darwin's *Origin of Species*, they transformed Western framings of nature by downplaying the organizing role of religious ideas and introducing new scientific notions. Furthermore, the awful loss of life during World War I promoted associations in the public mind between the senseless killing of humans and that of wildlife, further enhancing the legitimacy of collective action to govern the processes leading to extinction (Jepson and Whittaker 2002). Thus, by 1931, the influential American biweekly magazine, *Science News-Letter* (now *Science News*) was prompted to write the following with respect to the rapid demise of the thylacine (*Thylacinus cynocephalus*), also known as the Tasmanian tiger: "Australia, which has had a development more or less analogous to that of the American West, is now passing through a phase also experienced in America—the realization that reckless slaughter is threatening extermination of many of its unique animal species."[1] The article goes on to report how the government of Tasmania had recently taken steps to preserve the remaining thylacines by banning the exportation of pelts.[2] The last known thylacine died in Tasmania's Hobart Zoo on 7 September 1936 (Bulte et al. 2003).

Such high-profile extinctions and the fear that colonial expansion into Africa would result in the decimation and extinction of large mammal populations did much to galvanize the nascent global conservation movement. One notable result was the 1933 Convention Relative to the Preservation of Fauna and Flora in Their Natural State, which was signed in London and became known as the London Convention. This treaty, which established a set of African parks and wildlife sanctuaries (Hingston 1931; Jepson and Whittaker 2001), convinced Harold Coolidge, a leading conservationist of the time, that extinctions could no longer be treated as singular and unfortunate events. Coolidge mobilized the funds to commission the first two global reviews of extinction (Allen 1942; Greenway [1958] 1967), which played a significant role in promoting the view of individual extinctions and declines as indicators of a much larger pattern of global extermination.

A more catastrophic view of nature's trajectory came to the fore in the 1970s in the context of Malthusian worries over the population bomb (Ehrlich 1971), anxieties due to the oil crisis, and dire forecasts about the fate of tropical forests (see, e.g., Myers 1979). Arguably, it was during this period that the term 'crisis' became attached to the extinction concept and began to be used more systematically and strategically as an advocacy device by both scientists and conservationists—a trend that reflected the extent to which the concept had become embedded within Western societies.

The Transformative Power of the Extinction Concept

> Those who believe that the rate [of extinction] is several times higher than normal often talk of a biodiversity or extinction "crisis." Crisis talk has tremendous rhetorical value in the political terrain of the North. (Sarkar 2005: 6)

Due to a combination of inherent scientific uncertainties and deeply held but frequently undisclosed beliefs about the intrinsic value of other species, extinction rhetoric occupies a unique space in contemporary science discourse where estimates and forecasts of species loss can vary dramatically in magnitude and certainty, depending upon the intended audience and the communication media. Following Sarkar (2005), here we define the extinction crisis as the belief that the current rate of extinction is several times higher than the normal background rate (from the geological past). It is difficult to pinpoint the beginnings of the crisis framing of extinction, but some members of the scientific community were certainly making claims of this nature by the late 1960s and early 1970s. For example, Iltis (1970) quotes S. Dillon Ripley of the Smithsonian Institute as saying that the majority of the world's animal species will be extinct by the year 2000.

One of the first and most widely disseminated estimates of the current rate of global extinction was produced by Norman Myers (1979) in his influential book, *The Sinking Ark: A New*

Look at the Problem of Disappearing Species. In this book, Myers made the claim that, on average, 40,000 species per year would become extinct over the course of twenty-five years (ibid.: 5). Although this figure was clearly a first-cut estimate based on circular reasoning and very little hard information, it has frequently been quoted by environmental organizations, politicians, and other interested groups. Myers's claim has also been used as an illustration of misrepresentation and scaremongering and was a key example in Lomborg's (2001) *The Skeptical Environmentalist*, a highly publicized critique of the global environmental movement. Recently, Myers (2001) admitted that the estimate of 40,000 extinctions per year was "preliminary and exploratory, and advanced primarily to get the issue of extinction onto scientific and political agendas." In this respect, he certainly achieved his aim, as testified by the publication of *Global 2000 Report to the President of the United States*, in which it was estimated that 500,000 to 2,000,000 species could become extinct between the years 1980 and 2000 (Barney 1980).

During the 1990s, the notion of an extinction crisis was extended and given added gravity by dubbing it the 'sixth extinction' event (Leakey and Lewin 1995). This frame amplification constructs extinction as a major present-day phenomenon—one that has not happened for 45 million years and one that, in contrast to the last five extinction events, which were caused by physical changes, has been caused by anthropogenic influences (Eldredge 2001). The sixth extinction rhetoric has been widely deployed by leading scientists and is popular with the media. We suggest that in this guise the concept of extinction is constitutive of processes to adopt the term 'anthropocene' for the current geological epoch.[3]

Scientific Practice

The first systematic scientific account of extinction was made by Walter Rothschild, a scion of the Rothschild family who was a zoologist, as well as a banker and politician. His paper presented at the 4th International Ornithological Congress (Rothschild 1905) and his subsequent book (Rothschild 1907) compiled accounts of extinct species that had appeared in lists of the faunas of regions and islands since 1580. The book can be understood as part of the broader scientific effort to document forms of life, but Rothschild's treatment of the causes of extinction and his unequivocal statement—"the melancholy fact … that man and his satellite dogs, cats, rats and pigs are the worst and indeed only important agents of destruction of the native avifaunas wherever they go"—helped establish the idea in scientific circles that human expansion into new territories needed governing. Interestingly, there are three features of Rothschild's account that have virtually disappeared from contemporary scientific discourse on extinction.

The first is the use of the present continuous tense of the verb 'to vanish'. This was prominent in the title of early reviews of extinct and nearly extinct birds and mammals (e.g., Allen 1942; Greenway [1958] 1967; Harper 1945), but seems to have fallen out of vogue in the 1960s and 1970s, with the notable exception of Tim Halliday's (1978) book *Vanishing Birds*. The replacement of 'vanishing' with the terms 'threatened', 'endangered', and 'vulnerable' represents a shift in the logic of extinction rhetoric, a move toward representing extinction as a risk that could be managed through the proper execution of governance technologies and techniques.

The second change in authoritative accounts of extinction—nowadays framed as the assessment of threatened species—is the absence of anecdotal accounts provided by non-expert travelers and residents of the regions. Rothschild regularly quoted from letters that provide descriptions of field observations. Indeed, Greenway ([1958] 1967) included a list of 'hypothetical' extinct species known only from eyewitness accounts. However, over time, such local ecological knowledge has gradually been replaced by evidence published in scientific papers and reports or collated through networks of experts developed and managed by conservation institutions. This increasing 'scientification' of extinction risk accounts has presumably served to strengthen the authority of extinction-related information in official policy circles.

Finally, contemporary accounts of threatened species rarely mention the possibility that extinction might be the result of natural phenomena. By contrast, Rothschild (1907) talked about natural cataclysms and devastated populations losing their vitality when discussing the demise of the New Zealand moa, a large flightless bird, whose extinction coincided with the arrival of humans on the islands. Perhaps unsurprisingly, given the widespread transformations in ecosystems since Rothschild's day, contemporary extinction discourse focuses almost exclusively on the anthropogenic causes of extinction, thereby emphasizing the need for governance of human behaviors and societal systems.

A more distinct science of conservation started to take shape in the late 1970s and early 1980s when conservation biology was recognized as a sub-discipline 'worthy' of academic study in its own right, with university courses, textbooks, and peer-reviewed journals. The first international conference dedicated to conservation biology took place as recently as 1978 at the University of California, San Diego. However, the study of conservation biology noticeably accelerated with the founding of the US-based Society for Conservation Biology in 1986 and the publication of the first edition of its influential journal, *Conservation Biology*, in 1987. Founded and supported by some of the 'giants' of modern conservation and ecology—Edward O. Wilson, Ernst Mayr, Michael Soulé, Paul R. Ehrlich—the society has led the development of conservation science into the twenty-first century and has been instrumental in both legitimizing conservation as an academic discipline and providing the hard data and strong concepts that underlie modern conservation practice. From its beginnings, conservation biology was framed as a crisis discipline. Indeed, Soulé (1985) went so far as to state that this characteristic primarily distinguishes conservation biology from related disciplines.

One of the major components of this crisis was the belief, based on a variety of scientific evidence, that the current rate of extinction was many times greater than normal background rates (Sarkar 2005). Much of this evidence was strong, especially the rates of documented extinctions on oceanic islands, and there was a justifiable consensus among the scientific community about the reality of unusually high extinction rates (Pimm 2002). The evidence came from two main sources: (1) historically documented extinctions, and (2) models, simulations, and frameworks that relate environmental change (e.g., habitat loss and transformation) to probabilities of the extinction of individual species or to rates of extinction within specified areas and time frames (Lawton and May 1995). It is this second source of evidence, especially with respect to estimates and predictions of future global extinction rates, that has been the foundation for many of the most high-profile claims of the extinction crisis and the sixth extinction event discussed above.

There are many ways to forecast extinction (reviewed in Ladle 2009), but probably the most widely used (and misused) method is based on the observation that the relationship between the size of an oceanic island and the number of species it contains can be effectively captured by a simple mathematical relationship, known as the species-area curve. Scientists have used this relationship to calculate how many fewer species should be found in ecosystems such as tropical forests after large areas have been cleared. An example of this is Wilson's (1992: 268) 'conservative' prediction of approximately 27,000 species going extinct every year, based on the rate of tropical deforestation. There is nothing fundamentally wrong with such extrapolations, but without an appreciation of the underlying assumptions, such crude statements may give a false impression of certainty to non-scientists and—potentially more damaging to the conservation movement—may also result in scientists being accused of hyperbole. In the case of the species-area relationship, the key assumptions are that (1) the number of species and the proportion of endemic species prior to habitat destruction are known; (2) the slope of the species-area relationship has been correctly determined; (3) terrestrial islands such as fragments of rainforest act like oceanic islands; (4) the number of species in the original habitat was already in equilibrium; and (5) when the area of the habitat is reduced, the species do not go extinct immediately but are slowly lost due to a range of demographic, genetic, and environmental effects (Ladle 2009; Whittaker et al. 2005).

Of these assumptions, the estimated total number of species has the most scope for influencing the figure for global extinctions (Whittaker et al. 2005). There are anywhere between 1 and 100 million species that are yet to be formally described by science, many of which are represented by arthropods in tropical forests. Choosing a higher number gives a higher number of total extinctions and thus a higher rate (Pimm 2002). It also means that when scientists speak about extinction in terms of thousands or millions, they are mainly referring to species that are yet to be discovered. From a rhetorical perspective, calculations such as Wilson's gave the scientific community shockingly high numbers that were guaranteed to attract the attention of the world's polities and strongly supported the narrative of environmental crisis.

This type of reasoning and extrapolation has been taken up enthusiastically by science-based conservation non-governmental organizations (NGOs) such as the World Wide Fund for Nature (WWF), and Conservation International (CI). Indeed, CI has arguably transformed the Wilsonian extinction crisis into a modern multimedia interactive experience through its constantly running 'extinction clock', which is introduced in the following way: "Every 20 minutes, one species is pushed to extinction as more than 1,200 acres of forest are destroyed. At the same time, over 180,000 tons of carbon dioxide is released into the atmosphere. Protecting and restoring forests is a key solution to climate change and is vital to the survival of all life on Earth. Indeed, 2,000 species are at risk of disappearing forever each month."[4]

CI's extinction clock gives several interesting insights into the legitimization of extinction rhetoric through scientific reasoning and the strategic alignment of extinction with the currently dominant frame of climate change in public discourse. First, it should be noted that these extinctions are forecast on the impacts of habitat loss and climate change and assume that global species richness is 4 to 6 million (i.e., 2 to 4 million more species than have currently been documented). As a consequence, of the three species "pushed to extinction" every hour, two are currently unknown to science. In addition, the uncertainties involved with extinction estimates based on the reduction of a habitat's area and, even more so, with predictions based on climate change scenarios mean that any such estimates need to be treated with extreme caution (Ladle 2009; Whittaker et al. 2005). Second, the narrative strongly links extinction with the discharge of carbon dioxide and deforestation—two themes with arguably more relevance and traction in high-level policy circles. It is also interesting (and perhaps not coincidental) that the annual predicted rate of extinction from CI's clock is almost identical to Wilson's (1992) estimate of 27,000 species per year, based solely on habitat loss (but assuming a slightly higher global species richness). Such congruence can reinforce the validity of the statistics in the public consciousness. Finally, CI states that "[o]ur clock is based on the best science,"[5] a claim for scientific legitimacy that may not hold up to close scrutiny. Indeed, given the uncertainties involved, the word 'best' probably has little meaning when applied to extinction forecasting (Ladle 2009).

It would be wrong, however, to portray the scientific conservation community (academic, NGO, and governmental) as united in its perspective on current rates of extinction or how these should be communicated to the public. There is an increasing realization among some conservation scientists that (1) a serious lack of concordance exists between the robust language of early conservation scientists (e.g., phrases such as the 'sixth mass extinction' and the 'extinction crisis') and the slow (in human terms) rate of documented species loss, and (2) the very notion of an extinction crisis is becoming increasingly counterproductive. Reflecting on the successes and failures of conservation scientists in the global conservation movement, Redford and Sanjayan (2003: 1473) express the following lament: "Our focus on crisis has hampered conservation biology in achieving a scale of action required to match the world's environmental problems. Despite our best efforts to launch our cause into the mainstream culture, the world is suffering from crisis fatigue."

There are even some mainstream scientists who are challenging the consensus on the current and future rates of extinction. Two highly respected tropical forest scientists, S. Joseph Wright and Helene Muller-Landau, recently argued that current human demographic trends, including slower population growth and the migration of rural populations into urban centers, strongly suggest that tropical deforestation will slow, natural forest regeneration through secondary succession will accelerate, and that "the widely anticipated mass extinction of tropical forest species will be avoided" (Wright and Muller-Landau 2006: 287). This rather politically naive prognosis may be unlikely, especially given the spread of oil palm plantations in Southeast Asia, but it does signal an unusual break in scientific ranks.

The Media

The global media has always been a valuable source of publicity and rhetoric for conservation. This is beautifully illustrated by the publication of the special 'shock issue' of the *Daily Mirror* tabloid newspaper that coincided with the creation of WWF in 1961. The front-page headline, hovering over a photo of a black rhinoceros and its calf, boldly stated "DOOMED—to disappear from the face of the earth due to Man's FOLLY, GREED, NEGLECT."[6] This strategy reaped rich dividends: the *Daily Mirror* appeal generated an unprecedented public response and was able to raise the equivalent of about £2 million (Jepson and Ladle 2010). Shock stories such as these can help keep an issue in the public consciousness, aid fund-raising, and potentially influence public opinion and policy. They illustrate the ease with which the topic of extinction can be aligned with the broadened moral and societal concerns that sell newspapers. Moreover, increasing pressures on academics to garner attention for their work means that the media are fed on a diet of publicity releases from university press offices or, not infrequently, directly from the scientists involved.

The global news media has a strong agenda driven by economic imperatives that make it ill-suited to communicate the intrinsically complex and uncertain science of extinction. Consequently, when extinction science has appeared in the traditional print media, it has often been badly misrepresented. This is clearly illustrated by the coverage of a recent high-profile article that appeared in the journal *Nature* concerning the possible impacts of climate change on global biodiversity (Thomas et al. 2004). The results of the study suggested that, given a number of key assumptions and under "mid-range" climate change scenarios, 15–37 percent of the 1,103 species considered within the study would be "committed to extinction" by 2050 (ibid.: 145). The authors described these percentages as being an estimate of "proportions of species committed to future extinction as a consequence of climate change over the next 50 years, not the number of species that will become extinct during this period." Furthermore, they noted, "decades might elapse between area reduction (from habitat loss) [through climate change] and extinction" (ibid.).

After this article was published, Ladle et al. (2004) reviewed 29 reports in the UK's national and local newspapers and found a systematic pattern of errors in 26 of them. The most significant misrepresentation of the study's findings was the frequently repeated contention (in 21 reports) that over a million species would become extinct due to global warming by the year 2050, while two others went so far as to suggest that one-third of all the world's species would become extinct. Just two reports explained that only a few species would actually be extinct by 2050. Interestingly, the origin of the majority of the cruder generalizations and extrapolations in the media can be traced back to the original press release from the lead author of the study, which carried the headline "Climate Change Threatens a Million Species with Extinction." It is here that the million species reference first appears, along with the unattributed claim that a quarter of land animals and plants may go extinct.

The power of such widespread media representations to influence political debate soon became clear as some senior politicians commented on the study, apparently using the erroneous newspaper reports as their source material (Ladle et al. 2004; Ladle et al. 2005). For example, then EU Environment Commissioner Margot Wallström (2004) wrote in the *Guardian* newspaper: "Many people had a lot to say about the recently published study that suggests global warming could wipe out a third of the planet's species by 2050." The discussion of the subject in the UK House of Commons was similarly inaccurate.[7]

The sensationalist coverage of the Thomas et al. (2004) article in the UK news media is perhaps to be expected, given the short timescales and circulation-driven mentality of most newspapers. It is also clearly characteristic of much science reporting in the media. What is perhaps more surprising is that representatives of CI felt obliged to support such misrepresentation. Hannah and Phillips (2004), writing in *Nature*, argued that sensationalism is acceptable if it brings an environmental message (e.g., about the potential extinctions caused by climate change) to the attention of the public and polity. Similar motivations are clearly responsible for the massively unrealistic predictions for the imminent extinction of orangutans (*Pongo spp.*) used by Friends of the Earth to discredit the oil palm industry,[8] a strategy described as 'blackwashing' by Koh et al. (2010).

Scientists themselves are clearly important actors in the production of media-friendly stories about imminent extinctions, as are the well-known and influential journals that are willing to publish the articles on which the stories are based. The kudos associated with being responsible for research that gets extensive coverage in the global news media is clearly good for careers, tenure applications, research funding opportunities, and probably citations as well (Ladle 2004). At the time of writing, the article by Thomas et al. (2004) already had over 1,000 citations on the Web of Knowledge.[9] The benefits also extend to the departments that host the scientists and, ultimately, to the universities and research institutions where they work. The degree to which such pressures influence scientists' decisions to write press releases or to conceive research projects that will attract media attention is unknown, but this is clearly a fertile area for future research. Moreover, the perceived urgent need to mobilize conservation action may also draw acknowledged leaders of cutting-edge science out of their areas of expertise and encourage them to make poorly considered and inadequately researched statements about extinction, environmental crisis, and conservation imperatives.

Conservation Policy and Governance Institutions

Tarrow (1992) argued that when frames include collective actions that attract broad consent, they necessarily become institutionalized and thereby take a leading role in determining the future trajectory of policies and of scientific, management, and cultural practices. The institutionalization of extinction by the international conservation community began 50 years ago when the IUCN first attempted to codify different levels of extinction risk within their Red List of Threatened Species, published as the *Red Data Book*.[10] The first *Red Data Book*, a list of 135 endangered mammals, was published in 1960 (see Epstein 2006). By 2009, a total of 47,677 species from all the major plant and animal groups had been assessed, including all known birds and mammals (IUCN 2009). At first, a rather subjective system founded on expert judgment was used to categorize extinction risk, but this was replaced in 1994 by a much more quantitative approach, developed by Mace and Lande (1991), that is based on population and life history characteristics. At the heart of this system is a set of simple quantitative criteria—population sizes, population decline rates, range areas, and range declines—that are used to allocate species to one of several categories of extinction risk (e.g., endangered, critically endangered, extinct in the wild, etc.). It should be noted that the Red List employs different methods of assessing extinction risk, depending on the available data, and that some of the processes used to assess species status (e.g., population viability analysis) are also extinction risk forecasting methods. In this sense, it may be better to consider Red Lists as a framework for standardizing and communicating extinction risk (Ladle 2009). The act of defining and categorizing extinction risk has deeply affected what is measured, managed, discussed, and studied. In turn, this has fed back into endangered species policy and has influenced the degree to which it is meaningful to specialists and non-specialists.

This correlation is illustrated by the now widespread use of the rates of transition between IUCN categories as a means to forecast extinctions (Butchart et al. 2004). The key transition for extinction forecasting is between 'endangered' (where a population has a 'very high' risk of extinction in the wild) to 'critically endangered' (where the species is considered as having an 'extremely high' risk of extinction in the wild) (IUCN 2001). The criteria for inclusion in the latter category include very small populations and geographic ranges and a strong trend toward population decline. The category 'critically endangered' can thus be cautiously used as a surrogate for 'imminent extinction'. Brooke et al. (2008) tested this proposition by comparing the historical transition of bird species into the critically endangered category with verified extinctions at both a global level and within Australia. They concluded that species were actually going extinct at a rate that is 2 (Australia) to 10 (globally) times lower than predicted. The potential cause of this discrepancy was identified as the effectiveness of the global conservation community at rescuing bird species on the brink of extinction.

Red Lists are deployed and promoted as the global standard to (1) ensure consistency in conservation investment across taxa and regions; (2) inform national legislation and implementation; (3) guide the management of natural resources at multiple scales, including the identification of sites for conservation action; and (4) monitor changes in global biodiversity and the reasons for these changes (Rodrigues et al. 2006). Since 2004, Red Lists for birds (Butchart et al. 2004) and amphibians (Butchart et al. 2005) have been further reified into a Red List Index. The index is a quantification of the movement of sets of species (grouped taxonomically or geographically) through categories of the Red List (Butchart, Akçakaya et al. 2006). In essence, the indexing system conflates the complex realities of extinction into a single trend line. Such practices raise extinction (or extinction as an indicator of biodiversity) to a level of abstraction where it can be correlated with index data on the economy, population, and other major areas of policy concern. This helps maintain and elevate the international policy position of biodiversity conservation and the organizations producing such data.

The reification of continuous biological characteristics into extinction risk categories also provides an opportunity for manipulation to advance conservation aims. For example, the eminent Canadian zoologist and sea turtle expert, Nicholas Mrosovsky (1997), accused the IUCN's Marine Turtle Specialist Group of upgrading the listing of the Hawksbill sea turtle (*Eretmochelys imbricata*) without making available the scientific evidence for this change in status,[11] and then using it to influence proposals for the sustainable use of the species. Thus, it is possible that unconscious or conscious biases in the information accepted and used by specialist groups might influence the categorization of species and hence provide an unduly pessimistic prognosis of their future survival. This example also illustrates the perceived power of terms such as 'endangered' and 'extinct' to mobilize conservation action. More generally, it is clear that even the application of quantitative criteria can be subject to values-based biases with the ultimate aim of ensuring that a favored species maintains or achieves a label that reflects the highest possible level of extinction risk.

The extinction narratives exemplified by the pronouncements of the international conservation community demonstrate how, within this group of actors, extinction has become "locked within a more coolly constructed rationalistic argument" (Adams 2004: 25) that characterizes discourse about biodiversity. Moreover, like the rather technocratic term 'biodiversity', extinction has been increasingly subsumed into the dominating policy narrative of climate change. The shifting focus of the environmental polity has, inevitably, led to a certain amount of realignment and transformation of earlier agendas as a way of both maintaining their public visibility and accessing international funds focused on the 'new' threat of climate change. Thus, extinction narratives are increasingly being transformed and inserted into the climate change frame, providing a bridge between the two dominating conservation agendas of the last 30 years.

The impacts of this process of alignment pervade much of the current NGO and academic extinction rhetoric, such as CI's previously mentioned extinction clock or the high-profile Thomas et al. (2004) *Nature* article that included authors from the Royal Society for the Protection of Birds and CI. Another notable example among many is WWF's 2009 campaign to conserve the polar bear (*Ursus maritimus*), which clearly and effectively links traditional conservation themes of extinction to the threat of climate change: "The planet is rapidly moving towards a tipping point with climate change impacts on polar bears. Unless immediate action is taken by responsible governments, we may be relegating polar bears to extinction in the wild within the lifetime of our children."[12]

The above example indirectly illustrates one of the most interesting aspects of the insertion of extinction narratives into the climate change frame—the current lack of a conspicuous and unmistakable victim. Polar bears may or may not be threatened with extinction by anthropogenic climate change in the next 80 or so years, but for the time being they are still very much extant. Perhaps the best candidate, and a subject of some debate, is the Costa Rican golden toad (*Bufo periglenes*). This striking amphibian, which was discovered in 1964, had a known range of only a few square kilometers within the Monteverde Cloud Forest Reserve (Sarkar 1996). The toads were always elusive but could be reliably surveyed for a few weeks in April when they emerged from the dense undergrowth to mate in temporary pools. Their final demise was rapid and unambiguous: in 1987, more than 1,500 toads were observed, but in 1988 and 1989 only a single toad emerged (Crump et al. 1992), and there has been no verified sighting since then (Sarkar 1996).

Catastrophic declines in amphibian populations have been observed around the world since the 1970s, and the general consensus is that it is a global phenomenon with multiple local causes, including shifts in weather patterns (Alford and Richards 1999). So why did the golden toad become synonymous with climate change–induced extinction? This is a difficult question to answer, not least because there is a large degree of uncertainty in the science, which primarily relies on correlational inferences. The two most high-profile hypotheses and the probable source of the golden toad's status as a poster child for casualties of global warming are both complex and untestable. Pounds and Crump (1994) linked the toad's probable extinction to the 1986–1987 El Niño that was responsible for very low rainfall during several critical life history stages. In a far more ebullient and polemical article published in *Nature* in 2006, Pounds and his colleagues propose that climate change–induced outbreaks of pathogenic fungus were responsible for the extinction. The underlying message is hammered home in the final line of the abstract: "With climate change promoting infectious disease and eroding biodiversity, the urgency of reducing greenhouse-gas concentrations is now undeniable" (Pounds et al. 2006: 161).

This direct tone and strong linkage between an extinction event and the general threat of climate change to the natural environment are echoed by widely used NGO narratives concerning the probable extinction of the golden toad and another amphibian, the harlequin frog, which disappeared from the Monteverde reserve at around the same time. On its Web site, WWF puts it as follows: "The golden toad (*Bufo periglenes*) and the harlequin frog (*Atelopus varius*) of Costa Rica have disappeared as a direct result of global warming."[13] CI also makes the link: "For Ticos, as Costa Rican natives are known, reducing greenhouse gas emissions and stabilizing the climate is personally important, as the extinction of their emblematic golden toad (*Bufo periglenes*) due to climate change and altered weather patterns is still fresh."[14]

The convergence of scientific and NGO narratives is perhaps unsurprising, given the increasingly close ties between NGO and university researchers, funding sources that have targeted 'applied' research on the biodiversity consequences of climate change, and the increasing number of multi-partner research projects that aim to generate regional or global-scale databases. However, the willingness of major journals to publish articles with increasingly bombastic rhetoric in relation to extinction and climate change seems to be a relatively new trend. It carries with it the threat of undermining the legitimacy and credibility of scientists while at the same time playing into the hands of the powerful anti-environmentalist lobbyists, who are quick to exploit perceived weaknesses in the underlying science (see Ladle et al. 2005).

Conclusions

Adams (2004: 25) wryly commented that WWF's choice of the "cuddly" giant panda as an icon perfectly captures the "slightly mawkish public sentimentality about extinction," which he also characterizes as having "both an anger and a desperation about it" (ibid.). It is certainly true that extinction continues to have a special rhetorical power within natural resource policy, probably because its focus on species simultaneously constructs nature as a domain in need of governance, provides specific targets (species and places), and allows for the compilation of statistics that provide direction on how to become "the best possible governor" (Foucault 1991: 87).

However, the strong emotional response that can be provoked by the use of the term 'extinction' suggests that it should be used wisely by conservation advocates and scientists alike. In this article, we suggest that because the term resonates with deeply held human fears (e.g., extinction of a life, of a fire and warmth, of hope), the extinction of species provides a powerful organizing concept for frames relating to humanity's relationship with nature. The rhetoric of extinction as an expression of human failings (selfishness, greed, ignorance) and of human potential (compassion, morality, planning) creates a powerful rallying force for conservation action. The scientific act of systematically classifying extinction risks has resulted in a broadening of the frame so that extinction now refers to anything from the local disappearance of a cultural-valued species to the imagined extermination of masses of undiscovered species caused by global process of deforestation. This breadth creates the possibility for extinction frames to travel and align with a multitude of other frames—from deeply personal frames involving identity and place to big issue policy frames such as climate change and ecosystem services. This widening of the frame may, however, carry risks. Sarkar (2005) argues powerfully that the continued use of inflated rhetoric—some of which has been documented herein—is another way in which biodiversity conservationists may lose public credibility.

Given these comments, we argue for more research on the cultural agency of the concept of extinction. In particular, we suggest that it would be valuable to understand better where, when, and how the concept of species extinction has the capacity to produce a phenomenon or modify a state of affairs. On one level, this might involve, for example, surveys to test whether the concept of extinction is universal. Our opportunistic and unsystematic surveys among rural people in parts of Asia suggest that it is not. If this is the case, it could have important ramifications for communication strategies aimed at building local legitimacy for conservation projects. On another level, studies of extinction that adopt, for instance, actor-network perspectives (Latour 2005) could reveal important new insights on the assembly, persistence, and power of conservation institutions. We favor the development of a more balanced, nuanced, and culturally meaningful rhetoric of extinction that accurately reflects long-term temporal perspectives and the considerable uncertainties involved in the act of declaring a species extinct. In a previous article (Ladle and Jepson 2008), we have argued that one possible approach is to delineate carefully the different meanings of the term 'extinction', along with their varying powers and agency in society (see table 1).

The above typology of extinction recognizes that different meanings open up or enable entry to different governance spaces. For example, in the case of global extinction rates, derived from the species-area relationship in island biogeography (see table 1, Linnean extinctions), conservation scientists and NGOs have successfully used this meaning of extinction to attract the attention of the inter-state polity. The startling figures generated in support of the notion of a sixth mass extinction (see, e.g., Pimm and Brooks 1997; Thomas et al. 2004) bridge easily with political and bureaucratic worries about economic and social instability and the proposed macro-scale solutions, and they easily align with natural resource policy discourses favored by intergovernmental agencies. By contrast, 'ecological extinctions' (see table 1) clearly align with the interests of zoological gardens and seed banks because they provide an outlet and legitimization for their operations. In turn, these facilities bring conservation and endangered and exotic species into urban culture, creating links between cities and conservation sites around the world and thus effectively globalizing conservation culture.

We would argue that public trust in the legitimacy of conservation action can be restored and strengthened only through the development of a more culturally meaningful and sophisticated rhetoric of extinction, one that aligns scientific evidence in the form of documented and predicted extinctions with cultural frames, institutional frameworks, and organizational interests. To achieve such alignment, conservation practitioners and advocates must develop a more sophisticated understanding of the different meanings of extinction and the power that each has to influence individuals and societies. For example, the documented local extinction of a previously abundant bird species may have little power to mobilize funds for an international conservation NGO, but it may provide a focal point for the development of new community-led conservation initiatives. Likewise, the latest projection of global extinctions based on tropical deforestation may have some traction with environmental policy makers, but it may lead to apathy and compassion fatigue if overused for conservation fund-raising in the developed world.

Finally, it should also be noted that extinction rhetoric has been strongly founded on the use of the species as the key unit of classification and measurement. This is understandable, given the widespread recognition of the term among the public, but it has perhaps limited the discourse about the wider evolutionary significance of extinction. Nee and May (1997) demonstrate that approximately 80 percent of underlying evolutionary history would be preserved even if 95 percent of species should be lost. Moreover, choosing the 'survivors' based on algorithms that maximize the amount of evolutionary history preserved was only marginally better than choosing the survivors at random. As yet, little progress has been made in translating such scientific understandings into conservation policy and practices. However, there are signs that conservation science is starting to take the preservation of evolutionary history more seriously (e.g., Davies et al. 2008; Forest et al. 2007), a move that will undoubtedly provide a rhetorical challenge and may call for a whole new vocabulary of extinction.

Table 1: Typology, definitions, and potential agency of the term 'extinction'

Extinction Type	Definition	Potential Agency
Linnean extinction	Extinctions of undiscovered species inferred from the species-area relationship and estimates of species diversity for a given ecosystem or region. The assumed losses of these inferred species are termed 'centinelan extinctions' by Wilson (1992).	• Construct and make credible the notion of a 'sixth mass extinction event'. • The potential risks and opportunity costs associated with such an event make extinction relevant to a range of international policy areas (e.g., agriculture, forestry, and health), thereby contributing to the 'mainstreaming' of conservation/biodiversity in international policy. • Create a sense of urgency and crisis that can be used to raise funds and legitimize action.
Wallacean extinction	Species that have not been documented for many years, but for which extinction is uncertain because populations might survive in areas that have not been surveyed within the potential distributional range.	• Inspire local action. • Promote field surveys and expeditions to remote areas. • Attract funding to specific localities. • Support NGO communication strategies by providing a steady stream of 'good news' stories.
Phoenix extinction	Extinct in the wild, but genetic material is available in the form of stored material or a closely related conspecific or congeneric variety/breed/hybrid, allowing for the possibility of a future reintroduction of the same or a functionally equivalent form.	• Mobilize support for innovative conservation management strategies. • Create controversies by challenging fundamental precepts of the global conservation movement concerning issues of ecosystem composition, introduced species, and adaptive management. • Inspire innovative uses of technology. • Extend the range of actors participating in conservation (e.g., biotechnology laboratories).
Ecological extinction	Extinct in the wild but with extant captive-bred population, or present in the wild but at such low densities that it no longer interacts to a meaningful degree with other species in the community (i.e., it is functionally extinct).	• Support and legitimize the actions of botanical gardens, gene banks, and zoological parks. • Create momentum for habitat restoration and reintroductions. • Effectively communicate the reality of extinction to the urban citizenry.
Local extinction	Extinct in the wild within a clearly defined geographic area but with extant free-living populations outside that area.	• Mobilize local conservation action. • Reinforce local cultural identities. • Legitimize restoration and reintroduction programs.
True extinction 1: Contemporary extinction	Extinct since the birth of the international conservation movement (in the mid-nineteenth century). The last known population has been monitored and surveyed and is now considered globally extinct in the wild. No captive-bred population or genetic material is available.	• Create a moral imperative for conservation. • Legitimize and publicize actions aimed at specific drivers of extinction (e.g., climate change mitigation and the golden toad). • Mobilize action and support for global conservation initiatives.
True extinction 2: Historical extinction	Extinct prior to the birth of the international conservation movement. No authenticated record of an extant population. No captive-bred population or viable genetic material available.	• Create a moral imperative for conservation. • Legitimize the existence and actions of the global conservation movement. • Create a 'zero sum' threshold for conservation—embed the notion of the finality of extinction.

Source: Modified from Ladle and Jepson (2008).

RICHARD LADLE is a Brazilian-based conservationist and writer. He is a Visiting Professor at the Federal University of Alagoas and a Senior Research Associate at the School of Geography and the Environment, Oxford University, where he was formerly the Course Director of the MSc degree program in Biodiversity, Conservation, and Management (2003–2009). He has diverse and interdisciplinary research interests that span the public understanding of science, extinction theory, conservation biogeography, and theoretical ecology. Recent related publications include *Conservation Biogeography* (2011), with R. J. Whittaker; *Conservation: A Beginner's Guide* (2010), with P. J. Jepson; and *Critical Concepts in the Environment: Biodiversity and Conservation* (2008).

PAUL JEPSON is Course Director of Oxford University's Master of Science Degree program in Nature, Society, and Environmental Policy. He has consulted for a wide range of intergovernmental and non-governmental organizations and was Indonesia Programme Coordinator for BirdLife International (1991–1997). He leads an interdisciplinary Conservation Governance Lab and the communication work package for an EUF7 eco-informatics project. His research focuses on long-term interests in protected areas, wildlife trade, conservation history, media and technology, and the role, accountability, and legitimacy of conservation NGOs. Recent publications include conceptual pieces on conservation actors and interdisciplinary biogeography and a series on self-governance and bird keeping in Indonesia.

NOTES

1. "Australia Arousd [*sic*] to Preserve Wildlife," *Science News-Letter*, 16 May 1931, 307–308, here 307.
2. Ibid., 308.
3. Coined in 2000 by the atmospheric chemist Paul Crutzen, the term 'anthropocene' is used by some scientists to describe the current era in the earth's history in which humans have had a major impact on the earth's ecosystems.
4. http://getinvolved.conservation.org/site/PageServer?pagename=stoptheclock (accessed 5 November 2009).
5. http://www.conservation.org/act/get_involved/Pages/stop-the-clock-methodology.aspx (accessed 5 November 2009).
6. http://www.panda.org/who_we_are/history/wwf_conservation_1961_2006 (accessed 4 November 2009).
7. Reported in *Hansard*, the printed transcripts of UK parliamentary debates, on 8 January 2004.
8. http://www.foe.co.uk/resource/reports/oil_for_ape_full.pdf (accessed 3 July 2010).
9. Despite the limitations and uncertainties associated with Thomas et al. (2004), this article has been cited numerous times. See the Web of Knowledge's ISI Web of Science site at http://wokinfo.com/products_tools/multidisciplinary/webofscience/ (accessed 24 June 2010).
10. http://www.iucnredlist.org (accessed 18 January 2010).
11. Much of the 'scientific evidence' for Hawksbill sea turtles is actually derived from so-called gray literature (Mrosovsky and Godfrey 2008).
12. http://www.worldwildlife.org/species/finder/polarbear/rsm.html (accessed 20 January 2010).
13. http://www.panda.org/about_our_earth/aboutcc/problems/impacts (accessed 20 January 2010).
14. http://www.conservation.org/explore/north_america/costarica/Pages/costarica.aspx (accessed 20 February 2010).

REFERENCES

Adams, William M. 2004. *Against Extinction*. London: Earthscan.

Alford, Ross A., and Steven J. Richards. 1999. "Global Amphibian Declines: A Problem in Applied Ecology." *Annual Review of Ecology and Systematics* 30: 133–165.

Allen, Glover M. 1942. *Extinct and Vanishing Mammals of the Western Hemisphere with the Marine Species of All Oceans.* Cambridge, MA: American Committee for International Wildlife Protection.

Barney, Gerald O., ed. 1980. *Global 2000 Report to the President of the United States: The Summary Report*. New York: Elsevier.

Benford, Robert D., and David A. Snow. 2000. "Framing Processes and Social Movements: An Overview and Assessment." *Annual Review of Sociology* 26: 11–39.

Bengtson, Sven-Axel. 1984. "Breeding Ecology and Extinction of the Great Auk (*Pinguinus Impennis*): Anecdotal Evidence and Conjectures." *The Auk* 101 (1): 1–12.

Brooke, Michael, Stuart H. M. Butchart, Stephen T. Garnett, Gabriel M. Crowley, Natalia B. Mantilla-Beniers, and Alison J. Stattersfield. 2008. "Rates of Movement of Threatened Bird Species between IUCN Red List Categories and Toward Extinction." *Conservation Biology* 22 (2): 417–427.

Bulte, Erwin H., Richard D. Horan, and Jason F. Shogren. 2003. "Is the Tasmanian Tiger Extinct? A Biological-Economic Re-evaluation." *Ecological Economics* 45: 271–279.

Butchart, Stuart H. M., Alison J. Stattersfield, Jonathan E. M. Baillie, Leon A. Bennun, Simon N. Stewart, H. Resit Akçakaya, Craig Hilton-Taylor, and Georgina M. Mace. 2005. "Using Red List Indices to Measure Progress towards the 2010 Target and Beyond." *Philosophical Transactions of the Royal Society B* 360 (1454): 255–268.

Butchart, Stuart H. M., Alison J. Stattersfield, Leon A. Bennun, Sue M. Shutes, H. Resit Akçakaya, Jonathan E. M. Baillie, Simon N. Stuart, Craig Hilton-Taylor, and Georgina M. Mace. 2004. "Measuring Global Trends in the Status of Biodiversity: Red List Indices for Birds." *PLoS Biology* 2 (12): e383.

Butchart, Stuart H. M., H. Resit Akçakaya, Elizabeth Kennedy, and Craig Hilton-Taylor. 2006. "Biodiversity Indicators Based on Trends in Conservation Status: Strengths of the IUCN Red List Index." *Conservation Biology* 20 (2): 579–581.

Butchart, Stuart H. M., Alison J. Stattersfield, and Thomas M. Brooks. 2006. "Going or Gone: Defining 'Possibly Extinct' Species to give a Truer Picture of Recent Extinctions." *Bulletin of the British Ornithology Club* 126A: 7–24.

Callon, Michel. 1998. "An Essay on Framing and Overflowing: Economic Externalities Revisited by Sociology." Pp. 244–269 in *The Laws of the Markets*, ed. M. Callon. Oxford: Blackwell.

Channell, Rob, and Mark V. Lomolino. 2000. "Dynamic Biogeography and Conservation of Endangered Species." *Nature* 403: 84–86.

Courchamp, Franck, Tim Clutton-Brock, and Bryan Grenfell. 1999. "Inverse Density Dependence and the Allee Effect." *Trends in Ecology and Evolution* 14 (10): 405–410.

Crump, Martha L., Frank R. Hensley, and Kenneth L. Clark. 1992. "Apparent Decline of the Golden Toad: Underground or Extinct?" *Copeia* 2 (May): 413–420.

Davies, T. Jonathan, Susanne A. Fritz, Richard Grenyer, C. David L. Orme, Jon Bielby, Olaf R. P. Bininda-Emonds, Marcel Cardillo, Kate E. Jones, John L. Gittleman, Georgina M. Mace, and Andrew Purvis. 2008. "Phylogenetic Trees and the Future of Mammalian Diversity." *Proceedings of the National Academy of Sciences USA* 105 (s1): 11556–11563.

Davis, Joel J. 1995. "The Effects of Message Framing on Response to Environmental Communications." *Journalism & Mass Communication Quarterly* 72 (2): 285–299.

Doughty, Robin W. 1975. *Feather Fashions and Bird Preservation.* Berkeley: University of California Press.

Ehrlich, Paul. 1971. *The Population Bomb*. New York: Buccaneer.

Eldredge, Niles. 2001. "The Sixth Extinction." ActionBioScience.org. American Institute of Biological Sciences. http://www.actionbioscience.org/newfrontiers/eldredge2.html (accessed 29 June 2010).

Engeström, Yrjö. 2006. "Development, Movement and Agency: Breaking Away into Mycorrhizae Activities." *CHAT Technical Reports No. 01*. Center for Activity Theory and Developmental Work Research, University of Helsinki. http://www.chat.kansai-u.ac.jp/publications/tr/v1_1.pdf.

Epstein, Charlotte. 2006. "The Making of Global Environmental Norms: Endangered Species Protection." *Global Environmental Politics* 6 (2): 32–54.
Eskola, Antti. 1999. "Laws, Logics, and Human Activity." Pp. 107–114 in *Perspectives on Activity Theory*, ed. Y. Engeström, R. Miettinen, and R.-L. Punamäki. Cambridge: Cambridge University Press.
Forest, Félix, Richard Grenyer, Mathieu Rouget, T. Jonathan Davies, Richard M. Cowling, Daniel P. Faith, Andrew Balmford, John C. Manning, Şerban Procheş, Michelle van der Bank, Gail Reeves, Terry A. J. Hedderson, and Vincent Savolainen. 2007. "Preserving the Evolutionary Potential of Floras in Biodiversity Hotspots." *Nature* 445: 757–760.
Foucault, Michel. 1991. "Governmentality." Pp. 87–104 in *The Foucault Effect: Studies in Governmentality*, ed. G. Burchell, C. Gordon, and P. Miller. Chicago: University of Chicago Press.
Goffman, Erving. 1974. *Frame Analysis: An Essay on the Organization of Experience*. New York: Harper & Row.
Greenway, James C. [1958] 1967. *Extinct and Vanishing Birds of the World*. New York: Dover Publications.
Halliday, Tim R. 1978. *Vanishing Birds: Their Natural History and Conservation*. London: Sidgwick and Jackson.
Hannah, Lee, and Brad Phillips. 2004. "Extinction-Risk Coverage Is Worth Inaccuracies." *Nature* 430: 141.
Harper, Francis. 1945. *Extinct and Vanishing Mammals of the Old World*. New York: American Committee for International Wildlife Protection.
Hingston, Richard G. W. 1931. "Proposed British National Parks for Africa." *Geographical Journal* 74: 401–428.
Iltis, Hugh H. 1970. "Man First? Man Last? The Paradox of Human Ecology." *BioScience* 20: 820.
IUCN (International Union for Conservation of Nature). 2001. *Red List Categories and Criteria: Version 3.1*. Gland, Switzerland: IUCN.
______. 2009. "The IUCN Red List of Threatened Species. Version 2009.2." http://www.iucnredlist.org (accessed 20 January 2010).
Jablonski, David. 2001. "Lessons from the Past: Evolutionary Impacts of Mass Extinctions." *Proceedings of the National Academy of Sciences USA* 98 (10): 5393–5398.
Jepson, Paul, and Richard J. Ladle. 2010. *Conservation: A Beginner's Guide*. Oxford: One World.
Jepson, Paul, and Robert J. Whittaker. 2002. "Histories of Protected Areas: Internationalisation of Conservationist Values and Their Adoption in the Netherlands Indies (Indonesia)." *Environment and History* 8: 129–172.
Keller, Lukas F., and Donald M. Waller. 2002. "Inbreeding Effects in Wild Populations." *Trends in Ecology and Evolution* 17 (5): 230–241.
Koh, Lian P., Jaboury Ghazoul, Rhett A. Butler, William F. Laurance, Navjot S. Sodhi, Javier Mateo-Vega, and Corey J. A. Bradshaw. 2010. "Wash and Spin Cycle Threats to Tropical Diversity." *Biotropica* 42 (1): 67–71.
Ladle, Richard J. 2004. "Ecological Science Reporting: Another Victim of Global Warming?" *British Ecological Society Bulletin* 35 (4): 12–13.
______. 2009. "Forecasting Extinctions: Uncertainties and Limitations." *Diversity* 1 (2): 133–150.
Ladle, Richard J., and Paul Jepson. 2008. "Towards a Biocultural Theory of Avoided Extinction." *Conservation Letters* 1: 111–118.
Ladle, Richard J., Paul Jepson, Miguel B. Araújo, and Robert J. Whittaker. 2004. "Dangers of Crying Wolf over Risk of Extinctions." *Nature* 428: 799.
Ladle, Richard J., Paul Jepson, Steve Jennings, and Ana C. M. Malhado. 2009. "Caution with Claims That a Species Has Been Rediscovered." *Nature* 461: 723.
Ladle, Richard J., Paul Jepson, and Robert J. Whittaker. 2005. "Scientists and the Media: The Struggle for Legitimacy in Climate Change and Conservation Science." *Interdisciplinary Science Reviews* 30 (3): 231–240.
Latour, Bruno. 2005. *Reassembling the Social: An Introduction to Actor-Network-Theory*. Oxford: Oxford University Press.
Lawton, John H., and Robert M. May, eds. 1995. *Extinction Rates*. Oxford: Oxford University Press.

Leakey, Richard, and Richard Lewin. 1995. *The Sixth Extinction: Biodiversity and Its Survival.* New York: Doubleday.
Lomborg, Bjørn. 2001. *The Skeptical Environmentalist: Measuring the Real State of the World.* Cambridge: Cambridge University Press.
Lomolino, Mark V. 2004. "Conservation Biogeography." Pp. 293–296 in *Frontiers of Biogeography: New Directions in the Geography of Nature*, ed. M. V. Lomolino and L. R. Heaney. Sunderland, MA: Sinauer.
Lorimer, Jamie. 2006. "What about the Nematodes? Taxonomic Partialities in the Scope of UK Biodiversity Conservation." *Social and Cultural Geography* 7 (4): 539–558.
Lowe, Phillip D. 1983. "Values and Institutions in the History of British Nature Conservation." Pp. 329–352 in *Conservation in Perspective*, ed. A. Warren and F. B. Goldsmith. London: John Wiley & Sons.
Mace, Georgina M., and Russell Lande. 1991. "Assessing Extinction Threats: Toward a Reevaluation of IUCN Threatened Species Categories." *Conservation Biology* 5 (2): 148–157.
McKinney, Michael L. 1997. "Extinction Vulnerability and Selectivity: Combining Ecological and Paleontological Views." *Annual Review of Ecology and Systematics* 28: 495–516.
Mrosovsky, Nicholas. 1997. "IUCN's Credibility Critically Endangered." *Nature* 389: 436.
Mrosovsky, Nicholas, and Matthew H. Godfrey. 2008. "The Path from Grey Literature to Red Lists." *Endangered Species Research* 6: 185–191.
Myers, Norman. 1979. *The Sinking Ark: A New Look at the Problem of Disappearing Species.* New York: Permagon Press.
______. 2001. "On Bjorn Lomborg and Species Diversity." Grist online magazine. http://www.grist.org/article/specious/ (accessed 15 October 2009).
Nee, Sean, and Robert M. May. 1997. "Extinction and the Loss of Evolutionary History." *Science* 278 (5338): 692–694.
Orton, James. 1869. "The Great Auk." *American Naturalist* 3: 539–542.
Parejko, Ken. 2003. "Pliny the Elder's Silphium: First Recorded Species Extinction." *Conservation Biology* 17 (3): 925–927.
Pimm, Stuart L. 2002. "The Dodo Went Extinct (and Other Ecological Myths)." *Annals of the Missouri Botanical Gardens* 89 (2): 190–198.
Pimm, Stuart L., and Thomas M. Brooks. 1997. "The Sixth Extinction: How Large, how Soon, and Where?" Pp. 46–62 in *Nature and Human Society: The Quest for a Sustainable World*, ed. P. H. Raven. Washington, DC: National Academy Press.
Pounds, J. Alan, Martín R. Bustamante, Luis A. Coloma, Jamie A. Consuegra, Michael P. L. Fogden, Pru N. Foster, Enrique La Marca, Karen L. Masters, Andrés Merino-Viteri, Robert Puschendorf, Santiago R. Ron, G. Arturo Sánchez-Azofeifa, Christopher J. Still, and Bruce E. Young. 2006. "Widespread Amphibian Extinctions from Epidemic Disease Driven by Global Warming." *Nature* 439: 161–167.
Pounds, J. Alan, and Martha L. Crump. 1994. "Amphibian Declines and Climate Disturbance: The Case of the Golden Toad and the Harlequin Frog." *Conservation Biology* 8 (1): 72–85.
Redford, Kent, and M. A. Sanjayan. 2003. "Retiring Cassandra." *Conservation Biology* 17 (6): 1473–1474.
Roberts, David L. 2006. "Extinct or Possibly Extinct?" *Science* 312: 997–998.
Rodrigues, Ana S. L., John D. Pilgrim, John F. Lamoreux, Michael Hoffmann, and Thomas M. Brooks. 2006. "The Value of the IUCN Red List for Conservation." *Trends in Ecology and Evolution* 21 (2): 71–76.
Rothschild, Walter. 1905. "On Extinct and Vanishing Birds." Pp. 191–217 in *Proceedings of the 4th International Ornithological Conference*. London: Dulau.
______. 1907. *Extinct Birds: An Attempt to Unite in One Volume a Short Account of Those Birds Which Have Become Extinct in Historical Times: That is, within the Last Six or Seven Hundred Years: To Which Are Added a Few Which Still Exist, but Are on the Verge of Extinction*. London: Hutchinson.
Rudwick, Martin J. S. 1997. *Georges Cuvier, Fossil Bones, and Geological Catastrophes*. Chicago: University of Chicago Press.
Sarkar, Sahotra. 1996. "Ecological Theory and Anuran Declines." *BioScience* 46: 199–207.
______. 2005. *Biodiversity and Environmental Philosophy: An Introduction.* Cambridge: Cambridge University Press.

Soulé, Michael E. 1985. "What Is Conservation Biology?" *BioScience* 35: 727–734.

Tarrow, Sidney. 1992. "Mentalities, Political Cultures, and Collective Action Frames." Pp. 174–202 in *Frontiers in Social Movement Theory*, ed. A. D. Morris and C. M. Mueller. New Haven, CT: Yale University Press.

Thomas, Chris D., Alison Cameron, Rhys E. Green, Michel Bakkenes, Linda J. Beaumont, Yvonne C. Collingham, Barend F. N. Erasmus, Marinez Ferreira de Siqueira, Alan Grainger, Lee Hannah, Lesley Hughes, Brian Huntley, Albert S. van Jaarsveld, Guy F. Midgley, Lera Miles, Miguel A. Ortega-Huerta, A. Townsend Peterson, Oliver L. Phillips, and Stephen E. Williams. 2004. "Extinction Risk from Climate Change." *Nature* 427: 145–148.

Thomas, Keith. 1984. *Man and the Natural World: Changing Attitudes in England 1500–1800*. London: Penguin.

Trefethen, James B. 1961. *Crusade for Wildlife: Highlights in Conservation Progress*. Harrisburg, PA: Stackpole.

Triandafyllidou, Anna, and Anastasios Fotiou. 1998. "Sustainability and Modernity in the European Union: A Frame Theory Approach to Policy-Making." *Sociological Research Online* 3 (1). http://www.socresonline.org.uk/3/1/2.html.

Wallström, Margot. 2004. "Disproving Darwin." *Guardian*, 21 January. http://www.guardian.co.uk/society/2004/jan/21/environment.environment.

Ward, G. C. Tremaine. 1901. "The Passenger Pigeon." *The Auk* 18 (2): 191–192.

Whittaker, Robert J., Miguel B. Araújo, Paul Jepson, Richard J. Ladle, James E. M. Watson, and Katherine J. Willis. 2005. "Conservation Biogeography: Assessment and Prospect." *Diversity and Distributions* 11 (1): 3–23.

Wilson, Edward O. 1992. *The Diversity of Life*. Cambridge, MA: Belknap Press.

Wright, S. Joseph, and Helene C. Muller-Landau. 2006. "The Future of Tropical Forest Species." *Biotropica* 38 (3): 287–301.

Climate Changing Small Islands

Considering Social Science and the Production of Island Vulnerability and Opportunity

Amelia Moore

ABSTRACT: This article argues that climate change has influenced the way in which small island nations are viewed and understood by the international climate community. Climate change has become an internationally recognized and specific language of vulnerability that is deployed in requests for international aid to fund adaptation and mitigation measures in some small islands, for population relocation plans and human rights advocacy in other islands, and for overhauling the 'tourism product' and creating new markets for travel in others. Vulnerability is a powerful idiom, especially in the contemporary climate context that has come to imply crisis, change, uncertainty, and immediacy. Importantly, vulnerability also gestures unambiguously toward seemingly limitless scientific and even commercial opportunity. These developments come with new forms of expertise in the natural and social sciences and the travel industry, as well as with new or reinstated forms of inequity. As the areas of small island expertise increasingly overlap, they come to reproduce the very context and form of small islands themselves.

KEYWORDS: Bahamas, climate change, development, small islands, social science, tourism, vulnerability

This article is about small islands and small island nations and some of the ways that they have been situated at the nexus of climate change science and policy, vulnerability indexing, and sustainable economic development. Intended to contribute to ongoing discussions about the social science of climate change, this article also stands as an example of one of the many possible roles for social scientists in this milieu. The main thrust is that small islands themselves are being produced as an increasingly powerful and specific general category in international global climate change arenas. Social scientists have as much of a responsibility to examine these ongoing issues as they do the practices of islanders affected by climate change and climate policies. It is becoming especially important to understand the scientific and political processes of climate change as social movers of people and places, an undertaking that can both complement and complicate attempts to understand the biogeochemical nature of this planetary phenomenon and efforts to safeguard populations from this contemporary crisis.

Environment and Society: Advances in Research 1 (2010): 116–131 © Berghahn Books
doi:10.3167/ares.2010.010106

When it comes to climate change and small islands, the languages and institutionalizations of crisis, vulnerability, sustainability, opportunity, and human rights co-exist, although some of these languages and vocabularies are more explicit than others, depending on the island, region, or situation. The development of this nexus is due, in part, to the fact that climate change impacts have been framed as inevitable but also, importantly, as predictable by the Intergovernmental Panel on Climate Change (IPCC), said to be the highest scientific authority on the issue (Schneider et al. 2007). This predictability has allowed for the proliferation of planning and preparedness protocols and for an explicit linkage with development concerns. Sustainable development for small islands has been redefined in the context of climate change in terms of adaptation to climate impacts on ecosystems, food production, and economic growth. Small islands, already a recognized subset of small states, have become construed as less likely to meet sustainable development goals unless they incorporate climate change scenarios and adaptation measures into their national development plans. In light of these evolving circumstances, this article raises the question, what does it mean today to refer to a place as "a typical Small Island Developing State (SIDS)" (Ghina 2003: 145)?

As a general political category, it is likely that the small island owes its roots to international economic development forums as much as to the anthropological study of island peoples, which was itself rooted in biological notions of evolution and isolation.[1] Yet small islands are amazingly diverse in terms of population size, economic wealth, culture, and language (Julca and Paddison 2009). While there is no single, agreed upon global list of small islands or small island states and no single system of classification, there are currently 39 members and four observers of the Association of Small Island States (AOSIS).[2] Additionally, the Small Island Developing States (SIDS) Network,[3] a political category created in 1994 at the first Global Conference on the Sustainable Development of SIDS to address development concerns, has currently classified 51 states. Adrianto and Matsuda (2002: 394) define small islands (a separate category from small island states) as "islands with a population of 500,000 or less and a land area of approximately 10,000 km squared or less," although the SIDS Network does not follow this accounting. A common general characterization of small islands and small island states is that they have a small land area, a small population, limited natural resources, undiversified economies, and weak institutional capacities. In addition, they must contend with spatial remoteness, vulnerability to economic shock and natural disaster, and fragile ecosystems (van der Velde et al. 2007). This article refers at times to small islands as a general category and to small island states and SIDS as an institutionalized and politicized category, although one of the points of this piece is that even the general concept of small islands has been—and continues to be—institutionalized in specific ways.

In terms of small islands and the phenomenon of anthropogenic climate change, a prevalent view is that climate change is a global event that currently threatens the economic and social stability of small island states and demands coordinated international attention and research.[4] The predictions are dire: Hamilton (2003) says that two to three Commonwealth nations will disappear by the end of the century due to sea level rise; the Maldives are classified as under threat of complete submersion in the Indian Ocean; the Caribbean region may lose some low-lying coastal plains; some of the islands of The Bahamas are under threat of submersion; and in the Pacific, Tuvalu has begun population migration agreements with New Zealand. Furthermore, while the loss of territory is not currently predicted for all small islands, enormous economic costs are seen as the most severe threat to small island stability. This news is especially egregious when we consider that small islands have not directly contributed to the production of greenhouse gases to the extent that larger industrialized nations have. In other words, low-lying small island regions, governments, and people are set to suffer extensively from a climate situation that they had no substantive part in creating.

This inherent inequity between large industrialized states and small island or small developing coastal states has inspired block political organizing around climate debates within the United Nations negotiating system. AOSIS, mentioned above, was formed in the 1990s in order to provide a platform for small island concerns. It is important to note that, in this context, small islands have become an ethical orientation in the climate debates. They have been made to represent the unfair nature of the climate crisis, and identification as a small island state can become a tool for obtaining political leverage. Without detracting from these stakes and the seriousness of the issue, this article would like to suggest additionally that climate change is also a context and a problem that has helped to produce the small island state as a specific and newly recognizable political form based on notions of vulnerability. This article recommends an attention to the institutional language of science, policy, and even tourism, all of which tend to validate the international planning protocols that have linked both small island states and climate change as a particular sort of conjoined phenomenon.

In terms of the institutionalization of climate science, Dalmedico and Guillemot (2009) have compiled a brief and broad history of the genesis of the notion of climate change, rooting it in the post–World War II marriage of meteorology and new calculators for large-scale weather forecasting systems. In the 1970s, climate modeling converged with weather modeling, utilizing the development of new computers and satellites, and the scale of these scientific models became conceptualized and operationalized as global rather than regional. It became possible to imagine climate as an object with dynamic variables and physical processes based in the fluid mechanics of the atmosphere that could be calculated. Since its inception and prior to the 'discovery' of climate change, climate modeling has always been about societal needs such as the assumed relevance of weather forecasts and the dynamic effects of climate on agriculture. But in the 1980s, following developments in the earth sciences, such as geochemistry and paleoclimatology, combined with climate modeling, 'anthropogenic' climate change was born. The work of these earth scientists validated the models of the climate physicists, making them increasingly relevant as the exemplars for anthropogenic climate change, also known as 'global warming'. In the 1980s, the World Meteorological Organization, a specialized agency of the United Nations, designed its World Climate Programme, which hosted meetings that would lead to the creation of the international expert organization, the IPCC, whose scientists review policies related to greenhouse gas emissions and adaptation and mitigation strategies for the projected effects of global warming. Dalmedico and Guillemot (ibid.: 212) write that "climate change has been transformed from a complex scientific topic to a political issue with national, economic, social, and diplomatic ramifications involving conflicting economic interests, conceptions of law and equity, and perceptions of the future." They also note that the IPCC is therefore particularly interesting as an international body because its evaluation process is controlled by scientists (ibid.). Despite fears that their results are not adequately understood or appreciated, scientists are construed as essential experts having the appropriate knowledge to inform sound policy. Following in this vein, Mallard and Paradeise (2009: 31) say that in the IPCC, what "shifted, without apparent public polarization and controversy, [was] the identity of the problem that scientists were supposed to solve from 'climate' to 'impacts' and eventually to 'development.'" It is therefore crucial to note that science-based development tools involved in adaptation and mitigation measures and the creation of vulnerability indices are a major driver of the stakes when it comes to internationally recognized small island issues today.

The body of this article comprises three main sections. The first section explores the existing and possible roles for social science and anthropology in the nexus of climate change and small island issues. The following two sections are brief examples of the kind of work that social science can do within that nexus. The second section begins a discussion of the institutionalization of vulnerability discourse and its attendant paradigmatic vocabulary concerning small islands and the recent threat of the climate change crisis. The third section presents a climate conscious business opportunity case from The Bahamas in order to show how social science can attend to the various and unusual ways that climate science is made to matter in contemporary island life.

The Social Science of Climate Change

My own work in the archipelagic small island state of The Bahamas has led me to consider the role of social science in general when it comes to the analysis of processes and practices coalescing around the issue of climate change. How do social scientists follow these sorts of developments? What role has social science been asked to take in these problems? How can social science look at existing framings of small islands and climate change to open up new and relevant conversations? One possible answer is that social scientists can attend to the way that climate science and climate institutions attempt to reorient, familiarize, and reinscribe the very meaning of living in small island places. Social scientists can trace the development of scientific cosmologies that come to co-exist with and/or modify existing worldviews. They can also examine cosmologies, like climate change and vulnerability, that produce small island states as recognizable locations and research categories along with opportunities for related enterprise.

The existing role of social science in these arenas is far from typical or uniform, but there are some trends that are worth mentioning here. There is a loose body of research on small islands that is embedded in a 'vulnerability paradigm' (discussed below), which describes island social realities and histories through the lens of vulnerability.[5] The research construes many islands as existing in a state of social decline due to rapid population growth and the spread of inefficient economic development schemes. In this literature, small island states are construed as being in worse shape than they were before they became a part of the global economy, in that the 'traditional practices' of island people are seen as more sustainable than contemporary livelihood practices, especially in terms of agriculture and water use or 'living in harmony' with the environment (see, e.g., Ghina 2003; van der Velde et al. 2007). However, this article would like to stress that perhaps another way of looking at something like the revitalization of 'traditional' agricultural methods in the name of contemporary concerns about climate change is to see them not as older practices but as new forms that produce new kinds of knowledge and new experts in the name of the traditional. In this way, we can see that what has been produced in the present moment is a response to the current framing of climate-inspired problems.

Another trend involves social scientists advocating for the incorporation of local people into small island scientific research, modeling, and development planning in the form of human behavioral variables (Broad and Sanchirico 2008). For example, one area of international concern is small island state fisheries management, and, as a result of this focus, the role of fisheries social science is increasingly oriented around developing an understanding of fisher behavior and livelihood strategies. These are incorporated into the creation of more holistic models that include human dimensions and link individual decisions to state regulations and international recommendations (Colburn et al. 2006). This social science advocates for the revision of international means and methods of assessment and management to 'better fit' small island state characteristics (McConney 2002). This important work, done from within policy circles, is oriented toward making better policy decisions. Yet one of the additional advantages of social sciences such as anthropology is their vantage point from *outside* the management policy frame and their ability to produce knowledge that does not fit, a priori, into institutional molds, for instance, the small island state and its supposed management needs.

There are currently a number of social studies concerning local adaptation to climate change around the world (see, e.g., Adger and Kelly 1999; Berkes and Jolly 2001). Nelson et al. (2009) cite Julian Steward's (1955) cultural ecology as a scholarly precursor that has inspired decades of anthropological work, much of which has been ignored in international climate discussions that these authors claim are biased toward technological solutions and broad policy interventions at the expense of specificity and locatedness. Nelson et al. (2009) describe climate change adaptation as multi-scalar, linked to socio-economic inequalities, interrelated with natural and global stresses, and requiring community participation. Anthropology, they argue, is disciplinarily suited to engaging with these phenomena because it can contribute to an understanding of human dimensions and can focus attention on local vulnerabilities, that is, the 'social aspects' of adaptation that may or may not fit into current policy frameworks. These authors call for increased anthropological inclusion and for participation in the development of the Fifth Assessment Report of the IPCC.

Importantly, anthropology can also take the issue of climate change beyond the terms of adaptation, vulnerability, and resilience and can challenge the possibilities brought about by scientific research itself (Crate and Nuttall 2009). For example, Lahsen's (2005) work with climate scientists gives scientists as much of a human face and embodiment as anthropologists have attempted to give local and indigenous people affected by climate impacts and plans. Yearley (2009) advocates for more research on the social dimensions of the climate science community and a reconsideration of the way that social science participates and is represented in this field. Peterson and Broad (2009) attend to what they call 'climate narratives' created by climate scientists and the way that these narratives generate notions of globality and allow for the creation of new scientific productions. For these two authors, anthropology plays a checks and balances role, defending local priorities in climate change issues: "Anthropologists may likely find themselves arguing against the importance of global warming as a major risk factor versus more immediate (and longstanding) drivers of vulnerability including property rights, education, and access to water and health care. Climate change discourse has the potential to obfuscate unequal power relations, letting governments off the hook for poor environmental and social policies and practices" (ibid.: 80–81). Roncoli et al. (2009) argue that anthropological contributions to the field can go beyond structured surveys and analyses based on quantified results. The strength of anthropology still lies in fieldwork and the long-term situatedness of its research projects. For Roncoli et al., the areas of appropriate anthropological intervention in climate concerns center around perception, valuation, knowledge, and response that goes beyond cost benefit analyses and rational choice.

The above examples show that there are more opportunities and responsibilities for social science than are commonly stressed in the arena of small islands and climate change and even within climate-construed issues in general, although this is beginning to change. While policy-focused work certainly has its place, there is more for social science to do than to be 'policy appropriate'. The checks and balances role of social science is not always seen as relevant or helpful to the climate change institutions and organizations that deal in adaptation, mitigation, valuation, and policy frames. This is because some critiques cannot easily or ever be made feasible or actionable in policy programmatic ways. The relevance of some anthropological and social scientific work lies in its ability to describe other possibilities and to create an alternative space for thought. It can be, as described by Rabinow (2008), a purposefully untimely anthropology. The next two sections are examples of this sort of social science.

Small Island Vulnerability

Outside the milieu of climate change, small islands and vulnerability have already been institutionally linked through the discourse of economic development and environmental hazards (Cuny 1983; Cutter 1996; Pelling and Uitto 2001). Development projects in small islands set the precedent for small island states as a specific sort of development category that is linked to economic vulnerability indices. This use of vulnerability indices for planning purposes and prioritization seems to have taken off in the mid-1990s. Climate change has done nothing if not increase and refine what Lazarus (2009) refers to as the 'rubric of vulnerability'. This section is not about the quantification and development of indices themselves; rather, it is about their implications and the consequences of this policy-oriented vulnerability paradigm for small island nations.

The editors of the Third Assessment Report of the IPCC define vulnerability as "the degree to which a system is sensitive to, or unable to cope with, adverse effects of climate change, including climate variability and extremes. Vulnerability is a function of the character, magnitude, and rate of climate change and variation to which a system is exposed, its sensitivity, and its adaptive capacity" (McCarthy et al. 2001: 6). In the IPCC's Fourth Assessment Report, Schneider et al. (2007: 782) state: "Policy-makers and the scientific community have increasingly turned their attention to climate change impacts, vulnerabilities and associated risks that may be considered 'key' because of their magnitude, persistence and other characteristics. An impact describes a specific change in a system caused by its exposure to climate change. Impacts may be judged to be either harmful or beneficial. Vulnerability to climate change is the degree to which these systems are susceptible to, and unable to cope with, the adverse impacts." Briguglio (2000) has defined the term 'vulnerability' as the potential for aspects of a system to be damaged by external impacts. Vulnerability is therefore a vague term, but this vagueness allows for a great deal of work and organizing to be done in its name. Indexing and ranking are but two aspects of this orientation.

The current United Nations Framework Convention on Climate Change (UNFCCC) procedure for dealing with global SIDS planning includes processes of vulnerability assessment, along with information development, the creation of resource inventories, capacity-building priorities, systemic models, and the development of decision-making tools. Vulnerability is assessed by identifying the location and severity of short- and long-term climate impacts and the prioritization of the physical and economic impacts for concentrated action. One of the earliest discussions of economic vulnerability indices for small islands came from Briguglio (1995), following the call for increased vulnerability calculations from the Barbados Program of Action in 1994, and one of the latest institutionalizations of the current expanded notion of vulnerability indices considering climate change comes from Mimura et al. (2007). The desire was to generate appropriate indices for SIDS that represented more than the national GDP—indices that could delineate the complex relations between environmental resources, economies, and society and therefore pinpoint the institutional weaknesses of SIDS (Ghina 2003). The risk of natural hazards and the resilience/fragility of national economic and ecological systems became key components of vulnerability.

Vulnerability indices allow for comparison between entities and set categories, and there are currently multiple vulnerability indices and methods for the calculation and assessment of vulnerability. While some authors argue that small island vulnerabilities and climate-led disasters matter primarily in terms of their economic implications (Adrianto and Matsuda 2002), others have been attempting to integrate the analysis of ecology and economy (Costanza et al. 1996). No one index measures the vulnerability of all SIDS or all states. Some indices measure sea level rise vulnerability, while others measure economic and/or environmental vulnerability. There is even a Commonwealth Vulnerability Index (CVI)[6] and a major Environmental Vulnerability Index (EVI),[7] developed by the South Pacific Applied Geoscience Commission (SOPAC) and the United Nations Environment Programme (UNEP).

In terms of ranking the vulnerability of small islands, the Maldives has been shown by some calculations and indices to be the most vulnerable in terms of sea level rise, while all of the small islands in the Pacific are categorized as extremely vulnerable. The CVI calculated Vanuatu as the most vulnerable nation in the world, with 17 of the most vulnerable 25 being small island states. Small islands in the tropics in general are considered to be especially vulnerable to warming temperatures, despite the fact that the most extreme temperature rises are occurring at the poles. The calculated vulnerability of atoll islands places them near the top of most indices, and the Maldives and Tuvalu are thought to be the most vulnerable (Hamilton 2003). With regard to the Maldives, Ghina (2003: 144) has written that "the inherent nature of the islands predispose[s] them to frequent damage from storm and wave surges. This vulnerability has direct implications for a number of activities related to the economy and indeed livelihoods." For Julca and Paddison (2009), climate change is connected to the vulnerability of small islands, which they further link to the migration of the wealthier and skilled sectors of the population who can afford to leave when island life becomes difficult. This can lead to remittance dependency and a lack of qualified people remaining in the island country of origin.

In the Caribbean region, London (2004) points out that the impacts of climate change are now increasingly identified as erosion due to sea level rise, high water tables, property loss, threats to tourism infrastructure, and flooding of low-lying areas. Other results stemming from climate change are identified as the warming of coral reef habitats, shifts in rainfall affecting agriculture, landslides, and increased hurricane strength and frequency. These current and potential 'impacts' matter because the Caribbean region's systemic vulnerabilities have been identified, categorized, ranked, and indexed in order to develop regional and national 'targeted corrective action plans'. Therefore, key vulnerability issues for the Caribbean region include, among others, the intensification of water scarcity, unstable human settlement and infrastructure, threats to the viability of the tourism economy, increased fragility of coastal zone ecosystems and of coral reef systems, increasingly fragile fisheries and agricultural sectors, the growing prevalence of tropical diseases, and exposure to storms and hurricanes (R. Moore 2002). The Caribbean region has also been identified as the most economically volatile region in terms of consumption, as vulnerability is also linked to public debt (Julca and Paddison 2009). The broad calculation appears to be simple: climate impact + high island vulnerability = state disaster.

One result of these indexing exercises has been the development of the idea that the main commonalities of small islands consist of the possibility of disaster, the existence of multiple vulnerabilities, and the low 'capacity for resilience' in a number of sectors. Vulnerability has become generally recognized as an inherent condition of small islands: it is said to pre-exist the threat of natural hazards, making small islands extremely susceptible to all hazards. Indexing also shows that vulnerability has been grouped into social, economic, and environmental categories in terms of systemically identifying social inequities, the lack of economic diversification, and various environmental vulnerabilities to climate impacts. As a result, small islands are construed as inherently or naturally vulnerable, structurally vulnerable, and systemically vulnerable in terms of social, economic, and environmental systems.

Here I would like to point out that part of the very purpose of vulnerability assessment seems to be to influence public perception, to generate what is called 'political will' toward adaptive action, and to create economic incentives for adaptation measures. To this end, there are those who argue that small islands must be made into a global example for the development of effective climate change planning because they are characterized as some of the most vulnerable places in the world (London 2004). There are also examples of projects in the Maldives that are attempting to quantify and value the impacts of climate change to make them economically comprehensible for policy makers (Ghina 2003). London (2004) recommends looking to fund and carry out so-called no-regret projects whereby the environment and economic sectors both 'win', as well as developing a valuation scheme to represent the general national costs of climate change.

Vulnerability indices were developed along with the awareness of the climate crisis in order to make explicitly visible the disadvantages of small islands and small island regions. This form of policy programmatic assessment has now been institutionalized and repeatedly strengthened with each international meeting on the topic of small islands and climate change. As mentioned, the main benefits are seen to be attention to the issue, easy comprehension for policy makers (Adrianto and Matsuda 2002), and guidance for the allocation of scarce funds (Barnett et al. 2008). Small islands, in the institutionalized form of SIDS, have also become a model for coupled environmental and socio-economic analyses, and they are used to promote an articulation of vulnerability that is supposedly politically radical, as well as policy appropriate.

This article would like to introduce a healthy skepticism with regard to the generally popular use and proliferation of this vulnerability paradigm and to invite further inquiry into the implications of institutionalizing the so-called inherent vulnerability of small island places in terms and formats that appeal to, and are easily understood by, policy makers. Although climate change concerns have expanded the role of vulnerability assessments, they have as yet done little to alter the mainly economic focus and narrow quantitative analyses that make up the majority of the literature on vulnerability. That said, I would also like to invite inquiry into what I fear are the homogeneous ways that vulnerability, in the global milieu of climate change, has thus been linked to generalized economic, environmental, and social systems through the increasingly universal application of these systemic approaches in small island states. In addition, there must now be what we might call 'expert vulnerability practitioners' whose job it is to calculate, rank, and prioritize island vulnerabilities, among other things. There has yet been little research done with these populations of emergent experts and their expertise, especially as pertains to the codification and classification of the category of small island states.

There is already a nascent body of research critiquing the vulnerability paradigm. As mentioned above, Lazrus (2009) calls for the rethinking of the rubric of vulnerability, which she describes as top-down, denying local agency and meaning-making that proliferate around climate change and island issues. She asks us to recognize that climate change affects different people differently and that vulnerability is tied to histories of domination and social inequities that need more attention and that, I would like to point out, are not at all readily apparent in most vulnerability indices. For Lazrus (ibid.: 240), vulnerability is a "matter of representation to which questions of agency are central." Another danger is that small island communities become reified as lacking capacity or power (Campbell 1997). As a small island issue, vulnerability rankings can alter the decisions of aid donors and investors: the threat of climate change impacts can do more harm than the actually experienced impacts themselves (Barnett and Adger 2003). Further, Barnett et al. (2008) note that vulnerability indices presuppose what is valuable and meaningful for different groups. Indices assume that inhabitants of islands or regions share vulnerabilities in equal measure when in fact there may be infinite variations to susceptibility within regions and islands and between social groups and individuals. The supposed radicalism of vulnerability assessment and ranking is potentially profoundly exclusionary.[8]

The social science of vulnerability must consider who benefits from these calculations and identify what is made possible in this paradigm, along with what is made impossible. In the context of climate change, this sort of work is not merely deconstructive. It can be complementary to efforts to make ethical social decisions; it can also open the door to allow alternative and silenced realities to enter into the frame of debate. In terms of the stakes for small islands, this sort of orientation provides an attention to contemporary constructions that reinforce dangerous development notions of insufficiency, weakness, homogeneity, and isolation.

The next section, which discusses climate-inspired economic opportunities in travel markets, is another example of potential roles for social science in the milieu of small islands and climate change. This example also points to the creative aspects of the climate crisis and to the ways that various actors can make new forms with scientific information, such as the IPCC's regional predictions.

Construing Opportunities

Thus far, this article has briefly discussed the parameters of international small island categorization sparked by the climate crisis and the institutionalization of the vulnerability paradigm, especially vulnerability indexing. This is the categorical context in which small islands are able to lay claim to—or request adaptation and mitigation aid from—the international climate community, specifically the UN and those organizations concerned with issues of climate and justice between the developed and developing world. This is also the milieu in which it is possible for a crisis to become an unexpected opportunity for some actors. My own anthropological research and experiences in The Bahamas have added another example to this story concerning the potential for climate-conscious tourist development. What follows is an excerpt from my field experience in The Bahamas.

In 2008, I attended a regional workshop in Nassau on current climate change science and the threats and opportunities for the Caribbean tourism industry. The event was hosted by The Bahamas Ministry of Tourism, the Caribbean Tourism Organization (CTO), and CARIFORUM's Caribbean Regional Sustainable Tourism Development Programme, which is financed by the European Union. The workshop attendees consisted of representatives from regional tourism and hotel organizations, private business, Bahamian environmental governmental and non-governmental organizations, members of Parliament, and members of the press. The workshop was entitled "Climate Change-ing the Industry? The International Policy and Market Response to Global Warming and the Challenges and Opportunities that Climate Change Issues Present for the Caribbean Tourism Sector."

The CTO had invited a consultant from a British university to deliver a PowerPoint presentation on climate change adaptation and mitigation in the tourism industry. In it, the consultant made a number of connections, implicitly linking IPCC science with plans for the Caribbean and Bahamian "tourism product." This talk included a brief overview of current climate change science, showing charts and graphs published by the IPCC, which described the increase in global atmospheric temperatures and the projected increase in surface temperature over the course of the twenty-first century. In a segment of the presentation about the new realities for tourism in a world affected by climate change, the consultant identified the Caribbean region as a "tourism vulnerability hotspot," characterized by IPCC-predicted warmer summers, sea level rise, increased extreme weather events, fresh water scarcity issues, marine biodiversity loss, an increase in disease outbreaks, political destabilization, and travel cost increases due to mitigation policies. The Caribbean tourism industry was described as "crucially interdependent" on climate, and the connection between tourism and climate was identified as forming the basis for the economic development and maintenance of local livelihoods in most Caribbean nations. The development of "sustainable destinations" was touted as a crucial form of maintainable economic development in the region. Sea level rise, the increased rate and strength of Atlantic hurricanes, coral bleaching, coastal erosion, and changes in rainfall patterns were described as detractors from the natural resources of tourism, all of which would lead to the "loss of attractiveness of the region as a destination, the loss of employment in the industry, increased industry insurance costs, increased operating costs, and changes in patterns of tourist travel flows."

The presentation also included a discussion about the world tourism sector's contributions to global greenhouse gas emissions, 75 percent of which stem from the transportation of tourists around the world, which are calculated to be 5 percent of all world emissions. This places the world tourism industry as the fifth-largest emitter after the US, China, the EU, and Russia (UNWTO 2008). Tourism was construed as both a "vector and a victim" of climate change. The point was that although the Caribbean as a region has not significantly contributed to global warming through the production of industrial CO_2 emissions, the region's tourism industry most certainly has. Rather than being characterized in terms of a helpless small island region, the economies of The Bahamas and the Caribbean were described as complicit in producing high levels of travel-related emissions.

The explicit message of the presentation became quite clear when the consultant arrived at the notion of the "carbon neutral destination" as an adaptation and mitigation strategy. He stated that in order for a location to sell itself in a tourist market that has now been constrained by carbon, tourism officials must perform three steps: (1) they must measure and assess emissions produced by visitors' travel, accommodations, and activities and devise means to reduce them within the country; (2) they must "decarbonate" by switching to energy-efficient practices or alternative energy sources; and (3) they must offset their remaining emissions using offset purchasing options approved by Gold Standard Certified Emission Reductions registered with the UNFCCC. Tourists and other travelers, the consultant reasoned, would be willing to pay a small offsetting fee as part of their travel expenditures, knowing that they were contributing to the state of carbon neutrality that they were visiting. Along with these moves, the consultant noted the great "Caribbean potential" resulting from the fact that this is a "unique and groundbreaking time for tourism."

In this vision, the possibilities almost seem to outweigh the potential threats to the region, and the institutions organized around the production and protection of the Bahamian environment seem to agree that these opportunities come in the form of scientific and economic expansion under the aegis of tourism. The notion of the carbon neutral destination, partially based as it is in the projections and calculations of IPCC climate science, is particularly attractive for a tourism-dependent small island country like The Bahamas because it has the potential for what the consultant described as "greening the image of the country and the region." Institutions in The Bahamas are already quite savvy when it comes to branding and fashioning the country as a particular kind of destination, but the Ministry of Tourism was now presented with a novel opportunity to diversify the tourism product away from images associated with the 'sun, sand, and sea' brand to a marketing strategy that involved selling the very vulnerability of the islands themselves. This is a scheme where explaining how visitation negatively affects the country's fragile resources and asking visitors to pay a fee to protect these resources may actually result in the attraction of more visitors on a 'carbon free vacation'.[9] The 'threat' of climate change is not just an opportunity for the global scientific community to expand its horizons and conduct socially relevant projects on a massive scale. It is also viewed as a means to make the region a leading example of new, visitable approaches to climate change mitigation and adaptation measures if Caribbean countries like The Bahamas were to brand themselves in terms of offsets and alternative energy technologies.

The Bahamas, as a space of active experimentation, has become variously and generatively entangled in this climate change assemblage of international bodies, models, scientists, bureaucrats, consultancies, climate-articulated vulnerable natures, individuals, and groups (A. Moore 2010). However, the contingent history of the Bahamian tourism industry might very well allow climate to become part of a marketing package, and the contingent history of international climate science, briefly outlined at the outset of this article, might be all too easily oriented around touristic development in small island sites like this. Although it is a government body, the Ministry of Tourism is run as a business, as Strachan (2002) also notes, and this business, working with and within the global assemblage of climate change, has the potential to solidify particular forms. Tsing (2005) critiques climate science, stating that climate change creates a form of 'global nature' through experts' strict adherence to climate models on a global scale in which the local disappears. Yet within the framework of The Bahamas, climate change, and the Ministry of Tourism, the local does not disappear. Instead, it is already perceived, in the form of a carefully designed locality, as the Bahamian tourism product and is specifically branded as paradise. This potential to be rearticulated as a Bahamian climate conscious destination implies the selling of vulnerable and visitable nature, which depends on localization, not necessarily globalization, for its cachet.

These collaborations between climate science and the tourism industry could play into the productive tension between notions of The Bahamas and those of the Caribbean region. As Coombe (1998) notes, distinction is also necessarily the quality of being in a relationship. The Bahamas, as a national tourism product, is always in a relationship with the Caribbean as a regional tourism product, with the Ministry of Tourism constantly engineering distinction from it, positioning the country in a social system of difference. The brand of sun, sand, and sea is one example of the engineering of distinction that has lost its edge for the ministry, as the brand now represents the entire region and is no longer the exclusive domain of The Bahamas, if it ever was. Yet ministry officials recognize that marketing The Bahamas brings attention to the region and visa versa, as branding the Caribbean region also influences the collection of images and emotions that are associated with The Bahamas. Science-based climate conscious branding of the Caribbean as the world's first small island carbon neutral region might influence this tension in interesting ways that cannot yet be known.

What is important in this example is neither the expertise of the consultant nor the actual extent to which Caribbean governments will incorporate into their national policy frameworks the recommendations to mitigate, decarbonate, and offset. That outcome remains to be seen and, in any event, would depend on far more than a two-day workshop. Instead, this example shows that the idea of a carbon neutral destination has grown out of international climate change research and regional projections to the extent that it could feasibly be presented at a regional workshop with an audience of industry professionals, government bureaucrats, and NGO employees. The social science of these mid-level discussions illustrates the specific ways in which climate science is used to legitimate new business opportunities and allow for the creation of new forms of expertise and employment. In small island regions such as the Caribbean, dependent as many are on some degree of island tourism, science-inspired marketing strategies have the potential to carry real weight, even if the connections made here between climate change and the development of the tourism industry seem indirect or dubious at best. For social scientists interested in contemporary small island issues, cases like these can exemplify the problems of the day.

As we have seen, climate change itself has become a business model of a sort. This case demonstrates the economic development opportunities involved, as well as the possibility of creating new forms of market expertise, new research agendas, and new regional connections and distinctions based on the inevitability of global, regional, and local climate change processes. Tourism is but one arena in which this explicit mixing of climate science and the prospective generation of capital is happening. As stated in the previous section, it is appropriate for social scientists to ask, who benefits from situations like this and who is put at a disadvantage? In The Bahamas, unsurprisingly, there will likely be a perpetuation of the status quo in terms of social inequalities engendered by the tourism industry. At the same time, new institutional roles will continue to be created in order to accommodate climate-centered business plans. This article urges those concerned with small island issues everywhere and anywhere to pay attention to these developments, no matter how seemingly insignificant.

Conclusion

To conclude, I would like to reiterate the question asked at the outset of the article: what does it mean today to refer to any place as a 'typical SIDS'? What are the implications of this categorization? What are we perpetuating when we use this language, consciously or otherwise? Thinking about small islands and climate change requires that one consider the terminology involved and question the authoritative creation of typologies, whether or not one is a social scientist.

This article has attempted to demonstrate that more social scientific research can and should be done on the process of developing vulnerability indices and the classification of small island states and regions. Such work might open up for examination the practices of this interdisciplinary field and its experts, practitioners, and audiences, as well as the forms that they produce. Social science is capable of monitoring and describing those whose profession it is to observe the world and to make the world through observation, that is, to observe contemporary society *through* contemporary society (Luhmann 1998). Anthropology and social science are able to conceptualize ideas and research objects in ways that other disciplines cannot. As islands are reified through policies of protection and preservation concerning small island states and climate impacts, anthropologists can attend to and critique the stabilization of ideas that portray the small island as 'vulnerable'. Social scientists have the ability to recognize these and other terms as forms of intervention into human lives that can sometimes allow for the explicit targeting of populations in the name of development, adaptation, mitigation, vulnerability, and resilience.

To reiterate concerns with vulnerability indexing as currently understood, Kelman (2010) stresses the importance of recognizing the 'unnaturalness' of all potential island disasters. He argues that "disasters are social, not environmental phenomena … disasters are about people and their or their organization's reactions to environmental phenomena or potential environmental phenomena … Natural disasters do not exist." This is more than a semantic issue. It is a matter that involves the scientific and institutional ordering of the reality of island people who are more the subjects of generations of social and historical inequities than the victims of 'natural' bad luck.

Small islands and the people in them do not exist in isolation from each other or the rest of the world. I mention this here to make the point that thinking in categorical terms of small island states has the potential to obscure the constant and immediate relationships between people that have been forged over time and through space. As they stand now, climate change response policies—based primarily on nationally bounded frameworks and assessments, a standard set by the international community in the name of national development, sovereignty, and management—can do little to fully encompass island events. Issues such as climate-induced migration, investment, and climate conscious travel marketing are influenced in part by the very framing of the problem of island vulnerability, represented in indices for easy comprehension. Islands, even small ones, are not homogeneous. Nevertheless, one result of international classification and indexing is the creation of homogeneity in terms of the form of the international and national response to climate change and the understanding of the stakes involved.

Future international meetings of the climate community could result in the expansion or creation of international legal mechanisms involving climate impacts, the redefinition of international human rights standards in terms of island submersion, the loss of territory or state viability, and the designation of environmental refugee status. We may indeed see a renegotiation of the grounds for international asylum and migrant status, a redefinition of the rights of island citizens, a redetermination of when a state can be declared non-existent, a recalculation of the life expectancy of states, the reclassification of climate change as a potential state of public emergency, the incorporation of environmental law into human rights law, and an expanded definition of adaptation to include more radical options (Oliver 2009). These possibilities might inspire hope, but they should also prompt the aforementioned healthy skepticism concerning the ways in which new realities might be codified and new categories created.

Additionally, the failure to create binding resolutions on SIDS at the 2009 or 2010 UN Climate Change Conferences should not draw attention away from the ongoing dialogue of organizations and the re-enforcement of the framing of the problems of SIDS. The codification and delineation of vulnerabilities for small island states and the creation of frameworks for international and national responses and planning have become ends in themselves. These are powerfully creative tools. One contention of this article is that climate change is nothing if not political, from the realm of international negotiation to the way that people conceive of their own existence and their relationships with others. There are conceptual innovations running all through the discussion, especially in terms of how institutions think about possible and appropriate state actions and state orderings.

Scientific, political, and touristic realities are made to matter in small island settings, and they are particularly strong in that milieu—to the point that they can be said to have made the milieu itself. Social science, even applied, policy-oriented social science, can provide the necessary attention to idiomatic and material productions, an effort that is not currently reflected in many climate change and small island debates. Natural scientists and tourism officials could be more aware of their island-making practices, and international organizations could be more careful regarding the generative power of their language. Despite these observations, I want to stress that this article is not about delineating a potentially repressive regime. Instead, it is intended to draw attention to the way that problems, such as the relationship between small islands and climate change processes, continue to shape our contemporary world.

ACKNOWLEDGMENTS

The author would like to acknowledge Sean Vanaman, Cindy Chang, Denyse S. Dookie, Ilan Kelman, Joshua Wiese, Paige West, and Kenneth Broad for their assistance with the framing and content of this article.

AMELIA MOORE obtained her doctorate in Sociocultural Anthropology from the University of California, Berkeley in 2010. Her doctoral dissertation is on environmental politics in The Bahamas and the intersection of Caribbean anthropology, science and technology studies, and environmental anthropology. She is a Visiting Assistant Professor at the University of Miami, and her current research is an investigation of sustainable design principles and equitable decision making within community development projects in The Bahamas and southern Florida. She is currently working on two pending publications: an article on new developments in Bahamian fisheries and an article on educational field research as a form of tourism.

NOTES

1. For a historical discussion of social anthropology and island biogeography, see Kuklick (1996).
2. http://www.sidsnet.org/aosis/.
3. A project of the UN Department of Economic and Social Affairs (UN-DESA), the global network for small island developing states (SIDSnet) can be found at http://www.sidsnet.org/.
4. For a scientific discussion about envisioning the reality of global climate change, see Moss et al. (2010). In terms of human health, see Patz et al. (2005).
5. For a general discussion of development, disaster, and vulnerability, see Cuny (1983).
6. http://www.unescap.org/MCED2000/pacific/background/vulnerability.htm.
7. http://www.vulnerabilityindex.net/.
8. Barnett et al. (2008) also offer a thoughtful history and critique of the EVI.
9. See "Carbon Free Vacation," an informational pamphlet advertisement produced in 2008 for Paradise Bay Resort, Grenada.

REFERENCES

Adger, W. Neil, and P. Mick Kelly. 1999. "Social Vulnerability to Climate Change and the Architecture of Entitlements." *Mitigation and Adaptation Strategies for Global Change* 4 (3–4): 253–266.

Adrianto, Luky, and Yoshiaki Matsuda. 2002. "Developing Economic Vulnerability Indices of Environmental Disasters in Small Island Regions." *Environmental Impact Assessment Review* 22 (4): 393–414.

Barnett, Jon, and W. Neil Adger. 2003. "Climate Dangers and Atoll Countries." *Climatic Change* 61 (3): 321–337.

Barnett, Jon, Simon Lambert, and Ian Fry. 2008. "The Hazards of Indicators: Insights from the Environmental Vulnerability Index." *Annals of the Association of American Geographers* 98 (1): 102–119.

Berkes, Fikret, and Dyanna Jolly. 2001. "Adapting to Climate Change: Social-Ecological Resilience in a Canadian Western Arctic Community." *Conservation Ecology* 5 (2): 18.

Briguglio, Lino. 1995. "Small Island Developing States and their Economic Vulnerabilities." *World Development* 23 (9): 1615–1632.

______. 2000. "An Economic Vulnerability Index and Small Island Developing States: Recent Literatures." Working Paper Seminar on Islands Studies, Kagoshima University, Pacific Island Studies Center.

Broad, Kenneth, and James N. Sanchirico. 2008. "Local Perspectives on Marine Reserve Creation in the Bahamas." *Ocean and Coastal Management* 51 (11): 763–771.

Campbell, John. 1997. "Examining Pacific Island Vulnerability to Natural Hazards." Pp. 53–62 in *Proceedings, VIII Pacific Science Intercongress*, ed. A. Planitz and J. Chung. Suva, Fiji: United Nations Department for Humanitarian Affairs, South Pacific Programme Office.

Colburn, Lisa L., Susan Abbott-Jamieson, and Patricia M. Clay. 2006. "Anthropological Applications in the Management of Federally Managed Fisheries: Context, Institutional History, and Prospectus." *Human Organization* 65 (3): 231–239.

Coombe, Rosemary J. 1998. *The Cultural Life of Intellectual Properties: Authorship, Appropriation, and Law*. Durham, NC: Duke University Press.

Costanza, Robert, Olman Segura Bonilla, and Juan Martínez Alier, eds. 1996. *Getting Down to Earth: Practical Applications of Ecological Economics*. Washington, DC: Island Press.

Crate, Susan A., and Mark Nuttall. 2009. "Introduction: Anthropology and Climate Change." Pp. 9–36 in *Anthropology and Climate Change: From Encounters to Actions*, ed. S. A. Crate and M. Nuttall. Walnut Creek, CA: Left Coast Press.

Cuny, Frederick C. 1983. *Disasters and Development*. Ed. Susan Abrams. Oxford: Oxford University Press.

Cutter, Susan L. 1996. "Vulnerability to Environmental Hazards." *Progress in Human Geography* 20 (4): 529–539.

Dalmedico, Amy D., and Hélène Guillemot. 2009. "Climate Change: Scientific Dynamics, Expertise, and Geopolitical Challenges." Pp. 195–220 in *Global Science and National Sovereignty: Studies in Historical Sociology of Science*, ed. G. Mallard, C. Paradeise, and A. Peerbaye. New York: Routledge.

Ghina, Fathimath. 2003. "Sustainable Development in Small Island Developing States: The Case of the Maldives." *Environment, Development and Sustainability* 5 (1–2): 139–165.

Hamilton, Clive. 2003. "The Commonwealth and Sea-Level Rise." *Round Table* 371 (1): 517–531.

Julca, Alex, and Oliver Paddison. 2009. "Vulnerabilities and Migration in Small Island Developing States in the Context of Climate Change." *Natural Hazards* 55 (3): 717–728.

Kelman, Ilan. 2010. "Natural Disasters Do Not Exist (Natural Hazards Do Not Exist Either)." Version 3, 9 July 2010 (Version 1 was dated 26 July 2007). http://www.ilankelman.org/miscellany/NaturalDisasters.rtf.

Kuklick, Henrika. 1996. "Islands in the Pacific: Darwinian Biogeography and British Anthropology." *American Ethnologist* 23 (3): 611–638.

Lahsen, Myanna. 2005. "Seductive Simulations? Uncertainty Distribution around Climate Models." *Social Studies of Science* 35 (6): 895–922.

Lazrus, Heather. 2009. "The Governance of Vulnerability: Climate Change and Agency in Tuvalu, South Pacific." Pp. 240–249 in *Anthropology and Climate Change: From Encounters to Actions*, ed. S. A. Crate and M. Nuttall. Walnut Creek, CA: Left Coast Press.

London, James. 2004. "Implications of Climate Change in Small Island Developing States: Experience in the Caribbean Region." *Journal of Environmental Planning and Management* 47 (4): 491–501.

Luhmann, Niklas. 1998. *Observations on Modernity*. Trans. William Whobrey. Stanford, CA: Stanford University Press.

Mallard, Grégoire, and Catherine Paradeise. 2009. "Global Science and National Sovereignty: A New Terrain for the Historical Sociology of Science." Pp. 1–39 in *Global Science and National Sovereignty: Studies in Historical Sociology of Science*, ed. G. Mallard, C. Paradeise, and A. Peerbaye. New York: Routledge.

McCarthy, James J., Osvaldo F. Canziani, Neil A. Leary, David J. Dokken, and Kasey S. White, eds. 2001. *Climate Change 2001: Impacts, Adaptation, and Vulnerability. Contribution of Working Group II to the Third Assessment Report of the Intergovernmental Panel on Climate Change*. Cambridge: Cambridge University Press.

McConney, Patrick. 2002. "A Small Island Developing States (SIDS) and Social Science Perspective." Pp. 151–154 in *FAO Fisheries Report No. 672*. Report and Documentation of the International Workshop on Factors Contributing to Unsustainability and Overexploitation in Fisheries, 4–8 February, Bangkok. Rome: Food and Agriculture Organization of the United Nations.

Mimura, N., L. Nurse, R. F. McLean, J. Agard, L. Briguglio, P. Lefale, R. Payet, and G. Sem. 2007. "Small Islands." Pp. 687–716 in *Climate Change 2007: Impacts, Adaptation and Vulnerability. Contribution of Working Group II to the Fourth Assessment Report of the Intergovernmental Panel on Climate Change*, ed. M. L. Parry, O. F. Canziani, J. P. Palutikof, P. J. van der Linden, and C. E. Hanson. Cambridge: Cambridge University Press.

Moore, Amelia. 2010. "Life in the Living Laboratory: An Anthropological Investigation of Environmental Science, Tourism, and Design in the Contemporary Bahamas." PhD diss., University of California, Berkeley.

Moore, Rawleston. 2002. "Vulnerability and Adaptation: A Regional Synthesis of the Vulnerability and Adaptation Components of Caribbean National Communications." Caribbean Planning for Adaptation to Global Climate Change Project. http://www.caribbeanclimate.bz/accc/accc.html.

Moss, Richard H., Jae A. Edmonds, Kathy A. Hibbard, Martin R. Manning, Steven K. Rose, Detlef P. van Vuuren, Timothy R. Carter, Seita Emori, Mikiko Kainuma, Tom Kram, Gerald A. Meehl, John F. B. Mitchell, Nebojsa Nakicenovic, Keywan Riahi, Steven J. Smith, Ronald J. Stouffer, Allison M. Thomson, John P. Weyant, and Thomas J. Wilbanks. 2010. "The Next Generation of Scenarios for Climate Change Research and Assessment." *Nature* 463: 747–756.

Nelson, Donald R., Colin T. West, and Timothy J. Finan. 2009. "Introduction to "In Focus: Global Change and Adaptation in Local Places."" *American Anthropologist* 111 (3): 271–274.

Oliver, Selma. 2009. "A New Challenge to International Law: The Disappearance of the Entire Territory of a State." *International Journal on Minority and Group Rights* 16 (2): 209–243.

Patz, Jonathan A., Diarmid Campbell-Lendrum, Tracey Holloway, and Jonathan A. Foley. 2005. "Impact of Regional Climate Change on Human Health." *Nature* 438: 310–317.

Pelling, Mark, and Juha I. Uitto. 2001. "Small Island Developing States: Natural Disaster Vulnerability and Global Change." *Global Environmental Change Part B: Environmental Hazards* 3 (2): 49–62.

Peterson, Nicole, and Kenneth Broad. 2009. "Climate and Weather Discourse in Anthropology: From Determinism to Uncertain Futures." Pp. 70–86 in *Anthropology and Climate Change: From Encounters to Actions*, ed. S. A. Crate and M. Nuttall. Walnut Creek, CA: Left Coast Press.

Rabinow, Paul. 2008. *Marking Time: On the Anthropology of the Contemporary*. Princeton, NJ: Princeton University Press.

Roncoli, Carla, Todd Crane, and Ben Orlove. 2009. "Fielding Climate Change in Cultural Anthropology." Pp. 87–115 in *Anthropology and Climate Change: From Encounters to Actions*, ed. S. A. Crate and M. Nuttall. Walnut Creek, CA: Left Coast Press.

Schneider, S. H., S. Semenov, A. Patwardhan, I. Burton, C. H. D. Magadza, M. Oppenheimer, A. B. Pittock, A. Rahman, J. B. Smith, A. Suarez, and F. Yamin. 2007. "Assessing Key Vulnerabilities and the Risk from Climate Change." Pp. 779–810 in *Climate Change 2007: Impacts, Adaptation and Vulnerability. Contribution of Working Group II to the Fourth Assessment Report of the Intergovernmental Panel on Climate Change*, ed. M. L. Parry, O. F. Canziani, J. P. Palutikof, P. J. van der Linden, and C. E. Hanson. Cambridge: Cambridge University Press.

Steward, Julian H. 1955. *Theory of Culture Change: The Methodology of Multilinear Evolution*. Urbana: University of Illinois Press.

Strachan, Ian Gregory. 2002. *Paradise and Plantation: Tourism and Culture in the Anglophone Caribbean*. Charlottesville: University of Virginia Press.

Tsing, Anna L. 2005. *Friction: An Ethnography of Global Connection*. Princeton, NJ: Princeton University Press.

UNWTO (World Tourism Organization). 2008. *Climate Change and Tourism: Responding to Global Challenges*. Madrid: WTO and United Nations Environment Programme.

van der Velde, M., S. R. Green, M. Vanclooster, and B. E. Clothier. 2007. "Sustainable Development in Small Island Developing States: Agricultural Intensification, Economic Development, and Freshwater Resources Management on the Coral Atoll of Tongatapu." *Ecological Economics* 61 (2–3): 456–468.

Yearley, Steven. 2009. "Sociology and Climate Change after Kyoto: What Roles for Social Science in Understanding Climate Change?" *Current Sociology* 57 (3): 389–405.

Adaptation—Genuine and Spurious

Demystifying Adaptation Processes in Relation to Climate Change

Thomas F. Thornton and Nadia Manasfi

ABSTRACT: In climate change discourse and policy, adaptation has become a critical byword and frame of reference. An implicit assumption in much of the strategizing is the notion that adaptation can be rationally planned, funded, and governed largely through existing frameworks. But can adaptation really be managed or engineered, especially given the significant unpredictability and severe impacts that are forecast in a range of climate scenarios? Over millennia, successful societies have adapted to climate shifts, but evidence suggests that this was often accomplished only through wide-ranging reorganization or the institution of new measures in the face of extreme environmental stress. This essay critically examines the concept of human adaptation by dividing it into eight fundamental processes and viewing each in a broad cultural, ecological, and evolutionary context. We focus our assessment especially on northern indigenous peoples, who exist at the edges of present-day climate governance frameworks but at the center of increasingly acute climate stress.

KEYWORDS: adaptation, Arctic, climate change, culture, development, indigenous peoples, vulnerability

Adaptation has become a key watchword and action frame in climate change discourse and policy. In 2009, prior to the Copenhagen climate summit, United Nations Secretary-General Ban Ki-moon referred to climate change as "the greatest collective challenge we face as a human family," in response to which it is "absolutely crucial that the world agrees on a comprehensive framework for adaptation." Yet adaptation is not clearly defined in the text of the landmark 1992 United Nations Framework Convention on Climate Change (UNFCCC). It is mentioned only in relation to stabilizing greenhouse gas emissions at a level where adaptation of ecosystems, food production, and sustainable economic development are all still possible (Article 2), committing parties to develop adaptation programs (Article 4.1[b]), and encouraging cooperation in adapting vulnerable areas to climate change (Article 4.1[e]) (see Ford et al. 2007; Schipper and Burton 2008; Smit et al. 2000).

But can adaptation really be managed or engineered, especially given the significant unpredictability, stress, and impacts forecast in various climate scenarios? Can infrastructure and development projects simply be 'climate-proofed' as a matter of planning or 'mainstreaming' adaptation into existing frameworks? While successful societies have adapted to climate shifts for millennia, often adaptation was achieved only through radical reorganization or innovation in the face of destabilizing environmental stress. Or, as per Romer's rule, new adaptations were enabled by evolutionary changes that initially were selected in order better to maintain existing patterns of life (Allaby 2004). In short, adaptation is often a blind process that can be viewed as 'rational' only in hindsight. As a participant in the 2009 15th UNFCCC Conference of Parties (COP15) in Copenhagen suggested, "In order to say something about adaptation, you need to have lived for 10,000 years."

In the emerging mainstream climate change literature on adaptation, recognition of the power of so-called autonomous adaptation has often been neglected in favor of what may be termed 'planned' adaptation, that is, what humans must rationally do in order to reduce risk and vulnerability. This perspective, perhaps fueled by a progressivist, 'techno-fix' bias, often leads to the downplaying of ongoing processes of autonomous adaptation at the local level. This point is made forcefully in a recent Commission on Climate Change and Development report, produced by the Swedish Ministry of Foreign Affairs, which seeks to counter this bias (Christoplos et al. 2009: 3):

> [A]daptation should be built on efforts to more effectively support individuals, households, and businesses as they struggle to adapt to climate change and … this should be done with a deeper awareness of the social, economic, cultural, and political factors that frame their actions, incentives, opportunities, and limitations for action … The poor adapt in ways that are usually unnoticed, uncoordinated, and unaided by national governments, development agencies, or international agencies. People draw on resources and support from these sources, but they do it in ways that are rarely reflected in the formal mechanisms designed for poverty reduction and climate adaptation.

This realization is perhaps the beginning of a potentially constructive 'retrofitting' (Head 2009) of the concept of adaptation (long a part of biological and anthropological studies) to the considerable challenges that human societies face in the context of present and future climate change and other drivers of cultural change, such as development and globalization (Liverman 2008).

This essay seeks to contribute to this constructive retrofitting by examining critically the concept of adaptation and related ideas such as mitigation, vulnerability, and adaptive capacity in a broad cultural-evolutionary context, focusing especially on northern indigenous peoples, who are at the margins of current climate governance frameworks but at the center of increasingly profound climate stress. We review key themes in the adaptation literature with an eye toward sharpening the conceptualization of adaptation as a process and, in the spirit of Sapir's (1924) critique of the culture concept, highlight some genuine and spurious assumptions about the nature of adaptation. The case of northern indigenous peoples is explored in detail because these groups are on the front lines of climate change. Moreover, they are active not only in autonomously adapting to profound climate stress and to rapidly changing and less predictable environmental conditions, but also in shaping more just and efficacious national and international climate policies based on diverse but comparable local knowledge, cultural practices, and adaptation processes.

Defining Adaptation

Adaptation has been defined in many ways, and its meaning has evolved over time and as a result of consideration by scholars in different fields of study (Smithers and Smit 1997). In the biological sciences, adaptation refers to "genetic characteristics which allow individual organisms to survive and reproduce in the environment they inhabit" (Smithers and Smit 1997: 133, after Winterhalder 1980; see also Abercrombie et al. 1977; Lawrence 1995). Critically, the process of adaptation requires variation or diversity in the fundamental building blocks of the organism—genes in biological evolution—in order for environmental selection to operate successfully, favoring some characteristics over others. As an aspect of evolution, adaptation also requires a process of transmission through which those genetic characteristics that compete successfully within a selective environmental context are passed on to future generations.

In the social sciences, cultural adaptation typically refers to the processes by which individuals and groups of people adjust their behavior and organization in response to changes in their environment (Denevan 1983; Hardesty 1986; Smit et al. 2000). Like biological adaptation, cultural adaptation requires diversity in the fundamental building blocks of culture (ideas, practices, etc.),[1] as well as a system of reproduction (imitation, learning, etc.) and inheritance (cultural transmission). At another level, human adaptation is strongly linked to environmental extensity and diversity in that humans not only adapt to their existing environments but, when not ring-fenced by environmental or political barriers, also seek new environments and niches when their existing ones become too stressed or stressful.

Putting the biological and the cultural together, the Intergovernmental Panel on Climate Change (IPCC) defines adaptation as "the adjustment in natural or human systems in response to actual or expected climatic stimuli or their effects, which moderates harm or exploits beneficial opportunities" (IPCC 2007). This definition conveys the dynamism of adaptation in relation to both natural and human cultural systems. However, an awareness of the multi-scalar and coupled nature of human and biological networks as socio-ecological systems (Berkes and Folke 1998; Gallopín et al. 1989) is critical to understanding both stimuli and responses to processes such as climate change. When socio-ecological systems are linked through vast institutions (e.g., global markets) and environmental processes (e.g., climate change), the adage that 'all adaptation is local' proves false. Local adaptation may help a community-level socio-ecological system make minor improvements while neglecting major regional or global environmental forces and structural violence (Farmer 2003) that may be constraining or undermining it extra-locally. Although a useful heuristic, in reality few socio-ecological systems are discrete or simply joined; rather, they are nested and complexly linked. Young et al. (2006: 314) have particularly emphasized globalization as a phenomenon in which "linkages between biophysical and social systems across space and time produce surprising dynamics and novel emergent properties" across socio-ecological systems. Understanding these linkages presents a formidable challenge to researchers and policy makers in their efforts to assess resilience, reduce vulnerability, and build adaptive capacity.

Until recently, adaptation was neglected by the UNFCCC in international climate policy deliberations in favor of mitigation (Pielke et al. 2007; Schipper 2006). Reasons for this include (1) the fear that focusing on adaptation might give the impression of resignation and weaken the social will to mitigate; (2) the difficulty of incorporating and evaluating adaptation measures into negotiations, considering the uncertainty of climate change impacts (Burton 2008; Füssel 2007; Kates 2000; Pielke 1998); and (3) the notion that adaptation costs can be significantly compounded if mitigation is not attended to first (Stern 2007). However, the 2001 Marrakech Accords developed funding mechanisms to assist developing countries in adapting (Adger et al. 2003), and adaptation was included as one of the four pillars for any post-Kyoto Protocol agreement in the 2007 Bali Road Map and Action Plan (Schipper and Burton 2008). International financial commitments to adaptation continue to rise, most recently as a result of COP15 and COP16 agreements to boost funding for mitigation and adaptation by up to $100 billion a year by 2020, although which entities (including newly created ones) and initiatives will get the money has not yet been determined. To make the best use of adaptation funding mechanisms, a better understanding of contemporary and historical processes of adaptation in relation to climate change and other stressors in existing socio-ecological systems must be developed.

Climate change can impact the way that individuals interact with their natural and social environments, possibly resulting in irreversible losses to a community's cultural heritage (Adger et al. 2009) or even the extinction of communities or segments of society. Alternatively, over time, cultural adjustments, innovations, and the transmission of key ideas and artifacts to younger generations may result in an enriched cultural repertoire with well-designed adaptations (Boyd and Richerson 2005; O'Brien and Holland 1992). Significantly, there is no guarantee that adaptation will happen on any level, or that adaptation at one level (e.g., community environmental management) might not be maladaptive at another level (e.g., national security). In essence, adaptation is a blind, complex, dynamic, and contingent process. While politics emphasizes the short-term view, human adaptation must be judged from a long-term perspective. Further, given the inherent uncertainty with respect to the future, a long-term view of past socio-ecological systems that have adapted to climate change is a critical, yet underutilized, tool in adaptation assessment and planning. Dearing (2008), for example, analyzes some 3,000 years of erosion and land use, what he terms 'slow' environmental change processes, relative to 'fast' changes, such as monsoon intensity and flooding, in Yunnan, China. Such 'historical profiling' of landscapes requires a mixed methods approach that involves anthropological, geographical, and environmental science investigations in order to map the 'adaptive cycle' on to the millennial record of climate, land, and resource changes.

In a recent critique of the climate change literature, Nelson et al. (2009: 272) note that "climate-change debates have historically focused on technologies and the elusive search for large-scale, cookie-cutter solutions, leaving aside the important role that individuals, cultures, and societies play in constructing and living out an adaptation dynamic" (see also Finan and Nelson 2009). Similar critiques have been leveled at the development literature (Escobar 1995; Ferguson 1994; Shiva 1989), and, as in the development case, a remedy is often seen in the advent of 'community-based' approaches in which local stakeholders take part in the process of determining their own futures. Yet existing national plans, whether promoting development or adaptation, typically have not paid sufficient attention to the critical role that local institutions play in these processes, or how they link populations and policies in practice. As Agrawal (2008: 50) observes concerning adaptation, "As a result, despite a stated commitment to grassroots involvement, the actual focus in national adaptation plans is on technical and infrastructure options for adaptation, with little attention to their social and institutional context."

Just as it is a mistake to reify techno-fix as the only viable mode of adaptation, it can be similarly wrongheaded to regard existing polities or institutions, be they states or local communities or civic organizations, as the critical units of adaptation. Communities themselves can be adaptive or maladaptive and even multi-scalar, given modern communication, transportation, and identity networks. Community-based approaches often neglect these dynamics. Similarly, states have been reified as a political unit for adaptation. As the principal actors under the UNFCCC, states are seen as the critical agents in mitigating and adapting to climate change. If one takes the long view, however, states generally have evolved not so much to adapt to environmental constraints but rather to transcend them—through expansion, military conquest, or trade. As Carneiro (1970) hypothesized decades ago, states tend to develop in order to overcome conditions of environmental circumscription that limit organic processes of expansion and increase resource competition, stress, and conflict among expanding populations.

In the circumpolar North, most sedentary communities today are the products of expanding or developing states that actively sought to demobilize hunting-gathering peoples in order to exploit their territories and 'civilize' them through missionary work, education, collectivization, and other forms of economic rationalization and development. This process of state making enabled Arctic states to colonize the so-called Fourth World (Hall 1988) of marginal 'frontier' lands and indigenous peoples in order to exploit new sources of industrial wealth, such as minerals, oil, and gas, as well as to secure their borders during the Cold War through militarization in the interests of 'national security. Similar processes have enclosed mobile hunter-gatherers, pastoralists, and swidden agriculturalists in other parts of the world. As a result, these communities typically have become more environmentally vulnerable, more economically dependent on the state and global economy, and less diverse and resilient in terms of their livelihoods and cultural repertoires. While growing states have invested heavily in infrastructure to create certain kinds of peripheral communities in order to facilitate national goals, these marginal areas are often the first to suffer from divestment during economic downturns. In the Far North, as in other 'frontier environments', this is manifest in the classic 'boom and bust' cycles of the natural resource economies, which bring not only short-term socio-economic vulnerabilities but often long-term environmental and politico-economic problems, or what is sometimes referred to as the 'resource curse'. This pattern is especially acute in Russia due to the sudden withdrawal of the state from rural collectivization projects beginning in the 1980s, combined with the aggressive industrial development of selective regions based on their natural resources (Crate 2006; Ziker 2002).

Anthropological studies suggest that the most resilient and adaptive social unit over long periods may be the household rather than the community or state (Netting 1993; Netting et al. 1984). In the past, households and extended families have been the critical units for responding to crises spawned by climate variability and cyclical environmental stresses, such as food and water shortages. Households continue to adapt in patterned yet diverse ways to the stresses of modern climate, food, water, and economic crises. Analyzing data over a 30-year period, West (2009: 286) reports in his study of Mossi household organization in Burkina Faso that extended households "persist because they are better adapted to conditions of heightened agro-climatic risk brought on by regional desiccation. They are also better able to take advantage of opportunities to intensify agriculture." A key to their adaptive capacity is the unique ability of households to engage in processes of 'extension' and 'fragmentation' according to ecological constraints. Modern communities and states often lack this flexibility, with the result that responses to environmental change can be more stressful, especially on the most marginal settlements and segments of the population. In the next section we look more closely at central adaptation processes adopted by households and other social groupings.

Adaptation as Intersecting Processes

Cultural adaptation is an ongoing set of processes that must be viewed holistically over time. Smit et al. (2000) present a practical way of classifying aspects of cultural adaptation based on (1) who/what has to adapt (the system of interest), (2) what they have to adapt to (the stimulus), and (3) how they adapt (the processes and forms). The system of interest can vary from a whole system or country to individuals or species. It may be adapting to long-term mean climate variability, climate extremes, future climate change, or the risks and opportunities of climate stimuli, among other things. The adaptation process itself can vary in intent (autonomous, planned), timing (anticipatory, reactive), temporal and spatial scope, and form (technological, behavioral, institutional, etc.).

Given the dynamic and complex nature of adaptation and the bias of policy makers in favor of certain contemporary units (states, communities) and modes of adaptation (e.g., techno-fix) over others, we favor a processual approach that revolves around eight key dimensions of human adaptation without preference for a single social unit or mode. These dimensions are mobility, exchange, rationing, pooling, diversification, intensification, innovation, and revitalization. While functional overlap may occur between these processes, each has a unique motivational core that distinguishes it from the others. In addition, we emphasize a long-term view of human adaptation that analyzes historical and contemporary processes at work and the convergence and disjuncture between forces of environmental change and human livelihoods over long time periods and multiple spatial scales. In the literature to date, the conceptual approach that most closely mirrors ours is that developed by Halstead and O'Shea (1989), on how cultures manage risk, and further refined by Agrawal (2008), in the context of rural institutional adaptation. However, here we have expanded the discourse beyond modes of adaptation in primarily rural contexts to a wider, more basic discussion of fundamental human adaptation processes.

We first define and characterize these adaptation processes and then assess how they are playing out in autonomous and planned contexts in response to climate change in northern indigenous communities and states. Table 1 provides an overview of the eight processes.

Table 1: Adaptation processes: Overview

Adaptation Process	Description
Mobility	Seasonal movement or permanent migration to avoid risk or in search of better circumstances
Exchange	Flow of material and symbolic goods and services between people
Rationing	Controlling the circulation or consumption of limited or critical resources among members of a group
Pooling	Sharing or linking of assets (wealth, labor, knowledge) across social groups
Diversification	Increasing the variety of food, income production strategies, specialization, etc., to enhance livelihoods
Intensification	Increasing the availability of resources by boosting their yield within a certain space or time
Innovation	New, unplanned method or technique that arises to address a certain need
Revitalization	Organized reconfiguration of ideology and practices to reduce stress and create a more satisfying culture

1. *Mobility.* Whether through seasonal movements or permanent migrations, humans have sought to avoid environmental risk and obtain better circumstances. Despite the development of agricultural and industrial states and the decline of more nomadic modes of subsistence, such as foraging and pastoralism, human migration has increased, largely as a response to economic and environmental stresses and perceived opportunities in alternative environments. Richerson and Boyd (2008: 877) observe that "migration has a profound effect on how societies evolve culturally because it is selective. People move to societies that provide a more attractive way of life and, all other things being equal, this process spreads ideas and institutions that promote economic efficiency, social order and equality."

Rather than being a source of conflict and instability, migration may act as an engine of social change. Of course, this assumes voluntary migration rather than forced relocation or situations that involve so-called climate and environmental refugees, who may lack choices about whether or where to move. It also assumes that host societies have the ecological capacity and adequate ecosystem services to absorb migrants in sustainable ways, as well as the institutional means to adapt to potential socio-cultural 'organization of diversity' issues (Wallace 2009) posed by immigrants. Involuntary relocations especially can lead to incalculable cultural disruption, loss of livelihood, and increased dependency and stress (Oliver-Smith 2009). Presently, many such migrants are being absorbed into highly exploitive industries and urban slums, which are neither equitable nor sustainable. Unfortunately, under probabilistic scenarios generated in response to climate models, this is a process that is expected to intensify. As a result, conflict between migrants and residents in receiving areas may arise as a result of competition, ethnic tension, distrust, socio-economic fault lines, and political instability (Reuveny 2007).

In the modern industrial context, human mobility is intricately regulated by institutions and policies linked to immigration, transportation, and the enclosure of common lands into private space or property. A lack of sedentism can be confounding to states seeking to improve security or productivity through modernization, land use planning, and development. Nevertheless, it may be short-sighted to view mobility as maladaptation simply because it poses potential short-term social and political instabilities. As Agrawal (2008: 19) points out, mobility remains "a way of life for large groups of people in semi-arid regions, and a long standing mechanism to deal with spatio-temporal variations in rainfall and range productivity," especially among pastoralists (Niamir-Fuller 1999). Among northern indigenous pastoralists and foraging peoples, mobility remains a key dimension of economic production and adaptation (Rees et al. 2008).

2. *Exchange*. Like the flow of people, the flow of material and symbolic (knowledge) goods and services are a foundation not only of today's global economy but of every economy in history. Agrawal (2008: 21) observes that market exchange "is perhaps the most versatile of adaptation responses … to environmental risks but also … for specialization, trade, and welfare gains that result from specialization and trade at multiple scales." Agrawal cites a recent spate of insurance schemes to mitigate against weather-related damages among agricultural and pastoralist populations as an example of market exchange as a mode of adaptation.

A critique of the current dominant strategies of economic exchange is that the costs of exchange are being born disproportionately by the world's poor, underdeveloped, and thus most vulnerable communities, with negative consequences on their adaptive capacity. Their resources, natural and human, are being exploited at unsustainable rates in order to supply wealthier interests with resources, goods, and services. In its more extreme form, this inequitable, asymmetrical exchange (or negative reciprocity, as anthropologists refer to it) leads to a kind of 'eco-imperialist' expropriation, if not underdevelopment, of local environmental resources and ecosystem services (and even carbon) by well-financed (often distant) interests at the expense of the poorer local communities that depend on them for their livelihoods (Shiva 2008). The same is true of the knowledge economy, wherein intellectual property, even when based on local and traditional knowledge, can be captured as private property through patents and other legal regimes, or what Harvey (2005: 159) refers to as "accumulation by dispossession."

3. *Rationing*. All human societies have developed schemes and technologies for allocating provisions and other limited or critical resources among their members. The fundamental objective of rationing is to extend the supply of resources by controlling their circulation and consumption over time and space, and storage is among the most basic forms of rationing. The development of storage and preservation techniques has bolstered the resilience of households and communities enormously by allowing the consumption of otherwise perishable resources, such as fresh foods, over extended periods. Even among foraging societies with high mobility, these techniques have been instrumental in reducing risk due to climate and resource variability, as

evidenced by the widespread practice of caching foods among northern hunter-gatherers. The stockpiling of critical resources, such as energy, seeds, or even genetic material, is a modern means of utilizing storage to ration consumption (and conserve diversity) over time. In the case of genetic banking, it is a way of preserving biodiversity for the future by avoiding extinction due to present levels of consumption or destruction.

Rationing also can be applied to activities or elements that have a negative impact on critical resources or ecosystem services, such as pollution or overpopulation. Carbon capture and storage, for example, is seen as a potential technological solution to reducing the ill effects of carbon dioxide emissions into the atmosphere by rationing their release through sequestration. Cap-and-trade schemes are another potential means of controlling pollution (via the 'cap'), according to market-based principles. Rationing strategies are often productively linked with systems of exchange or pooling. Population control by limiting birth rates is among the most controversial of rationing programs, especially when it is applied coercively. Fortunately, there are many means available for promoting sustainable population growth at the family, ecosystem, regional, and national levels (Sachs 2008: 185).

Significantly, rationing does not by definition reduce consumption or pollution; rather, it may simply redistribute it. Similarly, rationing schemes do not necessarily result in greater equity among consumers or citizens. Population rationing programs, for example, may be nullified by increased per capita consumption of resources, or they may exacerbate existing inequalities due to differential values placed on children. Nevertheless, rationing remains among the most powerful adaptation tools that humans have developed for both limiting consumption and ensuring the equitable distribution of resources, including ecosystem services such as clean air and water.

4. *Pooling*. At base, pooling is a form of holding, sharing, or linking assets across social groups. Agrawal (2008: 20) defines it as "adaptation responses involving joint ownership of assets and resources; sharing of wealth, labor, or incomes from particular activities across households, or mobilization and use of resources that are held collectively during times of scarcity." Adaptive pooling assumes the ability to distribute or convert accumulated assets when needed in times of environmental stress (Moench 2007).

Perhaps the most important asset that can be pooled is credible knowledge. Credible knowledge is information that has been critically assessed by and rendered meaningful to members of a community. In the literature there is strong emphasis on improving the flows of information and scientific knowledge to communities. However, if the source of this information is not considered credible, or if the information is not considered relevant in terms of on-the-ground experience, such information pooling is unlikely to be successful. For example, regional weather forecasts or climate predictions may be judged as non-credible at a local level due to their higher incidence of error compared to local knowledge or monitoring techniques (cf. Strauss and Orlove 2003).

How communities pool resources is partly determined by tenure arrangements. Tenure systems are a means by which human societies hold prerogatives over key resources such as land, water, and knowledge. Tenure requires schemes for reckoning boundaries, access, and transfer. Such systems may facilitate or inhibit pooling, and globalization raises linkage potentials, as well as multi-level commons and tenure problems (Berkes 2007). Intellectual property tenure systems, for example, often seek to prevent pooling of information without formalized exchange or payment. As Brush (1999: 539) observes in the case of intellectual property patented through bio-prospecting in indigenous communities, "The monopoly privileges that one community can gain affect other communities that share the same knowledge and resources. If knowledge and genetic resources collected under contract lead to a patentable product, communities that are not part of the contract but have the same resources can be deprived of the opportunity to commercialize their knowledge." The monopolization problem also exists for material resources and services, as evidenced by strategies to privatize access to critical adaptation resources such as public land, water supplies, and utilities. Such schemes of accumulation by dispossession undermine the pooling of common resources and, according to neoliberal critics, lead to the "escalating depletion of the global environmental commons (land, air, water) and proliferating habitat degradations … [resulting] from the wholesale commodification of nature in all its forms" (Harvey 2005: 160).

5. *Diversification*. Diversification is a form of risk management that typically refers to food and income production strategies. But at a more basic level, diversification is adaptive because it either introduces or maintains variations upon which future cultural adaptation processes can act. Significantly, diversification does not negate specialization, except perhaps at the level of the individual organism. Depending on the social unit, organic specialization may occur at various scales. For example, in a household there may be a division of labor whereby a grandmother specializes in child care or food preparation, or young males specialize in hunting mobile prey, and so on. As with the other strategies, diversification is a hedge against uncertainty and a means of maintaining the availability of critical resources, including money.

Too much diversification can be limiting or even maladaptive, especially if it means a loss of knowledge and skills. Denevan (1983: 402) suspects that most diverse techniques of environmental manipulation are long present within a culture, perhaps as marginal or secondary strategies, before widespread adoption occurs, as opposed to being introduced 'Eureka style' through radical innovation. When minority strategies are perceived to respond to an identified need, more widespread use of these already available techniques (or concepts) results. The development of agriculture may very well have evolved along these lines, initially as a marginal or subsidiary diversification strategy, subordinate to foraging, which was not adopted wholesale until it was perceived to be more adaptive due to changes in socio-ecological or environmental conditions (see Bellwood 2005; Richerson et al. 2001; Rindos 1984; Winterhalder and Kennett 2006).

Niche theory holds that diversification in terms of livelihood may reduce competition and resource stress within or between species inhabiting the same area or ecosystem. Due to mobility and cultural adaptation, humans are able to dwell in an extremely broad range of habitats and are thus able to exploit a wide spectrum of niches. Niches and cultures may be co-evolutionary and co-dependent, shaping each other in fundamental ways to the point that one cannot truly exist without the other. Historical and anthropological studies suggest that ethnic and linguistic diversification may be associated with niche specialization. For example, Barth (1956) shows how niche specialization through partitioning (to reduce livelihood overlap) among the inhabitants of the rugged mountains and river valleys of northern Pakistan led to the formation and co-existence of three distinct ethnic groups. As a result of climate change, fundamental niches may also change, and there may be significant lags in the formation of realized niches, which adaptation policies could address (Pidwirny 2006).

6. *Intensification*. Intensification is a means of increasing the utilization of resources by boosting their yield within a certain space or time. Its inverse process is extensification, which typically implies mobility (and is thus partially treated under that rubric above, rather than separately). Agricultural intensification, through the Green Revolution and other technological and organizational innovations, has been a favored adaptive strategy for increasing food yields in conjunction with expanding populations. However, the dominant strategy of increased dependence

on engineered seeds, irrigation, chemical fertilizers, pesticides, and mechanization has been criticized as maladaptive for its deleterious ecological consequences, such as soil erosion and compaction, pollution, and loss of biodiversity, which have brought harm to a range of species through habitat loss or degradation (see, e.g., Birdlife International 2008).

Intensification can also lead to centralization and reduced diversity. In the realm of food production, this can be seen in the reliance on fewer varieties of agricultural seeds, developed and patented by large multinational agribusinesses, which are considered more efficient in maximizing yields and can be co-engineered with supplementary inputs such as fertilizers and pesticides. Dependency on fewer seed varieties and their specific inputs can lead not only to increased environmental impacts but also to greater vulnerability for local groups if the price, supply, or efficacy of these non-local products is adversely affected.

These criticisms notwithstanding, intensification as a broad adaptive strategy need not, in theory, be a zero-sum game for humans at the expense of other species or local diversity. Intensification may be achieved simply by improving efficiencies and making resources more usable, for example, by reducing waste through consuming rather than discarding vegetable or animal parts, or by using 'waste' to enhance environmental habitats (e.g., through fertilization). Among northern reindeer-herding peoples, a flexible dynamic between intensification and extensification of grazing in relation to land and labor conditions has proven highly adaptive over time. However, this flexibility has recently been compromised by reindeer 'rationalization' schemes that trade flexibility for more rigid territorial regimes and greater dependency on state subsidies or economic diversification (Beach et al. 1992).

7. *Innovation*. Innovation is perhaps the most cited but least understood of adaptation processes. As noted in the discussion above on diversification, the seeds of adaptation are often born not of true innovation but rather of shifts from dominant to minority strategies in the context of environmental change. True innovation, akin to mutation in biological evolution and adaptation, is typically random and therefore very difficult to predict or plan. Thus, it is problematic to rely upon innovation, whether technological, ideological, or organizational, to build adaptive capacity. Still, there is no doubt about the significance of innovation in the history of adaptation. Among Arctic indigenous peoples, for example, the invention of the toggle harpoon and float had a major effect on the adaptation and successful spread of Inuit culture across the northern Arctic coast of North America in conjunction with marine mammal populations, such as seals, which could not be reliably secured with simple spears. While modern tools have replaced many traditional technologies, the toggle harpoon persists because it retains numerous advantages for landing large sea mammals from breathing holes or open water. Moreover, the tool itself continues to be adapted and refined (Arnold 1989: 81).

Selective forces perfected the toggle harpoon to the extent that it was adopted (and mechanized) by the whaling (oil) industry in the mid-nineteenth century. However, as with many technological innovations applied on an industrial scale, the mechanized whaling harpoon was overdeployed, leading to a worldwide decline in whale populations, the eventual collapse of the industry, and the whale rationing system that exists today through the International Whaling Commission.

Innovation within a system may affect vulnerability and resilience in unanticipated ways. In the Middle East, Jordan illustrates a case of low system resilience that resulted in an inadvertent transformation. An expansion of land use (extensification) combined with advancements in agricultural technology (innovation) allowed agricultural output to increase. However, a lack of proper resource management eventually resulted in land degradation and a weakening of the socio-ecological system. It has been suggested that the agricultural system in Jordan ultimately collapsed due to a period of climatic stress (Nelson et al. 2007).

8. *Revitalization.* Revitalization is a society's capacity to adapt to environmental stress through a structured reconfiguration of its ideologies, practices, and organization in order to reduce stress and create a more satisfying culture (Wallace 1956). Revitalization theory as applied to social movements has been criticized for, among other things, its reliance on the metaphor of society as an adaptive 'organism' with a 'homeostatic' state (Harkin 2004), as opposed to a complex 'panarchic' system with multiple states (see Gunderson and Holling 2002). Yet there is no questioning the power of social movements to produce adaptive social change and more sustainable livelihoods by redefining human priorities and codes for living. Wallace (1956) emphasizes that such revitalization requires both reformulating old 'mazeway' patterns of cognition and routinizing new adaptive 'codes' for living. Often the new codes are defined by charismatic leaders at the margins of society who can see beyond current cultural models. Obviously, this new vision ultimately must be grounded in the material exigencies of life and successfully institutionalized if it is to survive.

From the perspective of adaptation, revitalization is important because it allows humans to achieve rapid social change in response to environmental shifts or stresses without recourse to violence or competitive conflict. Frequently, revitalization involves syncretism wherein traditional knowledge and lifeways are constructively realigned to respond to contemporary socio-ecological constraints. For example, in Wallace's (1956) prototypical case for revitalization—that is, the Code of Handsome Lake movement among the Seneca Iroquois of New York—the prophetic leader's vision included sanction for men to shift their productive roles in society from hunting to agriculture (formerly dominated by women). This was consistent with evolving material conditions that had resulted in declining game resources, decreasing land base and access to traditional hunting areas, and the concomitant growth of sedentism and agriculture in that area in the late eighteenth century.

Beyond social change, revitalization is an important means of restoring ecosystems and reducing stress on human health. Civic agricultural and local food movements are examples of efforts to revitalize selective aspects of traditional agriculture within a modern context of valuing diversity, quality, sustainability, and connections to land, place, and community (Lyson 2004). Similarly, efforts to restore urban pedestrian and communal spaces for congress, passage, and exchange may be seen as an attempt to revitalize healthy and sustainable aspects of urban culture while at the same time reducing human stress brought on by poor design, short-sighted planning, and mechanical obeisance to automotive transport. Revitalization is also evident in a variety of ecosystems that range from forests to deserts. Also included are marine contexts with efforts to restore important coastal habitats, such as wetlands, lagoons, reefs, and mangroves (Berkes 2008), and critical ecosystem foundation species, such as herring, often using local and traditional knowledge (Thornton et al. 2010). The Convention on Biological Diversity (CBD) calls for the adoption of measures to revitalize threatened species through reintroduction, combining *ex situ* sources with *in situ* measures for reducing the forces that drove them toward extinction in the first place.[2] Such repair efforts require social change and adaptive (co-)management (Berkes et al. 2007) in order to be successful in the long run.[3] To the extent that they can be effectively institutionalized and remain responsive to objective conditions, revitalization processes are critical to socio-ecological adaptation.

All of the above adaptive processes involve technologies, learning, and institutions at various scales, from the most basic households, using their members' embodied technologies of ambulation and speech to navigate and communicate the lay of the land, to the most complex states or multinational entities, which use supercomputers and other specialized tools to coordinate international transportation, model changes in climates, monitor trading, and optimize commodity exchange. The mixture of adaptive processes is complex and constantly changing, not only in relation to environmental constraints, but also in relation to other adaptation responses. Thus, adaptive processes must be understood as both drivers of change and responses to change.

Adaptation in Northern Indigenous Communities

The severity and disproportionate effects of climate change on communities of the Far North are well known. In Alaska alone, a recent report by the US Government Accountability Office concludes that 31 villages "face imminent threats" from climate change, and that 12 have decided to "explore relocation options" (GAO 2009: 12). However, as the report's own title indicates, only "limited progress has been made," in part due to the fact that government bureaucracies remain uncoordinated and unable to reassess their institutional priorities in relation to climate change and adaptation needs. Instead, "In the absence of a lead entity, federal agencies individually prioritize assistance to villages on the basis of their programs' criteria, which do not necessarily ensure that the villages in greatest peril get the highest priority for assistance" (ibid.: 36). The magnitude of accelerated climate change impacts on Arctic communities has made them the focus of intense scrutiny by national and global media, seeking to personalize and ground the effects of climate change in real life (Henshaw 2009; Marino and Schweitzer 2009). Many see the Arctic as being on the threshold or tipping point of major systemic transformations.

Arctic climate change impacts are driven by key physical and biological processes that are increasingly well-documented (ACIA 2005; IPCC 2007; UNESCO 2009). The most important of these are annual average temperature increases. In Alaska, the increase is 1.9 °C since the mid-twentieth century, about twice the rate of the rest of the US. Temperature increases, in turn, drive other environmental changes that include melting and retreating of snow cover and changing snowpack structure; melting of large glaciers and ice sheets; retreating sea ice; permafrost degradation; changing river and lake ice magnitude, timing, and stability; changes in precipitation; and increased exposure to ultraviolet radiation. Climatic changes and impacts of this magnitude have not been experienced in the Arctic for over 100,000 years.

As a result, Inuit and other northern peoples find themselves in the position of having to adapt to these unprecedented complex, interacting, and non-linear changes in short order. At COP15, we carried out interviews with 12 northern indigenous leaders and representatives, many of whom commented that their own households and communities were "already adapting," but that states and their governing institutions were not adapting or were not helping their communities to reduce climate vulnerabilities and build adaptive capacities rapidly enough.[4] The major literature on Arctic adaptation (e.g., Ford 2009; Ford and Furgal 2009; Furgal and Seguin 2006; Furgal and Prowse 2008; Krupnik and Jolly 2002; UNESCO 2009) supports this perspective, as do the major indigenous organizations, such as the Alaska Federation of Natives (AFN 2008), the Inuit Circumpolar Council (ICC 2005; Watt-Cloutier et al. 2004), and the Russian Association of Indigenous Peoples of the North (RAIPON). The following statements are representative of recent indigenous peoples' climate change adaptation and vulnerability assessments in the Far North.

> *Canada*
> Interviews conducted [for an Inuit Circumpolar Council] indicate that despite the increased difficulty in finding and harvesting big game and sea mammals due to thinning and less predictable sea ice, Inuit communities are persistent in maintaining their traditional diets. When asked whether changes in ice conditions were affecting their traditional diets, respondents spoke of having to travel farther or in a different month than usual; they spoke of dietary substitutions such as hunting more musk-oxen when the caribou migration shifted away from

their area, or they explained how melting permafrost has made the natural ice cellars used to age and store meat less effective. (Smith 2009: 17)
A buddy of mine is into making little sleds out of aluminum, which you can use as a little kayak or boat. If you're out on the ice and you have to cross an open lead or something you can use that … It's combined as a little sleigh and, if you have to, you can use it as a boat. That's one way I can adapt. (John Keogak, interview in Smith 2009: 19)

And because the permafrost is melting, infrastructure is starting to become damaged … The roads are starting to be … wobbly…

We're also starting to see different insects and blackberries that never came up to our communities. Polar bears are starting to wander closer into the communities [and] wolves …

[T]he fact that lakes are starting to dry up a little bit is causing flora that used to grow to not grow anymore, so different animals that used to come feed on that, like the geese, are going farther away from the communities, so hunters have to travel farther. The moss that the caribou feed on is starting to grow in different places, so the migration patterns are changing, and again, hunters have to travel farther …

There needs to be a bridge between that gap [between scientific studies and traditional knowledge] for mitigation and adaptation to be successful …

I'm only 20 years old but already the world that I was born into has changed so much … I'm really not sure what will happen in the future. (Janice Grey, interview, 2009)

Alaska
In later years, my father noticed how the climate seemed to be changing, and he would comment on how spring came earlier and warmed up faster, making our whaling activities more hazardous. We hunt from the edge of the shorefast ice in spring, and changes in the ice pack have a big impact on our ability to participate in this traditional hunt. As time went on, even us young folks could see the changes. The ice pack was shrinking, and shorefast ice was rotting earlier in the year, and the ice retreated farther out during the summer and stayed out longer. It was almost like a rug was being pulled from under us. (Itta 2009: 207)

There are changes in sea ice conditions, changes in weather, changes in coastal erosion problems, river system anomalies.

Eighty percent of our communities in Alaska are living along the coast, and there are examples of villages that are literally washing away because of the many storm surges and erosion … There used to be a wall of ice that protected the villages, but now, because the ice has changed, that is no longer there …

[I]t's not the same as it used to be when we were much more nomadic communities than now. Now we have our schools, our roads, and all these other things, so we can't pack up our tents and move quite as quickly. This has made a real difference in our ability to adapt in that way …

Some areas are looking at new [plants], new species, new food sources. They are relying on new sources that are now available. In Alaska there is a whole industry now growing up around shark fishing, and certainly this has not been the case historically in that area. You learn to deal with species that were not there before or were in much smaller numbers …

The modern technologies are now also looking back at how our people used to live. For example, in my community the house used to be subterranean, and now people are understanding why the homes were built in this fashion—they are the most ecologically friendly, they are more energy-efficient … and all the things that modern technology is now looking for. (Patricia Cochrane, interview, 2009)

Greenland
In Greenland, our Sila Inuk project focuses on Inuit hunters, asking them to document what climate change effects they have seen over their lives and what information they have gleaned

from their grandparents … Hunters speak of thinning sea ice that makes hunting much more dangerous, changes to permafrost that alter spring run-off patterns, a northward shift in seal and fish species, and rising sea levels with more extreme tidal fluctuations. One hunter told us, 'The sea must be getting warmer because it doesn't freeze where it used to, even when the air is very cold.' Another said that the snow melts so quickly in the spring now that 'it is as if the earth just swallowed it!' Many say their traditional knowledge is not as reliable as it was in the past for predicting safe ice conditions. This is a great source of anxiety for Inuit hunters. (Lynge 2009: 106)

Fennoscandia
The sensitive nature of the Arctic environment cannot sustain large numbers of people or the wide-scale exploitation of energy resources … Changes in the flora, fauna and climate, combined with the loss of entire living territories, will force indigenous people to seek new ways of adapting …

In addition to climate change, indigenous peoples must accommodate other competing land uses, such as oil fields, forest felling, tourism and mining … In reindeer husbandry, it may lead to a reduction in herd sizes, the need to acquire additional food for the animals, and changes in reindeer husbandry models and cultures. Continuous feeding will transform nomadic reindeer herding, bringing it closer to domestic animal husbandry … If reindeer herders have fewer reasons to travel within their environment, the use of environmental, snow and climate terminology will become less important, along with the ability to navigate within and read the environment … When reindeer no longer have some 300 plants in their diet, the taste and fat content of reindeer meat will change and the fat will become unhealthy. Due to climate change, the terminology and practical knowledge of nature and reindeer husbandry will decline and partly disappear. (Lemet-Klemetti Näkkäläjärvi 2009: 132, 140)

Russia
We have seen changes in nature and the climate, but it's very different from one region to another … [M]y region, Sakha Republic, is one of the coldest places in the world, where temperatures can go down to -50, -60, -70 degrees centigrade … In the tundra … we've seen that in that place where a few years ago there were no trees or crops, now there are …

We've also seen variability in the temperature in the fall. Before in November there was no rain, but now there is rain and the autumn is longer. After it rains, it can get very cold and an ice layer develops, which makes it very difficult for the reindeer to get to the food below. Also in winter it can rain and then it freezes—this is not good for reindeer health. We also see that because of the warmer temperature, the rivers and lakes are open longer. Because we are nomadic people, we are always moving with the reindeer, so we have to change our routes and change the time of moving …

But when you try to change the route, there will be other reindeer herders, and you will have the conflict between the reindeer herders and state enterprises of reindeer herding and private reindeer herding, so there is no clear regulation of land use …

We have new legislation about land now which states that, as of next year, reindeer herders need to buy or rent the pastures. But that is impossible because for one reindeer you need 300 hectares per year; so if you need to pay, you will need billions to rent …

One of the [adaptation] strategies is … when you get an ice layer, if you have a castrated bull in the herd, it will be strong enough [to dig through the ice], whereas the rutting males spend a lot of energy on getting the females and are weak and tired … [C]astrated males … are strong and they can dig through the ice layer, and the females then have access to the food. So you need to have castrates …

When there is a hot summer, we use the permafrost areas where the ice is open. This keeps the insects from attacking the reindeer … because it's colder there. But now with climate change, the number of these places is falling. Other strategies include building shelters for the

> reindeer to stay under in warm temperatures. We also burn moss to get the smoke that keeps insects away, but there are new kinds of insects. (Mikhail Pogodaev, interview, 2009)

> In the local context … pollutants take the form of thousands of empty metal drums, pesticides, radioisotope thermoelectric generators from lighthouses, and scrap metals abandoned by industry and the military. Waste burial sites remain in the permafrost, and with the permafrost thawing, the danger from this waste increases.
>
> The main result of all these problems is the depopulation of the North by the aborigines, with the loss of the unique northern gene pool and circumpolar culture.
>
> The most convenient 'solution' supported by the government is the displacement of aboriginal peoples from traditional spheres of life and economic activity, moving them into inhabited areas, and thus giving them the opportunity to receive unemployment benefits or to become workers in modern industries. This is a painful process for northern aboriginal peoples … In traditional economic activities reside the life and the future of the peoples of the North … With the destruction of this way of life certain aboriginal peoples will disappear. (Abryutina 2009: 168–169, 172)

Although not always couched in the same general terms, these perspectives emphasize the key adaptation processes examined above. Table 2 illustrates how the perspectives map on to these processes.

Table 2: Adaptation processes as applied to northern indigenous communities

Adaptation Process	Description
Mobility	• Traveling farther to hunt sea mammals, collect plants, feed reindeer • Relocating or migrating away from land that is vulnerable to storms or melting permafrost
Exchange	• Transfer of traditional knowledge (related to hunting, building shelters, etc.) across generations
Rationing	• Reduction in the size of reindeer herds to accommodate competing land uses, such as oil fields, forest felling, and mining • Caching food traditionally in ice cellar storage, but this faces disruptions because of melting permafrost
Pooling	• Sharing new scientific knowledge and skills
Diversification	• Altering prey choice and diet breadth, such as hunting musk oxen when caribou is not available, or shark fishing in Alaska • Use of castrates in reindeer herds
Intensification	• Feeding reindeer on smaller areas
Innovation	• Development of the hybrid aluminum sled-kayak • Invention of the toggle harpoon and float
Revitalization	• Return to traditional knowledge, organization, and technologies, such as semi-subterranean house structures and communal living arrangements for resilience, cost savings, energy efficiency, etc.

Mobility is discussed in terms of increased risk to travel; increased need to travel farther due to dispersals of wildlife populations (such as sea mammals), plants, or feed (reindeer); and increased circumscription due to competition from other users (herders), dependence on infrastructure (schools, roads, etc.), competing land uses (resource extraction or other herders), and even 'solutions' to displacement, such as relocation or depopulation through outmigration. Exchange and pooling are emphasized in terms of sharing new scientific knowledge, the reliability of traditional knowledge and skills and their transfer among generations, and sharing and tenure security over pools of subsistence resources. Rationing disruptions are caused by melting permafrost, which renders ice cellar storage ineffective. Diversification is highlighted in the need to alter prey choice and diet breadth and the use of castrates in reindeer herds. Intensification is discussed in terms of reindeer feeding and corralling, rather than pasturing, and also in the ill effects of intensified resource exploitation, particularly hydrocarbon extraction from Arctic lands and waters, in contradistinction to what is needed to reduce global warming. Innovation is highlighted in the hybrid sled-kayak that allows for better navigation, given the expanding leads between ice floes. Finally, revitalization is framed in terms of a return to traditional forms of knowledge, organization, and technologies. This is seen as part of an effort to re-empower northern indigenous peoples to determine their own cultural path of adaptation and development in the context of climate change, with appropriate support from larger socio-economic systems. Climate change thus becomes a human rights and cultural sovereignty issue.

As this brief sample of Arctic indigenous perspectives makes clear in light of the eight processes of human adaptation presented above, households and communities are coping and adapting in myriad ways. However, when these key processes are blocked, vulnerability and exposure to risk are potentially increased. Obviously, it is too early to tell, in the dynamic Arctic context, the extent to which some of the strategies, especially as concerns subsistence activities, will prove to be viable means of adaptation in the long term. Davies (2008) emphasizes the difference between short-term coping strategies and adaptation strategies. Coping strategies entail immediate measures that may help to survive an unusual decline in resources but may not be sustainable in the long run. In contrast, adaptation strategies tend to spawn new cultural configurations that have evolved in response to changed conditions. Even still, as Romer's rule suggests, pathways toward the maintenance of existing patterns may eventually serve as catalysts for adaptation.

Indigenous peoples emphasize the disconnect between their own adaptation needs and the responses and priorities of the state and other non-Native institutions, such as multinational energy corporations and conservation organizations. Adaptive capacity could be enhanced through further research into how key indigenous adaptation processes may be complicated or put at risk by socio-economic policies or adaptation processes occurring at other scales. In Arctic Alaska, this means assessing the foundational importance of sea ice habitats and Inuit adaptations to marine mammals against the potential risks of policies that support (1) the intensification of offshore oil and gas development, (2) shipping opportunities to cope with dwindling domestic energy supplies or to exploit reduced transport costs through the Arctic seas, and (3) restrictions on sea mammal hunting as a means of conserving populations through rationing kills (see Itta 2009). Analyzing the nexus of climate and other socio-ecological impacts and adaptation responses at the household, community, state, and other institutional levels over time offers the best opportunity for reducing the so-called adaptation deficit and identifying opportunities for coordinating adaptation policy across scales in the Arctic and elsewhere.

Conclusion

It has now become almost a cliché to point out that adaptation to climate change must happen in conjunction with mitigation in order to cope with increasing environmental change, and that adaptation should not be viewed as a stand-alone or single-sector, single-scale issue. Rather, adaptation must be considered in the context of long-standing and evolving multi-scalar ecological, social, economic, political, and institutional circumstances (Adger et al. 2003; Nelson et al. 2007; Smit et al. 2000; Smithers and Smit 1997). However, to date, adaptation in climate policy has been poorly theorized and plagued by conceptual biases. We have emphasized critical weaknesses in the present approach to adaptation that, unless addressed, may hamper even the most progressive, multi-scalar institutional efforts to deal with escalating climate change.

We have stressed that human adaptation is not a single strategy but rather a set of diverse, intersecting processes that may evolve autonomously or through planning in response to the panoply of climatic and non-climatic stressors. Up to the present, conceptual work on fundamental human adaptation processes in the climate change context has been underdeveloped with a few notable exceptions. As a consequence, analysts and planners often ignore key findings from long-term anthropological and geographical studies of adaptation, focusing too much on too few institutions (modern states and communities) and processes (market exchange and innovation in the form of techno-fix) at the expense of other units (households) and methods (rationing and revitalization). Successful adaptation strategies may entail interconnected aspects of all of the eight major adaptation processes we have identified—and potentially more. Even with a broader set of conceptual tools, putting adaptation theory into practice will remain challenging because of the uncertainty of future climate change and its impacts and due to the difficulty of evaluating and linking adaptation measures across scales and against short-term and long-term costs and opportunities, as interpreted locally within the 'limits' of existing socio-political entities (Adger et al. 2009).

By enlarging the typology of adaptation processes and creating a metalanguage to characterize and analyze diverse modes of reorganizing across cultures, we have sought to broaden the discussion on adaptation processes as sources of resilience and adaptive capacity. Hopefully, such an effort will serve to expand the scope of climate impact assessment and adaptation planning to address a range of processes that are currently neglected, taken for granted, or irreflexively viewed as maladaptive from a status quo perspective. Furthermore, adaptation policy and funding should address not only future vulnerabilities to climate change but also past and current vulnerabilities and adaptive strategies to both climate variability and non-climatic stressors. If adaptation funding (under UNFCCC and elsewhere) fails to enlarge its framing of risks and adaptation processes, vulnerable groups, such as indigenous communities in the fast-warming Arctic, are unlikely to be able to adapt successfully to future climate change or will be forced to do so in ways that serve dominant interests rather than their own diverse and long-term needs (cf. Adger et al. 2006; Barnett 2008; Burton 2008; Handmer 2008; Pielke 1998; Ribot et al. 2008; Smithers and Smit 1997; UNDP 2002).

Some institutional innovations suggest that awareness of these links is improving, which is an important first step. Newly formed 'boundary organizations' are attempting to coordinate adaptation processes across present sectors and scales. It is beyond the scope of this article to review these institutional efforts in detail, but one example may suffice to illustrate the developments and challenges faced. The United Kingdom Climate Impacts Programme (UKCIP), which was already established in 1997, is one of the earliest examples of such a boundary organization (see Metcalf 2008). To date, the organization's efforts have focused largely on building adaptive capacity as a prelude to specific actions, and some links to critical adaptation processes are already apparent. For example, a major focus of the first phase of the UK's national Adapting to Climate Change Programme (2008–2011) has been on building adaptive capacity by developing a comprehensive evidence base of climate change impacts, raising awareness of the need for and benefits of early action, and embedding the consideration of climate change into policies and programs. UKCIP is playing a leading role by liaising between scientists, policy makers, and various stakeholders. The organization is thus effectively pooling information from various sources and developing it into credible knowledge that is meaningful to businesses, local authorities, and other parties. Of course, it would be a mistake to assume that this capacity building will automatically lead to adaptation action. UKCIP's organizational and procedural approach shows potential for achieving adaptation action across scales, such as with strategic flood assessments, but it is still too early to tell how effective it will be in the long term.

Moreover, adaptation biases are evident. These include a reliance on bolstering existing infrastructure through techno-fix measures and a neglect of more socially transformative adaptation processes, such as revitalization and, to a certain extent, diversification. It can be argued that the greatest benefit would involve the introduction of new elements or significant restructurings to society, instead of merely 'shuffling around' with or within existing features or structures. Handmer and Dovers (1996) point out the human tendency to seek to maintain the status quo or to return to it after a disruption, as opposed to being more open to major changes. This is evident in the specific example of the UK's flood risk management option and in the country more generally, where alterations in physical structures (e.g., building the Thames Barrier to protect London and expanding the development ban in flood plains) are made rather than aiming for changes in human behavior.

It could be argued that in the UK the limiting factors to major adaptation reside in the willingness of the government to change its policies and the public's willingness to accept changes to its lifestyle or standard of living. This contrasts with northern indigenous people, whose resilience stems from being able and willing to change their lifestyle, depending on environmental conditions. In their case, the limiting factor is less their resistance to change than the barriers imposed by governments (e.g., development on herding or hunting lands or forced sedentism, etc.).

Climate change research reveals that the two groups' fates are inextricably linked, however, as a recent editorial in opposition to the expansion of the UK's Stansted airport, written by the president of Greenland's Inuit Circumpolar Council, Aqqaluk Lynge (2007), makes clear.

> What happens in Britain affects us in the north …
>
> Most flights from Stansted are not for an important purpose. They are mostly for holidays and leisure. Is it too much to ask for some moderation for the sake of my people today and your people tomorrow? For the sake also of our wildlife and everything else in the world's precious and fragile environment …
>
> Some might dismiss our concerns, saying: "The Arctic is far away and few people live there." That would be immensely short-sighted, as well as callous … [T]he Arctic is the barometer of the globe's environmental health. You can take the pulse of the world in the Arctic. Inuit, the people who live farther north than anyone else, are the canary in the global coal mine …
>
> We are not asking the world to take a backward economic step. All we are asking is that our neighbors in the south greatly reduce their emissions of greenhouse gases. This does not need big sacrifices, but it will need some change in people's lifestyles. Is that plane trip really necessary?

Lynge points out how the UK's short-term strategy for economic growth conflicts with its long-term strategy for greenhouse gas emissions reductions. His implicit suggestion that airplane travel must be rationed (or its deleterious effects as a mobility strategy otherwise reduced) is consistent with a framework that is willing to contemplate rationing or diversification as an adaptation strategy. Greenland itself is facing these contradictions in its own development and responses to climate change (Nuttall 2009).

Unfortunately, many national and local adaptation plans are not contemplating the full range of potential strategies or the implications of continued increases in consumption and emissions on others who may be disproportionally affected and whose vulnerability may be increased by such actions. Consequently, climate change is rapidly emerging as a human rights and environmental justice issue, as the statist framing of mitigation and adaptation measures often occludes crucial moral, livelihood, and entitlement considerations of minorities (Caney 2009), such as indigenous peoples of the Far North. However, gaining the moral and judicial high ground alone will not win the struggle against climate change impacts. Only genuine adaptation can do that. To transfer what Sapir (1924: 427–429) said about 'spurious' culture to adaptation:

> As long as [adaptation] is looked upon as a decorative appendage of large political units, one can plausibly argue that its [success] is bound up with the maintenance of the prestige of these units … The national-political unit tends to arrogate [adaptation] to itself and up to a certain point it succeeds in doing so, but only at the price of serious [adaptation] impoverishment of vast portions of its terrain …
>
> … The minute increment of individuality which alone makes [adaptation] in the self and eventually builds up [adaptation] in the community seems somehow overlooked. Canned [adaptation] is so much easier to administer.

ACKNOWLEDGMENTS

We would like to thank Patricia Cochrane, Janice Grey, and Mikhail Pogodaev for agreeing to be interviewed during the UNFCCC's 15th Conference of Parties in Copenhagen, Denmark, in December 2009. Thanks also go to two anonymous reviewers and the editors for their helpful comments on early drafts of the article.

THOMAS F. THORNTON is Senior Research Fellow at the Environmental Change Institute, School of Geography and the Environment, University of Oxford, where he also directs the Environmental Change and Management MSc course. An anthropologist, he has written widely on human ecology, adaptation, local and traditional ecological knowledge, conservation, coastal and marine environments, conceptualizations of space and place, and the political ecology of resource management among the indigenous peoples of the Pacific Northwest and the circumpolar North. His most recent publications include "A Tale of Three Parks: Tlingit Conservation, Representation, and Repatriation in Southeast Alaska's National Parks" (2010) and *Being and Place among the Tlingit* (2008).

NADIA MANASFI completed her masters degree in Environmental Change and Management at the University of Oxford in 2009. A paper based on the results of her dissertation, "Finding the Balance: Challenges and Opportunities for Climate Change Adaptation in Different Levels of English Local Government," co-authored with Elizabeth Greenhalgh, is in press. Following graduation, she conducted research on adaptation in indigenous communities as part of an internship at the University of Oxford's Environmental Change Institute. Her more recent work with the German International Cooperation (GIZ, formerly GTZ) has focused on climate change adaptation in developing countries, and she plays an active role in advising partner countries on climate-robust development planning.

NOTES

1. Dawkins (1989) proposes the term 'memes', an analogue of 'genes', as a fundamental unit of culture, although such discrete units have yet to be mapped (Benitez-Bribiesca 2001).
2. Initiated by the United Nations Environment Programme, the CBD entered into force on 29 December 1993. Its three main objectives are (1) the conservation of biological diversity, (2) the sustainable use of the components of biological diversity, and (3) the fair and equitable sharing of the benefits arising out of the utilization of genetic resources. See http://www.cbd.int/.
3. See Gorman (1999) about a failed attempt to revitalize a threatened species, the Arabian Oryx.
4. Seven northern indigenous representatives from Alaska, Canada, Siberia, and Norway were interviewed between 8–12 December 2009 at the COP15 in Copenhagen. Five of the interviews were held at the Bella Center and the other two at the World Wide Fund for Nature (WWF) Arctic Tent. The interviews were conducted in English and lasted approximately 15–30 minutes. With the exception of one interview that involved two interviewees, the interviews were conducted on a one-to-one basis, and five were audio-recorded. The interviews were intended to discover any changes in weather/climate that the interviewees have noticed in recent years, what the impact of those changes have been on their lifestyle and community, and what adaptation strategies, if any, the community has already put in place. In addition, the interviewees were asked about support or challenges they have faced in addressing climate change and what their hopes for COP15 were.
5. See Agrawal (2008) for recommendations regarding adapting institutions in poor, rural areas.

REFERENCES

Abercrombie, Michael, C. J. Hickman, and Minnie Louie Johnson. 1977. *A Dictionary of Biology*. Harmondsworth, UK: Penguin Books.

Abryutina, Larisa. 2009. "Indigenous Peoples of the Russian North: Social and Climatic Changes." Pp. 164–173 in *Climate Change and Arctic Sustainable Development*, UNESCO. Paris: UNESCO.

ACIA (Arctic Climate Impact Assessment). 2005. *Arctic Climate Impact Assessment.* Cambridge: Cambridge University Press.

Adger, W. Neil, Suraje Dessai, Marisa Goulden, Mike Hulme, Irene Lorenzoni, Donald R. Nelson, Lars Otto Naess, Johanna Wolf, and Anita Wreford. 2009. "Are There Social Limits to Adaptation to Climate Change?" *Climatic Change* 93 (3–4): 335–354.

Adger, W. Neil, Saleemul Huq, Katrina Brown, Declan Conway, and Mike Hulme. 2003. "Adaptation to Climate Change in the Developing World." *Progress in Development Studies* 3 (3): 179–195.

Adger, W. Neil, Jouni Paavola, Saleemul Huq, and M. J. Mace, eds. 2006. *Fairness in Adaptation to Climate Change*. Cambridge, MA: MIT Press.

AFN (Alaska Federation of Natives). 2008. "Alaska Federation of Natives 2008 Political Platform." http://www.nativefederation.org/documents/2008_AFN_Political_Platform.pdf (accessed 15 January 2009).

Agrawal, Arun. 2008. "The Role of Local Institutions in Adaptation to Climate Change." Paper prepared for the "Social Dimensions of Climate Change" workshop, Social Development Department, World Bank, Washington DC, 5–6 March.

Allaby, Michael. 2004. "Romer's Rule." *A Dictionary of Ecology*. http://www.encyclopedia.com/doc/1O14-Romersrule.html (accessed 20 January 2010).

Arnold, Charles. 1989. "Arctic Harpoons." *Arctic* 42 (1): 80–81.

Barnett, Jon. 2008. "The Effect of Aid on Capacity to Adapt to Climate Change: Insights from Niue." *Political Science* 60 (1): 31–45

Barth, Fredrik. 1956. "Ecologic Relationships of Ethnic Groups in Swat, North Pakistan." *American Anthropologist* 58: 1079–1089.

Beach, Hugh, Mydrene Anderson, and Pekka Aikio. 1992. "Dynamics of Saami Territoriality within the Nation-States of Norway, Sweden and Finland." Pp. 55–90 in *Mobility and Territoriality: Social and Spatial Boundaries among Foragers, Fishers, Pastoralists and Peripatetics*, ed. M. Casimir and A. Rao. Oxford: Berg.

Bellwood, Peter. 2005. *First Farmers: The Origins of Agricultural Societies*. Oxford: Blackwell.

Benitez-Bribiesca, Luis. 2001. "Memetics: A Dangerous Idea." *Interciencia* 26: 29–31.

Berkes, Fikret. 2007. "Community-Based Conservation in a Globalized World." *Proceedings of the National Academy of Sciences of the United States* 104 (39): 15188–15193.

_______. 2008. *Sacred Ecology*. 2nd ed. London: Routledge.

Berkes, Fikret, Derek Armitage, and Nancy Doubleday. 2007. "Synthesis: Adapting, Innovating, Evolving." Pp. 308–327 in *Adaptive Co-Management: Collaboration, Learning, and Multi-Level Governance*. Seattle: University of Washington Press.

Berkes, Fikret, and Carl Folke, eds. 1998. *Linking Social and Ecological Systems: Management Practices and Social Mechanisms for Building Resilience*. Cambridge: Cambridge University Press.

Birdlife International. 2008. "Agricultural Intensification Has Caused the Decline of Many Common Bird Species in Europe." http://www.birdlife.org/datazone/sowb/casestudy/141 (accessed 30 January 2011).

Boyd, Robert, and Peter J. Richerson. 2005. *The Origin and Evolution of Cultures*. Oxford: Oxford University Press.

Brush, Stephen B. 1999. "Bioprospecting the Public Domain." *Cultural Anthropology* 14 (4): 535–555.

Burton, Ian. 2008. "Climate Change and the Adaptation Deficit." Pp. 89–97 in *Earthscan Reader on Adaptation to Climate Change*, ed. E. L. F. Schipper and I. Burton. London: Earthscan. [Originally published in 2004.]

Caney, Simon. 2009. "Justice and the Distribution of Greenhouse Gas Emissions." *Journal of Global Ethics* 5 (2): 125–146.

Carneiro, Robert L. 1970. "A Theory of the Origin of the State." *Science* 169: 733–738.

Christoplos, I., S. Anderson, M. Arnold, V. Galaz, M. Hedger, R. J. T. Klein, and K. Le Goulven. 2009. *The Human Dimension of Climate Adaptation: The Importance of Local and Institutional Issues*. Report to the Commission on Climate Change and Development, Ministry for Foreign Affairs, Stockholm.

Crate, Susan A. 2006. *Cows, Kin, and Globalization: An Ethnography of Sustainability*. Plymouth, UK: AltaMira Press.

Davies, Susanna. 2008. "Are Coping Strategies a Cop-Out?" Pp. 99–116 in *Earthscan Reader on Adaptation to Climate Change*, ed. E. L. F. Schipper and I. Burton. London: Earthscan. [Originally published in 1993.]

Dawkins, Richard. 1989. *The Selfish Gene*. 2nd ed. Oxford: Oxford University Press.

Dearing, John A. 2008. "Landscape Change and Resilience Theory: A Palaeoenvironmental Assessment from Yunnan, SW China." *Holocene* 18 (1): 117–127.

Denevan, William M. 1983. "Adaptation, Variation, and Cultural Geography." *Professional Geographer* 35 (4): 399–407.

Escobar, Arturo. 1995. *Encountering Development: The Making and Unmaking of the Third World*. Princeton, NJ: Princeton University Press.

Farmer, Paul. 2003. *Pathologies of Power: Health, Human Rights, and the New War on the Poor*. Berkeley: University of California Press.

Ferguson, James. 1994. *The Anti-politics Machine: 'Development,' Depoliticization, and Bureaucratic Power in Lesotho*. Minneapolis, MN: University of Minnesota Press

Finan, Timothy J., and Donald R. Nelson. 2009. "Decentralized Planning and Climate Adaptation: Toward Transparent Governance." Pp. 335–349 in *Adapting to Climate Change: Thresholds, Values, Governance*, ed. W. N. Adger, I. Lorenzoni, and K. L. O'Brien. Cambridge: Cambridge University Press.

Ford, James D. 2009. "Dangerous Climate Change and the Importance of Adaptation for the Arctic's Inuit Population." *Environmental Research Letters* 4: 1–9.

Ford, James, and Chris Furgal. 2009. "Foreword to the Special Issue: Climate Change Impacts, Adaptation and Vulnerability in the Arctic." *Polar Research* 28: 1–9.

Ford, James, Tristan Pearce, Barry Smit, Johanna Wandel, Mishak Allurut, Kik Shappa, Harry Ittusujurat, and Kevin Qrunnut. 2007. "Reducing Vulnerability to Climate Change in the Arctic: The Case of Nunavut, Canada." *Arctic* 60 (2): 150–166.

Furgal, Chris, and Jacinthe Seguin. 2006. "Climate Change, Health, and Vulnerability in Canadian Northern Aboriginal Communities." *Environmental Health Perspectives* 114: 1964–1970.

Furgal, Chris, and Terry Prowse. 2008. "Northern Canada." Pp. 57–118 in *From Impacts to Adaptation: Canada in a Changing Climate 2007*, ed. D. Lemmen, F. Warren, E. Bush, and J. Lacroix. Ottawa: Natural Resources Canada.

Füssel, Hans-Martin. 2007. "Adaptation Planning for Climate Change: Concepts, Assessment Approaches, and Key Lessons." *Sustainability Science* 2 (2): 265–275.

Gallopín, Gilberto C., Pablo Gutman, and Hector Maletta. 1989. "Global Impoverishment, Sustainable Development and the Environment: A Conceptual Approach." *International Social Science Journal* 121: 375–397.

GAO (Government Accountability Office). 2009. *Alaska Native Villages: Limited Progress Has Been Made on Relocating Villages Threatened by Flooding and Erosion.* Report to Congressional Requesters. GAO-09-551. Washington, DC.

Gorman, Martyn. 1999. "Conservation: Oryx Go Back to the Brink." *Nature* 398: 190.

Gunderson, Lance H., and C. S. Holling, eds. 2002. *Panarchy: Understanding Transformations in Human and Natural Systems.* Covelo, CA: Island Press.

Hall, Sam. 1988. *The Fourth World: The Heritage of the Arctic and Its Destruction.* New York: Vintage Books.

Halstead, Paul, and John O'Shea. 1989. "Introduction: Cultural Responses to Risk and Uncertainty." Pp. 1–7 in *Bad Year Economics: Cultural Responses to Risk and Uncertainty*, ed. P. Halstead and J. O'Shea. Cambridge: Cambridge University Press.

Handmer, John W. 2008. "Adaptive Capacity: What Does It Mean in the Context of Natural Hazards?" Pp. 213–228 in *Earthscan Reader on Adaptation to Climate Change*, ed. E. L. F. Schipper and I. Burton. London: Earthscan. [Originally published in 2003.]

Handmer, John W., and Stephen R. Dovers. 1996. "A Typology of Resilience: Rethinking Institutions for Sustainable Development." *Organization and Environment* 9 (4): 482–511.

Hardesty, Donald L. 1986. "Rethinking Cultural Adaptation." *Professional Geographer* 38 (1): 11–18.

Harkin, Michael E., ed. 2004. *Reassessing Revitalization Movements: Perspectives from North America and the Pacific Islands.* Lincoln: University of Nebraska Press.

Harvey, David. 2005. *A Brief History of Neoliberalism.* Oxford: Oxford University Press.

Head, Lesley. 2009. "Cultural Ecology: Adaptation—Retrofitting a Concept?" *Progress in Human Geography* 34 (2): 234–242.

Henshaw, Anne. 2009. "Sea Ice: The Sociocultural Dimensions of a Melting Environment in the Arctic." Pp. 153–165 in *Anthropology and Climate Change: From Encounters to Actions*, ed. S. Crate and M. Nuttall. Walnut Creek, CA: Left Coast Press.

ICC (Inuit Circumpolar Conference). 2005. "Petition to the Inter-American Commission on Human Rights." Ottawa, Canada.

IPCC (Intergovernmental Panel on Climate Change). 2007. "Summary for Policymakers." Pp. 7–22 in *Climate Change 2007: Impacts, Adaptation and Vulnerability. Contribution of Working Group II to the Fourth Assessment Report of the Intergovernmental Panel on Climate Change*, ed. M. Parry, O. Canziani, J. Palutikof, P. van der Linden, and C. Hanson. Cambridge: Cambridge University Press.

Itta, Edward S. 2009. "Uses of the Arctic Ocean and the Future of Indigenous Cultures." Pp. 206–211 in *Climate Change and Arctic Sustainable Development*, UNESCO. Paris: UNESCO.

Kates, Robert W. 2000. "Cautionary Tales: Adaptation and the Global Poor." *Climatic Change* 45 (1): 5–17.

Krupnik, Igor, and Dyanna Jolly, eds. 2002. *The Earth Is Faster Now: Indigenous Observations of Climate Change.* Fairbanks, AK: Arctic Research Consortium of the United States.

Lawrence, Eleanor. 1995. *Henderson's Dictionary of Biological Terms.* Harlow, Essex: Longman Scientific and Technical.

Lemet-Klemetti Näkkäläjärvi, Juvvá. 2009. "Perspective of Sami Reindeer Herders on the Impact of Climate Change and Related Research." Pp. 131–143 in *Climate Change and Arctic Sustainable Development*, UNESCO. Paris: UNESCO.

Liverman, Diana. 2008. "Assessing Impacts, Adaptation and Vulnerability: Reflections on the Working Group II Report of the Intergovernmental Panel on Climate Change." *Global Environmental Change* 18: 4–7.

Lynge, Aqqaluk. 2007. "Global Warming Is Not Just a Theory to Us." *Independent*, 30 May. http://www.independent.co.uk/opinion/commentators/aqqaluk-lynge-global-warming-is-not-just-a-theory-to-us-450941.html (accessed 18 January 2010).

______. 2009. "Facing the Impact of Global Climate Change: Recommendations from the Arctic." Pp. 104–113 in *Climate Change and Arctic Sustainable Development*, UNESCO. Paris: UNESCO.

Lyson, Thomas A. 2004. *Civic Agriculture: Reconnecting Farm, Food, and Community*. Boston, MA: Tufts University Press.

Marino, Elizabeth, and Peter Schweitzer. 2009. "Talking and Not Talking about Climate Change in Northwestern Alaska." Pp. 197–208 in *Anthropology and Climate Change: From Encounters to Actions*, ed. S. Crate and M. Nuttall. Walnut Creek, CA: Left Coast Press

Metcalf, Gerry. 2008. "UKCIP Guidance Notes on NI188, etc." Presentation of Local Authority Support Programme, South West Regional Workshop on NI188, 9 October. http://www.oursouthwest.com/climate/registry/081009-ni188-workshop-gerry-metcalf-ukcip.pdf (accessed 28 January 2010).

Moench, Marcus. 2007. "Adapting to Climate Change and the Risks Associated with Other Natural Hazards: Methods for Moving from Concepts to Action." Pp. 13–48 in *Working with the Winds of Change*, 2nd ed., ed. M. Moench and A. Dixit. Kathmandu, Nepal: ProVention Consortium.

Nelson, Donald R., W. Neil Adger, and Katrina Brown. 2007. "Adaptation to Environmental Change: Contributions of a Resilience Framework." *Annual Review of Environment and Resources* 32: 395–419.

Nelson, Donald R., Colin T. West, and Timothy J. Finan. 2009. "Introduction to 'In Focus: Global Change and Adaptation in Local Places." *American Anthropologist* 111 (3): 271–274.

Netting, Robert. 1993. *Smallholders, Householders: Farm Families and the Ecology of Intensive Sustainable Agriculture*. Stanford, CA: Stanford University Press

Netting, Robert, Richard Wilk, and Eric J. Arnould. 1984. *Households: Comparative and Historical Studies of the Domestic Group*. Berkeley: University of California Press.

Niamir-Fuller, Maryam, ed. 1999. *Managing Mobility in African Rangelands: The Legitimization of Transhumance*. London: Intermediate Technology Publications.

Nuttall, Mark. 2009. "Living in a World of Movement: Human Resilience to Environmental Instability in Greenland." Pp. 292–310 in *Anthropology and Climate Change: From Encounters to Actions*, ed. S. Crate and M. Nuttall. Walnut Creek, CA: Left Coast Press.

O'Brien, Michael J., and Thomas D. Holland. 1992. "The Role of Adaptation in Archaeological Explanation." *American Antiquity* 57: 36–59

Oliver-Smith, Anthony. 2009. "Climate Change and Population Displacement: Disasters and Diasporas in the Twenty-First Century." Pp. 116–136 in *Anthropology and Climate Change: From Encounters to Actions*, ed. S. Crate and M. Nuttall. Walnut Creek, CA: Left Coast Press

Pidwirny, Michael. 2006. "Concept of Ecological Niche." *Fundamentals of Physical Geography*. 2nd ed. http://www.physicalgeography.net/fundamentals/9g.html (accessed 20 January 2010).

Pielke, Roger A., Jr. 1998. "Rethinking the Role of Adaptation in Climate Policy." *Global Environmental Change* 8 (2): 159–170.

Pielke, Roger A., Jr., Gwyn Prins, Steve Rayner, and Daniel Sarewitz. 2007. "Lifting the Taboo on Adaptation." *Nature* 445: 597–598.

Rees, W. Garreth, Florian M. Stammler, Fiona S. Danks, and Piers Vitebsky. 2008. "Vulnerability of European Reindeer Husbandry to Global Change." *Climatic Change* 87 (1–2): 199–217.

Reuveny, Rafael. 2007. Climate Change-Induced Migration and Violent Conflict." *Political Geography* 26 (6): 656–673.

Ribot, Jesse C., Adil Najam, and Gabrielle Watson. 2008. "Climate Variation, Vulnerability and Sustainable Development in the Semi-Arid Tropics." Pp. 117–160 in *Earthscan Reader on Adaptation to Climate Change*, ed. E. L. F. Schipper and I. Burton. London: Earthscan. [Originally published in 1996.]

Richerson, Peter J., and Robert Boyd. 2008. "Migration: An Engine for Social Change." *Nature* 456: 877.

Richerson, Peter J., Robert Boyd, and Robert L. Bettinger. 2001. "Was Agriculture Impossible during the Pleistocene but Mandatory during the Holocene? A Climate Change Hypothesis." *American Antiquity* 66 (3): 387–411.

Rindos, David. 1984. *The Origins of Agriculture: An Evolutionary Perspective*. New York: Academic Press.

Sachs, Jeffrey. 2008. *Common Wealth: Economics for a Crowded Planet*. London: Penguin Group.

Sapir, Edward. 1924. "Culture, Genuine and Spurious." *American Journal of Sociology* 29 (4): 401–429.

Schipper, E. Lisa F. 2006. "Conceptual History of Adaptation in the UNFCCC Process." *Review of European Community and International Environmental Law* 15 (1): 82–92.

Schipper, E. Lisa F., and Ian Burton, eds. 2008. *Earthscan Reader on Adaptation to Climate Change*. London: Earthscan.

Shiva, Vandana. 1989. *Staying Alive: Women, Ecology and Development*. London: Zed Books.

______. 2008. *Soil Not Oil: Climate Change, Peak Oil, and Food Insecurity*. London: Zed Books.

Smit, Barry, Ian Burton, Richard T. J. Klein, and Johanna Wandel. 2000. "An Anatomy of Adaptation to Climate Change and Variability." *Climatic Change* 45 (1): 223–251.

Smith, Duane. 2009. "The Sea Ice Is Our Highway: The Importance of Sea Ice to the Inuit Way of Life." Pp. 17–25 in *Climate Change and Arctic Sustainable Development*, UNESCO. Paris: UNESCO.

Smithers, John, and Barry Smit. 1997. "Human Adaptation to Climatic Variability and Change." *Global Environmental Change* 7 (2): 129–146.

Stern, Nicholas. 2007. *The Economics of Climate Change: The Stern Review*. Cambridge, Cambridge University Press.

Strauss, Sarah, and Benjamin S. Orlove. 2003. *Weather, Climate, Culture*. Oxford: Berg.

Thornton, Thomas F., Virginia Butler, Fritz Funk, Madonna Moss, Jamie Hebert, J. Tait Elder, Robi Craig, Shingo Hamada, and Adela Maciejewski Scheer. 2010. *Herring Synthesis: Documenting and Modeling Herring Spawning Areas within Socio-Ecological Systems over Time in Southeast Alaska*. Report to North Pacific Research Board, Anchorage, Alaska.

UNDP (United Nations Development Programme). 2002. "A Climate Risk Management Approach to Disaster Reduction and Adaptation to Climate Change." UNDP Expert Group Meeting, Integrating Disaster Reduction with Adaptation to Climate Change, Havana, Cuba.

UNESCO (United Nations Educational, Scientific and Cultural Organization). 2009. *Climate Change and Arctic Sustainable Development: Scientific, Social, Cultural and Educational Challenges*. Paris: UNESCO.

Wallace, Anthony F. C. 1956. "Revitalization Movements." *American Anthropologist* 58: 264–281.

______. 2009. "Epilogue: On the Organization of Diversity." *Ethos* 37 (2): 251–255.

Watt-Cloutier, Sheila, Terry Fenge, and Paul Crowley. 2004. "Responding to Global Climate Change: The Perspective of the Inuit Circumpolar Conference on The Arctic Climate Impact Assessment." Paper presented at the Inuit Circumpolar Conference. Ontario, Canada.

West, Colin T. 2009. "Domestic Transitions, Desiccation, Agricultural Intensification, and Livelihood Diversification among Rural Households on the Central Plateau, Burkina Faso." *American Anthropologist* 111 (3): 275–288.

Winterhalder, Bruce. 1980. "Environmental Analysis in Human Evolution and Adaptation Research." *Human Ecology* 8 (2): 135–170.

Winterhalder, Bruce, and Douglas J. Kennett. 2006. "Behavioral Ecology and the Transition from Hunting and Gathering to Agriculture." Pp. 1–21 in *Behavioral Ecology and the Transition to Agriculture*, ed. D. J. Kennett and B. Winterhalder. Berkeley: University of California Press.

Young, Oran R., Frans Berkhout, Gilberto C. Gallopín, Marco A. Janssen, Elinor Ostrom, and Sander van der Leeuw. 2006. "The Globalization of Socio-ecological Systems: An Agenda for Scientific Research." *Global Environmental Change* 16 (3): 304–316.

Ziker, John P. 2002. *Peoples of the Tundra Northern Siberians in the Post-Communist Transition*. Long Grove, IL: Waveland Press.

Climate Change Resilience and Adaptation

Perspectives from a Century of Water Resources Development

Clive Agnew and Philip Woodhouse

ABSTRACT: The Fourth Assessment Report of the Intergovernmental Panel on Climate Change and the influential Stern Report both reinforce the warming of the earth's climate system. The alarming environmental, social, and economic consequences of this trend call for immediate action from individuals, institutions, and governments. This article identifies parallels between the problem of adaptive management presented by climate change and an earlier 'global water crisis'. It explores how adaptive strategies have successively emphasized three different principles, based on science, economics, and politics/institutions. The article contends that the close association between climate change and water resources development enables a comparative analysis to be made between the strategies that have been adopted for the latter over the last 100 years. It argues that the experience of water resources development suggests a strong interdependence between the three principles and concludes that conceptualizing them as different dimensions of a single governance framework is necessary to meet the challenge of climate change adaptation.

KEYWORDS: climate change, governance, ecological principle, institutional principle, instrument principle, resilience, water resources

The IPCC has produced four key reports (IPCC 1990, 1995, 2001, 2007b), providing increasing evidence that twentieth-century changes in the earth's climate are unprecedented in historical times. In this body of work, physical models and empirical evidence have in recent decades established links between the atmospheric concentration of greenhouse gases and temperature rises, and the IPCC (2007b: 30) has argued strongly that "[w]arming of the climate system is unequivocal, as is now evident from observations of increases in global average air and ocean temperatures, widespread melting of snow and ice and rising global average sea level." Although this represents a broad scientific consensus, skeptics remain, and Hulme (2009: 1) notes that, in the UK, "only 41% believed humans are causing climate change, 32% remained unsure and 15% were convinced we aren't." This explains the recent vigorous defense of the science of global warming that has been conducted, both online and in academic publications, by institutions such as the UK Meteorological Office[1] and the Royal Society (2008). The science forms the basis

Environment and Society: Advances in Research 1 (2010): 156–183 © Berghahn Books
doi:10.3167/ares.2010.010108

of startling predictions of social and economic consequences, such as those in the influential analysis by Stern (2006) (hereafter, the Stern Report), which assumed a 2 °C rise in global temperature. Yet Anderson and Bows (2008) argue that this understates the likely scale of impacts because past accumulation of greenhouse gases has already reached levels that will see temperatures rising inexorably past the 2 °C 'tipping point', beyond which many believe that warming will result in serious disruptions to human activities and natural ecosystems. They also argue that warming will most likely exceed a 4 °C rise during the twenty-first century.

Predicted consequences are so great that most political leaders support the need for action to curb future emissions. However, the system of financial incentives agreed upon in the 1997 Kyoto Protocol and at conferences thereafter is widely regarded as being ineffective, and agreement on coordinated action to replace it appears elusive, as exemplified by widespread disappointment with the 2009 climate conference in Copenhagen (Vidal and Watts 2009). Dimitrov (2010: 796) describes the Copenhagen conference as "a failure whose magnitude exceeded our worst fears, and the resulting Copenhagen Accord was a desperate attempt to mask this failure." This depressing analysis and a renewal of public doubt about the IPCC's scientific methodology form the backdrop against which individuals, institutions, and governments must act to respond to the threats posed by anticipated future global warming.

Two broad strategies that have emerged from UN deliberations and have been promoted by the IPCC to tackle global warming are 'mitigation' and 'adaptation'. Mitigation seeks to reduce atmospheric greenhouse gas emissions through emissions control and technological changes (e.g., by achieving greater environmental absorption and storage), whereas adaptation is seen to involve changing human behavior in response to predicted threats. Mitigation strategies are not without criticism, but there is strong support for the economic measures associated with them, such as carbon trading and carbon offsets, if only because poorer nations hope that this will provide much needed financial support, while governments of richer countries hope that it will minimize the economic (and political) cost of reducing emissions. Adaptation has been promoted as necessary to increase the ability of people and ecosystems to survive the 'shocks' associated with climate change. However, there is also some confusion, not least because mitigation includes elements of adaptation, such as changes in energy consumption, waste reduction, and the like, alongside economic incentives for investment in technological change. Similarly, adaptation strategies also utilize technical solutions that have to be funded; hence, economic tools are part of the adaptation debate.

In discussions on mitigation and adaptation, it is clear that favored strategies are concerned both with technological change and with influencing human behavior through a variety of instruments that include both financial and regulatory measures. In this article, we argue that similar conclusions can be drawn from an examination of water resources development, which, during the twentieth century, sought to manage the hydrological cycle through solutions based initially on science and engineering. More recently, and into the twenty-first century, the effort to maintain this cycle has been made through (re)designing systems of water governance (Plummer and Slaymaker 2007). The aim of this article is to assess, from a variety of perspectives, the experiences that have been gained during the last 100 years in the process of tackling water resources problems and to apply and contrast these to concerns about the impacts of global warming that have been emerging over recent decades. Water resources and climate are obviously linked through hydrological cycles and energy exchanges between the earth and the atmosphere, but more than this, we argue, they share common management approaches based around the notions of resilience and governance of ecosystem resources that can help us gain critical insights into strategies to combat climate change.

The article initially reviews predicted impacts of climate change on water resources in order to explore the continuum between the atmosphere and hydrology and to demonstrate why strategies developed to tackle flood risk and water scarcity may be relevant to current attempts to respond to global warming. We then turn to a discussion of resilience and governance, which is used to present three principles of environmental management: science, embracing an 'ecological principle'; economics, embracing an 'instrument principle'; and politics/institutions, embracing an 'institutional principle'. We use these three principles to structure the rest of the article, which systematically considers first the lessons learned from water resources development and then contrasts these with current approaches to climate change adaptation.

Climate Change and Water Resources

The IPCC has concluded in both its Third Assessment Report (IPCC 2001) and its Fourth Assessment Report (IPCC 2007b) that freshwater resources are being affected by climate change. Future predictions include a decline in glacier storage and increases in the occurrence of precipitation extremes, leading to more droughts and floods. Increased runoff of 10–40 percent is predicted in higher latitudes and wet tropics, with a 10–30 percent increase in dry tropics. The greatest regional changes are likely to occur in Africa, where water stress is expected to rise, and in Australia, which is likely to experience water security problems through reduced precipitation. In Asia, decreased meltwater flows are likely to offset climate conditions that would be otherwise favorable to increased crop yields. The three key manifestations of climate change in relation to freshwater resources are rising sea levels, increased flood hazards, and more frequent incidence of drought.

Sea Level Rise

Freshwater resources are threatened by sea level rises, including both surface flooding in coastal areas and saline intrusion into groundwater. Gregory and Lowe (2000) forecast a rise of around 40 centimeters from 1990 to 2100, with 15 centimeters already observed during the twentieth century (Dessler and Parson 2006). More controversially, Meir et al. (2007) suggest that previously unaccounted for contributions through glacial melting will add a further 10 to 25 centimeters of sea level rise by 2100, or 50 percent more than the IPCC's (2007b) overall projections of 20 to 50 centimeters (see also Oppenheimer et al. 2007; Solomon et al. 2008). As 100 million people live less than 1 meter above sea level and 40 percent of the world's population lives within 60 miles of the coast, this is a serious threat, especially in areas of Southeast Asia (Dow and Downing 2006). For this hazard, as for those arising from increased water stress, the actual impact will depend upon the human capacity to respond and adapt to these changing threats.

Flood Hazard

According to the IPCC's (2007b) Fourth Assessment Report, it is likely that up to 20 percent of the world's population will inhabit areas where flood risk has risen due to climate change. There is evidence that precipitation has increased in eastern areas of the Americas, in Northern Europe, and in Central Asia, and that it has decreased in Southern Asia, Southern Africa, and the Mediterranean. It is also highly likely that for glacier- and snow-fed areas, runoff has increased, while rivers have warmed. In general, higher storm flows can be expected to result from global warming, as explained by Arnell (2002, 2003) and Kerr (2007). However, even where increased precipitation causes higher runoff, this water resource benefit is likely to be moderated by increasing variability and seasonal changes. Considerable uncertainty is attached to predictions of river flow, since the conversion of precipitation changes into river flood regimes involves extremes of two variables: intensity and duration (Bell et al. 2007; Prudhomme et al. 2002). There are also spatial differences: in the UK, for example, upland catchments have proved easier to model than lowland catchments, where groundwater input is more significant. Similarly, rural and urban climates may be distinct (Arnfield 2003; Roth 2007). Impacts in urban areas are a significant challenge because of the complex nature of the urban hydrological system (Marsalek et al. 2006), with combined threats of fluvial flooding from river channels and pluvial floods due to higher rainfall intensity and poor urban drainage infrastructure.

Drought

The Fourth Assessment Report (IPCC 2007b) has identified the risk of drought as rising during the twenty-first century, principally in tropical regions. However, with more extremes in precipitation and drier summers predicted for the northern hemisphere, the incidence of droughts is likely to increase across the globe (Lloyd-Hughes and Saunders 2002). Drought needs to be differentiated from desiccation. The former is a short-term and abnormal water shortage due to an imbalance between water supply and demand (Agnew 2002; Wilhite 2000). It can result from a reduction in precipitation but can also arise from changes in human activity that raise demand for water above the available supply. Desiccation refers to increasing aridity as determined through 30-year climate averages. Climate change threatens both increased drought and desiccation. However, adaptation strategies will differ according to whether one is dealing with drought, which needs an immediate response, or desiccation, which requires longer-term strategies.

The regions predicted by the IPCC (2007b) to be most threatened by increasing drought and aridity are areas of high precipitation variability where agriculture is the dominant land use, in particular, sub-Saharan Africa. Southern Africa could lose almost a third of maize production, while losses of rice and millet could be over 10 percent in South Asia. Impacts and possible responses are not straightforward, however, and Challinor et al. (2009) and Soussana et al. (2010) both argue the need for improvements in agro-climate modeling. Drought effects on crop growth may be tackled through irrigation and other technologies, if they can be afforded (Parry et al. 2005). However, crop choice has a significant effect on drought hazard, and a warming world with higher carbon dioxide levels also heralds other changes. For example, crops such as wheat, soybean, and rice (known as 'C_3 plants' because CO_2 is fixed by a 3-carbon compound, phosphoglyceric acid) are well placed to respond positively, because they have lower water use efficiencies compared to typical dryland crops such as millet or sorghum. Thus, a CO_2 richer world means greater growth potential for some plants in temperate areas, but only if there is sufficient water to offset the impacts of higher temperatures. A further factor is that the vulnerability of poorer areas of sub-Saharan Africa to water scarcity may be increased where production systems are geared to global economic linkages rather than to local risk factors (Franke and Chasin 1980; Glantz 1994). The interaction of socio-economic and biophysical factors in determining overall vulnerability prompted Thornton et al. (2006) to argue for a move beyond considering only water supply when deciding upon drought management strategies. Similarly, the Energy and Resources Institute (TERI 2003) suggests three forms of drought vulnerability: biophysical, social, and technological.

These concerns about climate change intersect with perceptions of a 'global water crisis' (UNESCO 2009) that is identified both with the failure to meet the demand for water in a more populated and urbanized world (WWC 2010) and the commitment included in the UN Millennium Development Goals (MDGs)[2] to reduce by half the proportion of the global population lacking access to clean water and adequate sanitation, numbering some 1.1 billion and 2.6 billion people, respectively (UNDP 2006). This 'crisis' narrative is problematic, however, for two reasons.

First, the focus on domestic water needs, as in the MDGs, raises questions of what standard to use. In a report for the World Health Organization (WHO), Howard and Bartram (2003) state that a water supply of 20 lcd (liters per capita per day) will provide water sufficient only for drinking and basic hygiene, whereas a supply of at least 50 lcd, from a source no more than 1 kilometer away, is required to meet most essential water needs. In practice, measured water consumption values are greater, typically around 100 lcd from urban street standpipes (Twort et al. 2000), rising to 300–500 lcd for developed countries and even up to 800 lcd (in the US). The range of values poses questions about how those who are already facing water scarcity can possibly adapt to a warmer, drier world, and how the higher levels of demand that are typical of urban households can be met. The latter is significant because most of the future global population growth is expected to take place in cities, and there are already 21 'megacities' with populations that exceed 10 million. Against this, Potts's (2009) work concludes that urbanization trends in Africa are not as pronounced as they are commonly represented, especially in medium-sized towns where most of the observed increases are due primarily to local population growth and not to in-migration.

A second problem with the water crisis narrative expressed in terms of domestic use, as in the MDGs, is that, in terms of total water withdrawal, agriculture dominates: 70 percent of global extraction is for that purpose in rural areas, compared to 22 percent for industry and 8 percent for domestic household use. Moreover, the pattern of rising withdrawals is not reproduced everywhere. For example, in the US, total water withdrawals reached a plateau in the 1980s,[3] while household water consumption in the UK has remained around 150 lcd since the mid-1990s (Ofwat 2007).

Nonetheless, since the eighteenth century, global demand for water has grown 35 fold, whereas population has increased 8 fold (Altinbilek 2002).There are undoubtedly problems of water scarcity in some parts of the world, most obviously in arid regions. Twenty-nine countries are classified as being 'water stressed' in terms of the amount of water that is annually available for all economic activity (less than 1,700 m^3 c^{-1} y^{-1}). This list includes India and Denmark but also many of the poorest developing countries. A further 17 countries, mainly from the Middle East and North Africa, are classified as facing 'water scarcity' (less than 1,000 m^3 c^{-1} y^{-1}), based on data from the World Resources Institute.[4] As a consequence, a major concern emanating from future climate scenarios is that changes to the global water supply, along with increases in demand for food production and a growing urban population, will be among the biggest challenges. While concern over global warming is relatively recent, the need to manage and adapt to water resources constraints is not new, and this presents an opportunity to learn from past methods of adaptation to different environmental conditions. In the next section, we outline a framework to identify different dimensions of adaptation.

Governance, Resilience, and Water Development Principles

Governance is a key concept in adaptation strategies: essentially, it is about how decisions are made. Reed and Bruyneel (2010: 647) observe that the meaning of governance varies between, on the one hand, an emphasis upon government as a centralized authority that deploys financial and regulatory instruments and, on the other hand, concepts of devolved decision making by people and communities. They further note a distinction between governance and management, which are presented as operational procedures. Figure 1 depicts three elements of water governance as presented in the UN's Second World Water Development Report (UNESCO 2006). Recent reports on water resources development have stressed the importance of effective governance, as exemplified in the EU Water Framework Directive of 2000[5] and repeated in the UN's Third World Water Development Report (UNESCO 2009). It should be noted that figure 1 presents a normative view of environmental management, whereas in practice markets are not perfect, environments degrade, and equity is rarely achieved. Nevertheless, it illustrates an integrated approach that combines environmental, economic, and socio-political dimensions. To the extent that these dimensions are combined successfully, we might draw the same diagram with 'sustainability' in the center, or even 'resilience'. This is to not say that governance is the same as sustainability or resilience; rather, good and effective governance is a requisite for those features.

Figure 1: Dimensions of Water Governance

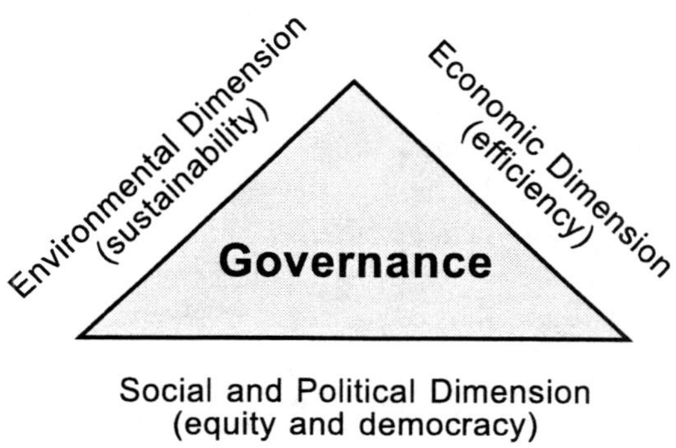

Source: UNESCO (2006).

The need to integrate these dimensions for effective environmental management has formed the basis for thinking about water development since a recasting of water development priorities emerged from the International Conference on Water and the Environment (ICWE), held in Dublin in 1992. These key issues were subsequently incorporated into the Earth Summit held in Rio de Janeiro in the same year (Lundqvist 2000; UNCED 1992). The three principles emerging from the ICWE have been reworked, notably through the incorporation of gender issues into the third principle (WWC 2000). They typically include the following:

1. *Ecological principle*—recognizes the need for holistic management of the water resources, usually interpreted as managing water in an 'integrated' way, according to its hydrological units, such as river basins or subterranean aquifers. Here scientific management and expert decision making are the norm, generally invoking the agency of a Weberian state model to ensure that individual, smaller-scale actions are subordinated to a 'greater good', defined variously in environmental, political/security, or economic terms.
2. *Instrument principle*—recognizes water as an economic good whose efficient use and conservation should be promoted through charges payable by users. Here neoliberal ideas, whose resurgence in the 1980s coincided with the concept of 'sustainable development', are embraced, notably through the use of water pricing as a means of determining allocation priorities and the valuation of ecological services.
3. *Institutional principle*—calls for decision making on water resources to be decentralized to the smallest scale feasible, following the criterion of subsidiarity, and to be representative

of all water users. Here importance is attributed to decentralized and participatory decision-making processes.

It is possible to map these principles onto the development of water resources. Within the environmental or biophysical dimension (ecological principle), we can locate the engineering approaches that dominated water resources development, from early water treatment some 4,000 years ago (Symons 2006) to the first municipal water treatment plant in Scotland in the early nineteenth century. Walski (2006) charts this technological development of water supply, noting that 2,000 years ago cities in Asia had functioning water systems long before the development of modern hydraulic engineering in the nineteenth century. Today, a plethora of water control technologies are available to tackle environmental constraints, including mega-dams (Altinbilek 2002), desalination (Burbano et al. 2007), and even cloud seeding (Silverman 2003). However, this emphasis upon hydrology and engineering has become substantially counterbalanced, if not replaced, by the economic dimension (instrument principle) and the socio-political dimension (institutional principle). This has shifted the focus to water allocation through both pricing mechanisms and 'deliberative' decision making in which all water users or 'stakeholders' are represented. The category of stakeholders is typically drawn widely to include all those whose interests, such as amenity value, may be affected by the water use of others. Women have been explicitly identified as needing stronger representation in water management, in recognition of the gendered nature of the labor that provides household water supplies in many societies (Peter 2006). Together, the instrument and institutional principles have shifted attention from technology to governance. This shift, which Renzetti (2002) traces to the middle of the twentieth century, is evident in the historical development of water supply as identified by Swyngedouw (2006):

- Pre-1850s: Small-scale supply enhancement through private investors for profit in Western countries
- 1850s–1920s: Large-scale municipal sanitation and potable supply development
- 1920s–1970s: National-scale public provision of basic needs and economic development
- Post-1970s: Privatization, with water viewed as an economic good

For the first three of these stages, the financing and organization of water resources development was driven primarily by the increasing scale of engineering works needed to meet growing urban demand during the nineteenth and early twentieth centuries. Only since the last quarter of the twentieth century has investment in water become guided by neoliberal ideas of pricing and cost recovery, indicated as the final stage in the above chronology. Over the past two decades, this primacy of neoliberal models of market-based management has, in turn, become increasingly contested on the ground that it does not satisfy social goals for universal access to adequate water, as reviewed by Agnew and Woodhouse (2011).

We argue that there are strong parallels between water resources development and responses to climate change: both were initially focused upon scientific evidence leading to technological solutions, followed by economic instruments to provide incentives for change via 'commodification', with finally a move to embrace political mechanisms based on considerations of social inclusion and equity. In figure 2, we have presented these three principles as key instruments for maintaining resilience in the face of climate change impacts. In contrast to figure 1, we have introduced a fourth element, 'hazards', in figure 2 to acknowledge that not all environmental or climate change necessarily has a destructive nature. Environmental change is thus not synonymous with impacts that require an adaptive response.

Figure 2: Framework for understanding responses to climate change using ecological, instrument, and institutional principles

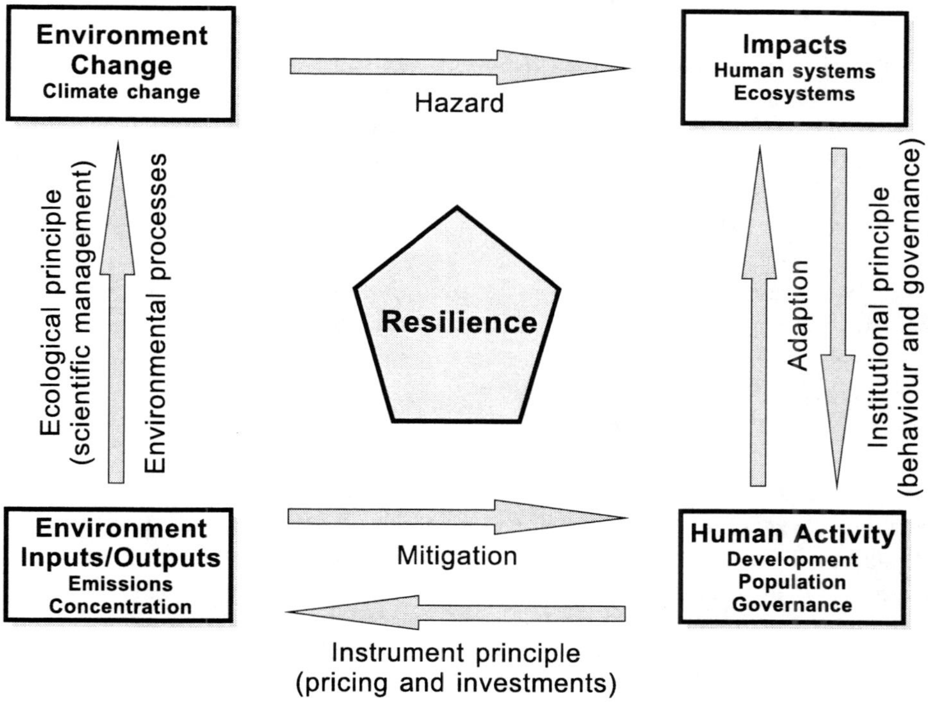

Resilience

The concept of resilience—the ability to respond to environmental shocks—is attributed to the ecologist C. S. Holling (1973). It is defined by the IPCC (2007a: 880) as "[t]he ability of a social or ecological system to absorb disturbances while retaining the same basic structure and ways of functioning, the capacity for self-organisation, and the capacity to adapt to stress and change." It has been found useful in work on climate change because it recognizes that the same climate change might produce different impacts due to variations in the stability and vulnerability of different social or ecological systems. Resilience has, for example, been incorporated into work by the International Livestock Research Institute on mapping climate vulnerability and poverty in Africa (Thornton et al. 2006).

Pelling's (2011: 42) recent review notes that resilience "is not simply synonymous with adaptation." This is illustrated by an example where (short-term) risk management can lead to (longer-term) institutional inertia, which highlights the need to understand the "social processes shaping resilience" (ibid.: 43). Figure 2 has the merit of illustrating that adaptation and mitigation measures are part of a continuum linking human impacts and actions to resilience. Perhaps more controversial is our use of three water development principles (ecological, instrument, and institutional) to label particular types of engagement. While oversimplified, as discussed below, this framework provides an opportunity to identify key management dimensions underpinning resilience in the face of climate change and to show how different climate change responses can be mapped onto different paradigms for natural resources management. The article now deals with the three principles in turn, each section commencing with a discussion of the lessons learned from water resources development before turning to consider the possible implications for climate change adaptation.

Ecological Principle—Scientific Management

Some have challenged the 'global' nature of the water crisis (Agnew and Woodhouse 2011; Rogers et al. 2006). They suggest that this focus ignores the successful application of science and technology in many places, which has resulted in water being brought to areas of potable need, allowing for massive increases in food supply through irrigation, as well as health improvements through improved sanitation and water treatment. This progress has been underpinned by advances in hydrological science, moving from understanding channel flow characteristics to catchment-wide linkages between land surfaces and the atmosphere (Ward and Robinson 1999) and predictive modeling of hydrological variables, most recently based upon artificial neural networks (Maier et al. 2010). Developments in the science of water treatment and sanitation, using filtration and chlorination from its early origins in nineteenth-century Scotland (McGuire 2006; Symons 2006; Twort et al. 2000), have been credited with a 50 percent increase in life expectancy in the US during the twentieth century (Christman 1998). Perhaps most emblematic of these engineering achievements is the construction of large dams, which has allowed for a fivefold increase in irrigated area during the twentieth century (presently covering about 300 million hectares),[6] a growing supply of drinking water for expanding urban centers, and the generation of 19 percent of the world's electric power (Altinbilek 2002; World Bank 2009). This remains an important development model,[7] and although three countries (China, the US, and India) dominate, by 2000 some 45,000 large dams had been constructed in over 150 countries. As a result, in less industrialized economies, irrigation is often the largest use of water, typically 80 percent or more (Anand 2007). However, in the face of increasingly uncertain rainfall, access to irrigation technology, along with fertilizers and new seed varieties, is an important asset in adaptive strategies (Parry et al. 2005).

The success of scientific and engineering-led water development initiatives has provided a model that has been promoted by international development agencies, such UNESCO's International Hydrological Decade (1965–1974) and the UN International Drinking Water Supply and Sanitation Decade (1981–1990). There has nonetheless been increasing acknowledgment of failures and of the unpredictability of outcomes from water development projects in which a top-down management style has given priority to scientific and engineering considerations. Adverse impacts of large-scale water developments—dams in particular—have been documented in detailed case studies (Goldsmith and Hildyard 1984; WCD 2000) that identified many dam projects as failing to reach their goals of hydroelectric power production (HEP). Altinbilek (2002) and others[8] have also warned of hazards associated with dams and their potential failure, while Ortolano and Cushing's (2002) study of one of the first mega-dams, the Grand Coulee in the US, concludes that the adverse impact upon the indigenous population continues some 70 years later. Questions about the sustainability of irrigation have also been raised, particularly with regard to problems of soil salinization and groundwater contamination, leading to a decline in irrigated area per capita of population (Postel 1992). Even California, one of the richest regions on the earth, has struggled to meet its insatiable appetite for water through technology. In February 2009, Governor Arnold Schwarzenegger proclaimed a state of emergency as California entered a third consecutive year of low rainfall, with consequent reductions of up to 50 percent in deliveries of water to irrigated agriculture, notably in the San Joaquin Valley, resulting in losses in agricultural production estimated at almost $1 billion (CDWR 2009). Local water shortages are compounded by widespread deterioration in water quality, with only modest improvements achieved between 1993 and 2003 in the most heavily affected streams in the US (Sprague et al. 2009).

Evidence that irrigation did not necessarily benefit all of those involved resulted in a shift in focus of irrigation design, particularly in South Asia, from technical considerations of water storage, transmission, and application to issues of poverty, social organization, and participation in decision making (Chambers 1994). More generally, declining confidence in the ability of engineering design to deliver predictable economic benefits led to growing political opposition to large-scale water projects and a decline in investment. Bank support for development dropped from $4.4 billion per annum in the 1980s to $2.6 billion per annum by the late 1990s (Bird 2002). Moreover, despite a renewal of World Bank interest in investment in dams—notably in Africa since 2000 (Pearce 2006)—loans for irrigation and drainage in Africa were lower in 2002–2005 than they had been in 1978–1981 (CAWMA 2007: 73).

Large-scale engineering remains an important element of water resources development, especially when HEP is central to cheaper energy production (cf. the Three Gorges Dam in China and the Grand Anatolia Project in Turkey). More broadly, science and engineering play a central role in programs to 're-engineer' rivers in industrialized countries in order to improve their ecological or flood management functions. However, engineering has come to be seen as part of a continuum in which human organization and behavior play a much larger role. This is exemplified in proposals for flood responses in the Netherlands (de Graaf et al. 2007) that list four types of adaptive capacity:

1. *threshold capacity*—engineering to prevent damage, which requires a high degree of social organization;
2. *coping capacity*—reducing immediate impacts with a focus upon relief, requiring good communication and clear responsibilities;
3. *recovery capacity*—short-term return to previous state, requiring insurance and disaster funds); and
4. *adaptive capacity*—the ability of a community to cope and adjust in the future, which may require significant changes in lifestyle and land use.

A further key function of science and engineering is evident here: the provision of information relevant to environmental decision making. As we shall discuss below, 'asymmetries of information' (i.e., situations in which one party in a transaction has more—or possibly more reliable—information than the other) have a major impact on political and institutional processes. Thus, environmental 'threats' can become 'institutional facts' that are propagated to support specific land use policies. Lemos (2003), for example, describes how the threat of drought in Brazil is used to 'insulate' policy makers from public accountability for the policy choices that they have made. For our purposes here, we need to note that the levels of uncertainty attached to many types of environmental data have a major influence on such decision-making processes. This has been found to be true of water resources management (e.g., with respect to the availability of records of river flows), and it is of primary significance when we turn to the science of predicting climate change and designing responses to it.

Scientific Uncertainties and Adaptation to Climate Change

Successful adaptation requires stakeholder engagement to have access to the necessary resources, including adequate information to support the mobilization of investment by those developing and implementing climate change policy. Much of this depends upon an understanding that climate change is a threat that will materialize during the twenty-first century. It is therefore important to understand the remaining scientific uncertainties and the impact that they may have on stakeholder engagement The areas of uncertainty can be exemplified by considering the predictions of the impacts of climate change on water resources. These may be categorized as (1) a lack of agreement over how best (which measurements, which variables) to measure and predict climate change; (2) a more specific problem associated with 'downscaling', that is, forecasting change at the scale of a region or sub-region; and (3) an unstable database that is inadequate and deteriorating in certain critical respects. We shall review each of these briefly in turn.

1. *Measuring and predicting climate change.* Although it has long been recognized that the earth's climate has undergone major changes over geological timescales, current concern centers on the observation that mean global temperatures have risen at 0.13 °C per decade from 1956 to 2005, or nearly twice the rate observed (0.74 °C rise in 10 decades) for the twentieth century as a whole. The interpretation of this fact as a sign of accelerating global warming is disputed, and it is worth recalling that only three decades ago climatologists were writing about global cooling (Ponte 1976). Skeptics' objections (Oldfield 2005; Schiermeier 2010) include challenges to historical interpretations of temperature records and claims that warming trends are much smaller than normal inter-annual temperature variations; that existing climate models are too limited and fail to take feedback mechanisms sufficiently into account; that observed temperature trends are due to natural causes, such as solar changes; and that the climate system will be resilient to any changes being observed. Conversely, Oppenheimer et al. (2007) caution that the climate change consensus may hide important feedbacks that could exacerbate warming effects, such as increased contribution from melting glaciers on sea level rise, the amplification of feedbacks between climate and carbon cycles (e.g., the release of carbon from thawing permafrost), and possible interactions between sea surface temperatures and tropical circulation. Taken together, these arguments define a 'chain of uncertainty' that starts with predictions of future human activities, including energy scenarios and greenhouse gas emissions, to which are added variations in impacts on atmospheric composition before introducing questions of scale, both temporal and spatial, over climate and hydrological responses.

Much of the climate change literature focuses on where there is the greatest consensus, that is, on temperature and precipitation. However, predicting flood and drought hazards requires data not only on rainfall frequency and intensity but also on variables such as evaporation, soil moisture, and groundwater, which are much less commonly measured. This deficiency is particularly marked for evaporation—a key meteorological variable because it reflects the distribution of energy between sensible and latent heat fluxes, is required for predictions of soil moisture deficits that have implications for groundwater recharge and surface runoff, and is the basis for agro-climatological forecasts of food production. However, measuring evaporation is complex (Allen et al. 1998), and recent reviews of methods for calculating potential evaporation for climate change studies note that it may be an important source of uncertainty (Kay and Davies 2008; Kingston et al. 2009).

2. *The problem of downscaling.* The IPCC's (2007b) Fourth Assessment Report contains broad generalizations about future conditions, but the literature identifies severe limitations for predicting localized and short-term patterns of change. Some predictions—for example, forecasts of future European summers that are drier and winters that are wetter—are fairly consistent. In contrast, predictions for South Africa diverge, ranging from more days with heavy rain, to changes in the number of consecutive rain days to no trend at all for annual total rainfall (Hulme 1996; Kruger 2006). Others predict local spatial variations, with increased summer rainfall over central and

eastern plateaus in the Drakensberg mountain range and little change in the Western Cape (Hewitson and Crane 2006). This illustrates the uncertainty involved in interpreting results of regional GCM (global circulation model) forecasts (for scale units of 100 kilometers or more) in order to create local and catchment-scale hydrological forecasts. This was termed 'climate inversion' by Kim et al. (1984) and recognized by the IPCC (2001) as 'downscaling'. Of the two downscaling methods commonly used, the statistical approach is faster but produces uncertain predictions for extremes of both precipitation and temperature than the more computationally demanding dynamic (synoptic) approach (Hundecha and Bardossy 2008). Prudhomme et al. (2002) believe that at present there is little confidence in precipitation predictions at timescales of less than a month. They state that "there exists no universal downscaling method for all situations" (ibid.: 1139), a view that is supported by Hewitson and Crane (2006). The problem is not merely one of choosing an appropriate method, as the statistical approach is dependent upon having a reliable database to capture conditions. This presents problems for assessing the impacts of climate change upon water resources, above all, at a local level where they are 'visible' to the majority of people.

3. *An unstable database.* A further complication for climate analysis is that the actual networks of stations have changed during the observation period, creating the possibility of an added artifact in the climate signal (Chappell and Agnew 2008). In West Africa, for example, the number of stations decreased in the wetter coastal area and increased in the drier continental area, while the total number of stations with complete records decreased dramatically after the late 1970s. The same observation was recently made for Africa as a whole, and Kenya in particular, at the Second Africa Water Week in Johannesburg: "You can only get reliable rainfall data in Kenya up to 1980 as the existing infrastructure, like rain gauges and stream gauges, are no longer working" (Ogodo 2009). There are, then, a number of uncertainties from the basic observational networks through to predictions of global economic activity that need to be understood when giving advice about adaptation strategies. Taking these uncertainties into account, it can be understood why resilience—the ability to withstand (unspecified) shocks—and adaptive capacity have become such important concepts.

Technology and Adaptation to Climate Change

As with water resources management, many of the initial steps in climate change adaptation have centered on technological change. Adaptation to water scarcity has focused on infrastructure to secure supplies (Parry et al. 2009). As Page (2005: 302) notes: "It is imperative that more capital should be fixed in the form of pipes, taps and reservoirs in the developing world." The wide range of available technologies (Agnew and Woodhouse 2011) also addresses the priority of conserving water through more efficient use, including reducing pipe leakage that causes a 'normal' loss of 10–20 percent (Lahlou 2001) or that averages 36 percent (within a range of 8 to 62 percent) in Asia. While the problem of such water loss is mainly associated with urban areas, many rural irrigation systems have efficiencies well below 50 percent. Heavily promoted water conservation measures include rainwater storage, water audits, and water-efficient household appliances (Sharp 2006), many of which may be 'retrofitted'. Deoreo et al. (2001) noted that such measures in the US enabled household consumption to be significantly reduced to around 150 lcd. Similarly, Kolokytha and Mylopoulos (2004) estimated a 20 percent decrease in consumption through residential retrofitting and calculated that using gray water (e.g., rainwater) saved up to 39 percent. The reuse of waste water is well established (Symons 2006), with techniques that are suitable for water-scarce regions and developing countries (Al Baz et al. 2008; Sánchez et al. 2006), including the treatment of contaminated water through desalination (Burbano et al. 2007).

More strategic technological adaptation to climate change is evident in the concept of the 'absorbent city' (White 2008), according to which the design of urban areas can be used to adapt and even mitigate the expected increase in floods, while retaining water in the city can reduce scarcity. Similar goals inform 'green landscapes', in which water is conserved and temperatures are mitigated (Gill et al. 2007). Comparable holistic approaches are evident in the use of rainfall water harvesting to improve resilience to water scarcity in Africa (Rockstrom 2003). There are, then, many promising attempts to use technology to adapt to future climate extremes, but it is a challenge to introduce these ideas in congested and impoverished areas of cities in the South. Therkildsen (1988) noted failed attempts in Africa to supply water in the 1960s and 1970s, and Thompson et al. (2001) found in East Africa that mean water use had actually declined between the 1960s and 1990s, due to the reduced reliability of the piped water supply, as municipal water services could not maintain the infrastructure.

As with our review of water resource infrastructure at the beginning of this section, these failures mark the limits of engineering approaches, and responses have invariably highlighted a need for greater emphasis on economic, social, and political aspects, such as the need for greater partnership with communities (Rietveld et al. 2008), management at the lowest appropriate level, community ownership of schemes, full cost recovery for operation and maintenance, recognizing key roles of women, and inclusion of the poor (Gine and Perez-Foguet 2008). These conclusions are consistent with the lessons learned by the UK's Department for International Development regarding water provision: recognize water as an economic good, respond to demand, and put people at the center (DFID 2001). However, there is also evidence that such formulations may fall short of what is required. Examining failures in the supply of water in Mumbai, for example, Gandy (2008: 122) observes that an "array of technological, scientific and architectural innovations … enable wealthy households to insulate themselves from the environmental conditions of the poor." He concludes: "The recent history of Mumbai has militated against the kind of progressive political movements that galvanized processes of sanitary reform in, for example, European cities during the second half of the nineteenth century and the early decades of the twentieth century" (ibid.).

We suggest, therefore, that while science and engineering lie at the heart of the ecological principle and offer the potential to deliver fresh impetus in adaptive behavior at a range of scales, the environmental dimension is unlikely to be successful in isolation from the economic and socio-political dimensions. This is illustrated by integrated water resources management (IWRM), a key application of the ecological principle. IWRM has been promoted widely (UNESCO 2006) as it enables different water users and different types of water to be managed within a single unit—normally, the watershed defined by topography and surface hydrology (Lenton and Muller 2009). Mollinga et al. (2007) argue that this hydrological catchment may impose a false boundary and that 'problemsheds', which are socio-political constructs, are more appropriate as a unit of management and hence regulatory organization. The ecological principle, then, as with engineering-led adaptive strategies, may be imposing false boundaries, which require people to 'fit' the data, and institutional arrangements that are likely to be problematic. We now turn to the economic dimension and institutional dimensions of governance.

Instrument Principle—Economic Strategies

We noted above that, over the last 100 years, the financing of water management has changed in industrialized societies from the municipal to the national scale and has increasingly been linked to strategic economic planning (Swyngedouw 2006). The broader scale has required more extensive investment and greater emphasis upon cost recovery, associated with intensifying contestation of how costs should be distributed. The need to raise investment from private sources and the adoption of the instrument principle for water services and supply have moved the debate to address whether water is a basic right or can be privately owned. The Organisation for Economic Co-operation and Development (OECD 2003), for example, notes that water can be regarded as a 'public good', to which everyone has a right in a non-competitive manner; as a 'private good', which is owned and to which there are access restrictions; and as a 'common good', to which there is non-competitive access only until the resource becomes scarce. It is argued that, for adaptation to be successful, local resources will have to be harnessed alongside external injections of capital. The MDGs also recognize the need for adaptation to be based around economic and behavioral changes (Moriarty and Butterworth 2003).

The Fourth World Water Forum of the World Water Council (WWC), held in Mexico City in 2006, noted the necessity to develop local financing capacity and markets. The forum concluded: "This necessity is founded on the recognition that users and taxpayers are in the end the main financiers, and on the associated shift from full cost recovery to a solidarity system of fair tariffs combined with targeted subsidies" (WWC 2006: 108). This is echoed in UNESCO's Third World Water Development Report, published at the Fifth World Water Forum in Istanbul, which states, in relation to financing: "While there may appear to be many financing options for water resources development, governments still have only three basic means of financing them: tariffs, taxes and transfers through external aid and philanthropy" (UNESCO 2009: xxi). This makes clear that the cost of borrowed capital, if it is not written off through international aid, must be paid by users or by taxes. Here is a perspective in which the instrument principle remains since access to water is defined as having a cost, but the distribution of that cost is governed by the social criteria of need and the ability to pay. The same debate will need to take place over interventions concerning climate change adaptation.

One of the key financial mechanisms to raise revenues has been water pricing. However, pricing is also proposed as a means of adapting to greater scarcity through its inclusion in water demand management (WDM), which has become a key strategy across the world, including Europe (Sharp 2006), the Middle East (Magiera et al. 2006), and China (Chen et al. 2005). WDM contains three elements that are reflected in adaptation strategies to climate change: economic (water pricing); technical (infrastructure, including metering, recycling, and retrofitting); and social (education, legislation, and regulation). Most attention has focused on a range of socio-economic strategies that seek to influence water demand (Babel et al. 2007). Kolokytha and Mylopoulos (2004: 263) are emphatic about the need for a change in direction, asserting that "demand management is considered to be the best potential solution to meet future [water] needs" because, they argue, supply enhancement does not effectively deal with growing competition between consumers. Treating water as a commodity rather than a basic human right is not without criticism, and many factors can influence responses to water prices (Jeffrey and Geary 2005; Kenney et al. 2008; Martinez-Espineira 2002). Renzetti (2002: 157, 158) concludes: "In general, the demand for water is a function of its price, the prices for other goods, the scale of activity … and the nature of the preferences or technology of the decision maker … [E]xcept in very unusual circumstances, the value of water is neither zero nor infinite." However, he notes that economic theory does not provide guidance on the actual decision making by the consumer. Sharp (2006) echoes this perspective, criticizing previous studies of demand management for their aggregation of data while failing to understand household decisions sufficiently. Jansen and Schulz (2006) note a lack of micro-level studies and cite a household study from Sri Lanka, which showed that demand was actually price-inelastic and income-inelastic, so that price increases may not conserve water. Their study of demand in South Africa similarly found that water use among low-income groups showed only modest responses to pricing changes, while cheap water for poorer groups did not lead to greater wastage.

The South Africa case is of particular interest because the 1997 Water Services Act explicitly recognized a "right of access to basic water supply and basic sanitation,"[9] to be funded in a variety of ways. In urban municipalities with substantial income from metered water users, a rising block tariff recouped the cost of the initial 'free' water, while rural water supply was capped at a 'free basic' level by 'appropriate service levels' provided through standpipes with restricted flow rates. It is relevant that there is a history of non-payment for services in South Africa, although J. Brown's (2005) study in Nelspruit (renamed Mbombela in 2009) led her to argue that a culture of non-payment is not simply a legacy of the historical boycotting of rates, although this is still within living memory. Rather, it is made up of many contributing factors that have resulted in an environment where it is the norm not to pay service bills, regardless of the ability to pay. This highlights the importance of legitimacy in determining the viability of pricing policy. The association of pricing policies with a transfer of ownership (privatization) has significantly undermined legitimacy in some cases. However, less than 5 percent of the world's population are supplied water by the private sector (Anand 2007: 78), and it is important to note that privatization is not necessary for the operation of a pricing system for water. Anand finds that the notion of fairness is paramount in attitudes toward privatization and whether water should be treated as a public good (ibid.: 9). The degree to which people trust the government to ensure that everyone is treated in a similar manner may explain why privatization in England and Wales was accepted, while in India, Tanzania, and Ghana it produced protests (ibid.: 252).

Turning from investing in water to funding responses to climate change adaptation and mitigation, we find that the sums required are staggering. Drought is often cited as the world's costliest natural disaster, causing annual damage of US$6 to $8 billion (Wilhite 2000). During the UN Drinking Water Supply and Sanitation Decade (1981–1990), a total of US$13.5 billion was invested in water supply and sanitation (Tebbutt 1998: 259), and at that time it was estimated that future investment would have to increase fivefold to US$5 billion a year. All this is dwarfed, however, by the sums required to tackle climate change. The Stern Report (2006) predicted climate change impacts would cost 5 to 20 percent of global GDP, whereas mitigation costs would be less than 1 percent of GDP if acted upon now. At the 2008 World Water Congress of the International Water Association (IWA) in Vienna, it was reported that World Bank investment to combat climate change would need to more than double the current annual rate of $80 billion, just to cope with the present situation through mitigation and adaptation (Jowit 2008). Parry et al. (2009: 9) have noted a convergence of predictions of the expected costs of adaptation for developing countries, with figures falling between US$10 to $100 billion per annum: "The UNFCCC report concluded that total funding need for adaptation by 2030 could amount to $49–171 billion per annum globally, of which $27–66 billion would accrue [mostly for infrastructure] in developing countries." Unfortunately, Parry et al. (ibid.: 41) also conclude that the UNFCCC forecasts may be an underestimate. Dimitrov (2010: 815) reports that the 2009 Copenhagen Accord contains a commitment to additional funding for mitigation and adaptation of US$30 billion for 2010–2012, rising to US$100 billion by 2020, from public and private sources, but notes that the accord is non-binding and does not provide details on institutional arrangements.

This is not just a problem of funding adaptation for the rural poor. Since 2008, some 3.3 billion people are estimated to be living in urban areas, and this number is expected to rise to 5 billion by 2030. Much of this growth is taking place in Africa and Asia, so that by 2030, 81 percent of the urban population will reside in developing countries. We mention here the caveat that such approximations may overestimate urban growth rates (Potts 2009), but existing numbers of urban poor mean that many now live in squalor with limited ability to respond to climate change

(Bicknell et al. 2009; Moser and Satterthwaite 2008). De Bruijne et al. (2007) note that in many respects the situation facing urban areas in the South is far worse than that of Europe 150 years ago. Poverty is deeper, there are fewer opportunities to access land and water resources, and existing resources are being degraded through pollution and over-abstraction.

Responding to climate change includes a combination of market-based mechanisms, technology transfers, and human lifestyle changes (Dessler and Parson 2006). Investment is required, and the question of who should pay for it is key. We have already presented a growing emphasis on cost recovery with regard to water resources investment, moving from state intervention to private capital. Swyngedouw (2006) notes that much of the investment required to meet the water MDGs is expected from the private sector, and this is reinforced in a recent UN analysis (Griffith-Jones et al. 2009). Discussions of climate change mitigation focus upon the need to recognize the responsibility (both current and historical) of the industrialized nations of the North for high greenhouse gas emissions, while the costs of impacts fall upon the most vulnerable and poorest communities of the South. The Stern Report (2006) was a major contribution to the argument that the global cost of immediate action to mitigate climate change outweighed the eventual cost of dealing with climate change impacts. However, the size of the costs involved have arguably only served to intensify the argument over who should pay. There is also controversy over the effectiveness of the economic tools available for mitigation and the potential conflict between mitigation and economic growth (Cosbey 2009), but the major problem is the expected cost of sharing the burden. Ahmad and Opschoor (2009: 1) note that there is presently limited funding, especially for current adaptation measures, never mind those extending well into the future: "Given the nature of adaptive capacity and its linkage with development in general, strategizing development planning and raising development finance is essential in meeting adaptation requirements."

We are, then, at an interesting turning point in the management of climate change impacts, where the need to raise capital for investment and the adoption of economic principles for management have promoted private, non-state engagement. Similar moves in the management of water resources two decades ago have been followed by promotion of greater public engagement, for example, in the EU's 2009 Water Framework Directive (Carter 2007). Dungumaro (2007) cites public participation as a core of integrated water resources management. Hence, questions of governance and stakeholder involvement have come to the fore, as illustrated in figure 1. We will expand upon these next, not least because it has become clear that economic tools need to be informed by a detailed political understanding of how behavior can and will be changed.

Institutional Principle—Politics and Participatory Decision Making

By the 1970s, the effectiveness of state-led top-down development was being questioned, and two alternatives were proposed. One, which we discussed above, sought to make resource allocation decisions subject to the 'discipline' of markets, as understood in classical economics. The second, associated with Jürgen Habermas, promoted ideas of a more active, 'participatory' citizenship, with decisions being subjected to 'deliberative democracy' in which technical experts would engage in discussion with 'ordinary' citizens to examine the rationale behind and consequences of new developments. In this section we consider increased participation in adaptation strategies, which can broadly be seen in 'upscaling', dissemination, and efforts to raise awareness and strengthen stakeholder engagement.

The goal of involving citizens as stakeholders in planning activities has been fostered by many reforms to water management institutions and by efforts to use spatial planning (Howe and White

2004) to contribute to sustainable development. Carter (2007) argues that spatial planning has become key to promoting participatory measures: increasing democratic legitimacy, building consensus, and strengthening decision making are some of the derived benefits. Equally, some countries, such as South Africa (J. Brown 2007), and the EU (Carter 2007) have incorporated stakeholder participation in water development policy. Gaining experience in implementing such initiatives has seen the concept of 'social learning'—described by van Slobbe et al. (2008) as a move away from linear planning and the reliance on expert knowledge—come to the fore. Blackmore et al. (2004: 6) explain social learning as "a process of knowing based upon experience and practice that is developed collectively and interactively among multiple interdependent stakeholders."

Increased participation needs to recognize the importance of gender, which was included in the institutional principle and incorporated into Goal 3 of the MDGs.[10] On the UNDP's Web site for women's empowerment, UNDP administrator Helen Clark states: "Development cannot be achieved if fifty percent of the population is excluded from the opportunities it brings."[11] In the past, water resources development assumed that households would alter their practices in a predictable manner to take advantage of improvements in water supply. Such assumptions have been criticized for failing to take adequately into account intra-household decision making and the gendered outcomes of changing natural resources management (Hunter 2006; Nyong and Kanaroglou 2001), including cultural restrictions placed on women who attempt to move from private (household) spaces into the public realm (Sultana 2009). In urban areas, constraints still exist, although space may be constructed differently and women in urban households may have greater access to assets (Moser and Dani 2008). Efforts to overcome gendered limitations to participation are reinforced by research that aims to identify positive instances involving women in water management (Fonjong 2008; Were et al. 2008; Wirfa et al. 2008).

An appreciation of the importance of stakeholder participation has been associated with a reinterpretation of the reasons for the failure of water pricing to deliver expected changes. In particular, a reappraisal of the role of 'communication', sometimes referred to as 'education', has identified the danger that this becomes top-down, reinforcing the position of the 'expert'. As Page (2005: 298) observes: "The engineers and politicians who now manage water supplies in Cameroon claim that most consumers are reluctant to pay for their water because they are ignorant; their attitudes need to be modified." The messages from water companies through Web sites and leaflets, plus occasional media campaigns (e.g., during droughts), are criticized as being ineffective in changing behavior (Kenny et al. 2008; Sharp 2006). Conversely, Fenemor et al. (2008) studied community resilience in New Zealand that developed through active partnerships involving the knowledge base of resource users and stakeholder participation in management. It appears to be important that conversations are initiated as early as possible so that both agendas and language are shared. Lundqvist (2000: 264) also notes the significance of water user engagement, as well as the need for better training and capacity building: "Education, training and research must transcend borders, between disciplines and between sectors and between cultures and countries."

Blackmore et al. (2004) conclude that obstacles to stakeholder participation in water resources development include overly complex institutional contexts and lack of experience with this strategy. Van Slobbe et al. (2008) note that in Sri Lanka there has been frustration over 'too much' participatory engagement: participatory practices require an investment in time and social spaces in order for interactions and negotiations to take place. Also needed are patience and tolerance, since results are not instantaneous. J. Brown (2007) has cautioned about the limitations that participatory approaches may offer in practice and the frequent absence of evaluation of the outcomes of stakeholder involvement. Her study in South Africa suggests that, where stakeholders have strongly competing interests in water use and where there are major asymmetries of information, it is unlikely that consensus will be reached through consultation among local stakeholders. Another South Africa study (H. Brown 2009) broadly supports this conclusion but also wonders whether social learning raises too many expectations, for example, that stakeholders' collective interest in collaborative management would be greater than their individual (and potentially competing) benefits from non-collective water use.

In seeking to develop more effective stakeholder engagement in adaptive water resources management, a persistent challenge is the upscaling of participation by single individuals to the level of communities. A similar step may be needed to enable collective management of the impacts of climate change. Ryan (2004) notes that adaptation is constrained by poverty due to limited resources (funds, human capital, institutional capacity), lack of knowledge or shared understanding, and resistance to change. Community-driven approaches, such as community-led total sanitation (CLTS),[12] offer some promise of improvements (Deak 2008), but there is no consensus over the best approach to scaling up adaptive action of individuals and households, either spatially or vertically (social hierarchy). Ryan's (2004) review of the literature on upscaling underlines a lack of rigor in defining project objectives and an overemphasis upon anecdotal reporting of project achievements. The review describes the following key roles for different stakeholders:

- Local community: managing the process and being in control
- Central government: institutional support through financial and legal provisions
- NGOs: facilitation, support, and training
- INGOs: advocacy and support
- Private sector: supply chain development
- International community: long-term commitment without intervention

Large disparities in socio-economic conditions present problems for the consistent engagement of stakeholders across a water catchment. In South Africa, stakeholder preferences varied so widely within studied catchments that consensus and effective decision making were blocked (J. Brown 2007). In Bangladesh, there have been positive experiences involving CLTS projects, and Deak (2008) notes here the importance of 'self-spread', with enthusiastic supporters helping to promote programs. Other critical factors that have been identified include (1) planning that assesses coverage (spatial) together with sustainability (timescale); (2) institutions, partnerships, and policy support for community engagement; (3) capacity building and financial stability; and (4) a shift from 'project implementation' to 'service delivery'. We have, then, identified from the water development literature a number of obstacles to greater subsidiarity and devolved decision making, ranging from questions of scale to the need for social learning and capacity building. There is strong evidence that, for technical and scientific support to be fully effective, engagement with stakeholders should be undertaken at the earliest opportunity, while consultation must be more than simply seeking responses to written reports.

Strategies of greater engagement with stakeholders have also gained strong support in discussions of climate change adaptation, promoted by the IPCC (2001) through the notion of a community's adaptive capacity. Similar ideas at the household scale have sought to estimate vulnerability in terms of household assets and capital, which are identified as key components of an individual's ability to make environmental responses to climate change (de Graaf et al. 2007; Nicol 2000). This emphasis upon household adaptive capacity is also found in Wamsler's (2007) analysis of disaster risk reduction in El Salvador. She lists three types of coping strategies: (1) reducing risks through technical measures (e.g., construction), environmental controls (e.g., runoff reduction), economic measures (e.g., diversifying income), and organizational change (e.g., social networks, community engagement); (2) ensuring against risks through self-insurance (e.g., education, ownership, family or institutional insurance); and (3) recovering from risks (e.g., through loss financing and community action). A similar approach is evident in a recent change in attitude to flood risk management in the UK (Johnson et al. 2007), moving from a purely engineering approach of flood defense measures to management of risks in much the same way as noted by Pereira (2007) for drought management. Rather than seeing nature as something to be controlled and dominated, the new attitude ('living with floods') places more emphasis upon adaptation that ensures ecological integrity.

In his review of adaptive management, Pelling (2011) argues that the concept emerged three decades ago to support decision making under conditions of uncertainty with an emphasis upon social learning. He goes on to argue that adaptive management has a role to play in adaptation to climate change, where there is much uncertainty and a need for multi-stakeholder engagement for social learning. However, Pelling cautions that, to be effective, institutions and organizations need to be receptive to local viewpoints, that sustaining local engagement is a challenge, and that, for community-led efforts to be credible, they need to secure technical and scientific support (ibid.: 32). This last point brings us full circle to the first principle, while reinforcing that all three (science/ecological, economics/instrument, and politics/institutional) are not meant to be implemented in isolation; rather, it is a matter of emphasis.

Conclusions

In this article we have explored the growing consensus on water development strategies in terms of three principles—ecological, instrument, and institutional—and we have argued that the same consensus is required for responses to climate change. In particular, we advocate the need to embrace the institutional principle, with an emphasis upon governance and stakeholder engagement, while noting the significance of the instrument principle, that is, that investments need to be financed in an equitable and legitimate manner. However, recent efforts to implement the institutional principle, which fosters subsidiarity and the engagement of citizens and local communities in water governance, have not been without difficulties. This principle does not offer a panacea, and we have reached the same conclusion for strategies that focus on science or economics.

Tackling environmental constraints through science and technology (e.g., by increasing the supply of water or by conservation measures to maintain ecosystem functions) can have major impacts, for instance, when large dams are involved. In addition, problems frequently arise over the distribution of benefits and the subordination of individual or local community interests to a 'greater common good'. We have argued that science and technology are necessary but that they are insufficient, on their own, to foster adaptive capacity, since they do not provide the means with which to manage the uncertainty that is inherent in scientific measurement and that arises from economic and social factors. If science therefore does not hold all of the answers, the (Weberian) scientific arguments, based on a greater common (public) good, may not find acceptance.

Two alternative and opposing views have gained importance as the efficacy of science has been questioned:

1. *Neoliberal*—pricing to ensure efficient resource use. Understanding the costs associated with resource use is a necessary dimension of adaptive capacity, but considerations of economic efficiency will be conditioned by questions of equity (distribution of costs and

benefits) and, linked to this, political perceptions of fairness (e.g., whether some people have unmet needs that may be regarded as 'human rights') and legitimacy.
2. *Institutional*—deliberative and accountable democracy to decide who gets allocated what. This view rests on a belief in the power of rational argument to ensure that adaptive behavior is equitable and politically legitimate, but it is undermined by asymmetric power relations, through which more powerful stakeholders prevail over others, irrespective of questions of equitable resource use. A related key factor involves asymmetries of information, particularly with regard to scientific information and an understanding of environmental processes and change.

These viewpoints suggest the need to re-evaluate the role of the state and the political processes through which development goals and priorities are determined and implemented.

A number of parallel but more recent discussions have developed concerning adaptation to climate change. Science has arguably led the dialogue and dominated policy formulation, yet it is beset with uncertainty, especially at the local scale where adaptation will have to take place. Technology will be important for various forms of adaptation, from water conservation measures to flood storage engineering and water supply enhancement—but it has to be paid for. The sums involved globally are staggering, and this has moved us increasingly toward discussions of investment in mitigation and financial support to developing countries for adaptation measures. The experience of carbon trading, tradable emissions quotas, and so forth suggests, however, that wider issues of ownership and equity need to be addressed, alongside the capacity for private investment. In practice, the lack of binding commitments resulting from the 2009 Copenhagen conference has highlighted the problems of adopting the instrument principle to combat climate change. As with water resources, the lack of funds in many parts of the world that are most at need means that adaptation will have to rely heavily upon social and political organization. In addition, one lesson learned from nearly half a decade of water pricing and demand management is that people do not always respond in a predictable manner to economic measures.

This has contributed to more a recent emphasis upon governance, including community-based actions and social learning, in order to meet the challenges presented by climate change. The experiences of water development, however, are not entirely promising. They have exposed the limitations of local participation in bringing about change, particularly in circumstances when asymmetries of power and conflicts of interest translate into competition to control scientific information and its interpretation. Reliance on local decision making has proved difficult, not only because of these asymmetries of power, but also due to the need for upscaling. We will thus find all three principles advocated, with different degrees of emphasis, in support of adaptation measures. Assessing which measure is most appropriate becomes a question of the purpose of the adaptation: is it a short-term response to an immediate threat, or is it intended to support the community in the long term through fundamental changes in lifestyle and land use? As with the adaptive management of water resources, the question essentially requires that the development goals of climate change adaptation be determined and agreed upon, for it is only against these objectives that measures to adapt to climate change can be effectively evaluated.

■ ACKNOWLEDGMENTS

The diagrams have been produced by the drawing office in the School of Environment and Development at the University of Manchester, and we thank Graham and Nick for their patience and skill.

CLIVE AGNEW holds a Chair in Physical Geography in the School of Environment and Development at the University of Manchester. He is an applied climatologist and hydrometeorologist working on problems of environmental degradation and environmental status assessment. He has worked extensively in both the drylands and the wetlands of Africa, Europe, and the Middle East, developing strategies to improve the management of water shortages at local and regional levels. His current research is on climate change impacts on cities in the tropics and the possibilities of flood alleviation through UK upland reclamation. Over the last 10 years, he has been Head of Geography and Head of the School of Environment and Development at the University of Manchester.

PHILIP WOODHOUSE is Senior Lecturer in Environment and Rural Development in the Institute for Development Policy and Management, University of Manchester. Trained as an agricultural scientist, he has worked for more than 30 years on land and water management in developing countries. His recent research has focused on the interaction between political, economic, and technological factors in changing land and water use, particularly in sub-Saharan Africa. Recent publications include *Water Resources and Development* (2011), co-authored with Clive Agnew.

NOTES

1. "Climate Change," Met Office, UK National Weather Service, http://www.metoffice.gov.uk/climate-change (accessed January 2011).
2. Adopted in 2000 and set to be achieved by 2015, the MDGs are intended to deal with extreme poverty around the world. See "What Are the Millennium Development Goals?" United Nations Development Programme, http://www.undp.org/mdg/basics.shtml (accessed January 2011).
3. "Water Resources of the United States," Web page of the US Geological Survey (USGS), http://water.usgs.gov/ (accessed January 2011).
4. See "EarthTrends: Environmental Information," World Resources Institute, Washington, DC, http://earthtrends.wri.org (accessed January 2011).
5. "The EU Water Framework Directive: Integrated River Basin Management for Europe," European Commission, http://ec.europa.eu/ENVIRONMENT/water/water-framework/ (accessed January 2011).
6. See FAO-AQUASTAT data tables on irrigation area at http://www.fao.org/nr/water/aquastat/dbase/AquastatWorldDataEng_20101129.pdf (accessed November 2010).
7. For an analysis of economic development and water-related issues in the Ganges-Brahmaputra-Meghna basin, see Bandyopadhyay (2002).
8. "A Scramble for Africa: Large Dams Benefit Contractors and Corrupt Governments More Than They Aid the African People," *Nature* 440 (2006): 383–384; "China's Dams in the Danger Zone," *New Scientist* 196 (2008): 5.
9. See the 1997 Water Services Act at http://www.dwaf.gov.za/Documents/Legislature/a108-97.pdf (accessed January 2011).
10. For more on Goal 3 of the MDGs, see http://www.undp.org/mdg/goal3.shtml (accessed January 2011).
11. The UNDP addresses women's empowerment and gender equality in its main areas of work. See http://www.undp.org/women/ (accessed January 2011).
12. For more information on CLTS, see http://www.communityledtotalsanitation.org/ (accessed January 2011).

REFERENCES

Agnew, Clive T. 2002. "Drought, Desertification and Desiccation: The Need for Further Analysis." *Geography Journal* 87 (3): 256–267.

Agnew, Clive T., and Philip Woodhouse. 2011. *Water Resources and Development*. London: Routledge.

Ahmad, Imram H., and Johannes B. Opschoor. 2009. "Reaching a Climate Deal in Copenhagen." UN-DESA Policy Brief No. 17, Department of Economic and Social Affairs, New York.

Al Baz, Ismail, Ralf Otterphohl, and Claudia Wendland, eds. 2008. *Efficient Management of Wastewater: Its Treatment and Reuse in Water-Scarce Countries*. Berlin: Springer-Verlag.

Allen, Richard G., Luis S. Pereira, Dirk Raes, and Martin Smith. 1998. "Crop Evapotranspiration: Guidelines for Computing Crop Water Requirements." FAO Irrigation and Drainage Paper No. 56. Rome. http://www.fao.org/docrep/X0490E/X0490E00.htm (accessed January 2011).

Altinbilek, Dogan. 2002. "The Role of Dams in Development." *International Journal of Water Resources Development* 18 (1): 9–24.

Anand, P. B. 2007. *Scarcity, Entitlements and the Economics of Water in Developing Countries*. Cheltenham, UK: Edward Elgar.

Anderson, Kevin, and Alice Bows. 2008. "Reframing the Climate Change Challenge in Light of Post-2000 Emission Trends." *Philosophical Transactions of the Royal Society A* 366: 3863–3882.

Arnell, Nigel W. 2002. *Hydrology and Global Environmental Change*. Harlow, UK: Pearson.

______. 2003. "Relative Effects of Multi-decadal Climatic Variability and Changes in the Mean and Variability of Climate Due to Global Warming: Future Stream Flows in Britain." *Journal of Hydrology* 270: 195–213.

Arnfield, A. John. 2003. "Two Decades of Urban Climate Research: A Review of Turbulence, Exchanges of Energy and Water, and the Urban Heat Island." *International Journal of Climatology* 23 (1): 1–26.

Babel, M. S., A. Das Gupta, and P. Pradhan. 2007. "A Multivariate Econometric Approach for Domestic Water Demand Modeling: An Application to Kathmandu, Nepal." *Water Resource Management* 21 (3): 573–589.

Bandyopadhyay, Jayanta. 2002. "A Critical Look at the Report of the World Commission on Dams in the Context of the Debate on Large Dams on the Himalayan Rivers." *International Journal of Water Resources Development* 18 (1): 127–145.

Bell, V. A., A. L. Kay, R. J. Jones, and R. J. Moore. 2007. "Use of a Grid-Based Hydrological Model and Regional Climate Model Outputs to Assess Changing Flood Risk." *International Journal of Climatology* 27 (12): 1657–1671.

Bicknell, Jane, David Dodman, and David Satterthwaite, eds. 2009. *Adapting Cities to Climate Change*. London: Earthscan.

Bird, Jeremy. 2002. "Nine Months after the Launch of the World Commission on Dams Report." *International Journal of Water Resources Development* 18 (1): 111–126.

Blackmore, Chris, Kevin Collins, Pam Furniss, Dick Morris, and Martin Reynolds. 2004. "The UK Policy Context for Water Management: I—the English and Welsh Policy Context." SLIM Case Study Monograph 12A. Open University. http://slim.open.ac.uk (accessed January 2011).

Brown, Julia. 2005. "Water Services Subsidies and the Poor: A Case Study of the Greater Nelspruit Utility Company, Mbombela Municipality, South Africa." CRC Working Paper No. 112. http://www.competition-regulation.org.uk/publications/working_papers/WP112.pdf/ (accessed January 2011).

______. 2007. "Participatory Processes and Outcomes of South Africa's Water Reforms." PhD diss., School of Environment and Development, University of Manchester.

Brown, Helen. 2009. "Social Learning within Participatory, Catchment-Based Water Management Processes in South Africa and Namibia." PhD diss., University of Sheffield.

Burbano, Arturo A., Samer S. Adham, and William R. Pearce. 2007. "The State of Full-Scale RO/NF Desalination: Results from a Worldwide Survey." *Journal of the American Water Works Association* 99 (4): 116–127.

Carter, Jeremy G. 2007. "Spatial Planning, Water and the Water Framework Directive: Insights from Theory and Practice." *Geographical Journal* 173 (4): 330–342.

CAWMA (Comprehensive Assessment of Water Management in Agriculture). 2007. *Water for Food, Water for Life: A Comprehensive Assessment of Water Management in Agriculture*. Ed. D. Molden. London: Earthscan; Colombo: International Water Management Institute.

CDWR (California Department of Water Resources). 2009. *California Drought. An Update: June 2009*. Report of the California Department of Water Resources, Sacramento. http://www.water.ca.gov/drought/docs/Drought_report_30june2009_web.pdf (accessed January 2011).

Challinor, Andrew. J., Frank Ewert, Steve Arnold, Elisabeth Simelton, and Evan Fraser. 2009. "Crops and Climate Change: Progress, Trends, and Challenges in Simulating Impacts and Informing Adaptation." *Journal of Experimental Botany* 60: 2775–2789.

Chambers, Robert. 1994. "The Origins and Practice of Participatory Rural Appraisal." *World Development* 22 (7): 953–969.

Chappell, Adrian, and Clive Agnew. 2008. "How Certain Is Desiccation in West African Sahel Rainfall (1930–1990)?" *Journal of Geophysical Research* 113, D07111, doi:10.1029/2007JD009233.

Chen, Y., D. Zhang, Y. Sun, X. Liu, N. Wang, and H. G. Savenije. 2005. "Water Demand Management: A Case Study of the Heihe River Basin in China." *Physics and Chemistry of the Earth* 30 (6–7): 408–419.

Christman, Keith. 1998. "The History of Chlorine." Water Quality and Health Council. http://www.waterandhealth.org/drinkingwater/history.html (accessed January 2011).

Cosbey, Aaron. 2009. "Developing Country Interests in Climate Change Action and the Implications for a Post-2012 Climate Change Regime." Paper for the UN Conference on Trade and Development. http://www.unctad.org/en/docs/ditcbcc20092_en.pdf (accessed January 2011).

Deak, Andrew. 2008. "Taking Community-Led Total Sanitation to Scale: Movement, Spread and Adaptation." Institute of Development Studies Working Paper No. 298, University of Sussex.

de Bruijne, Gert, Mirjam Geurts, and Brian Appleton. 2007. "Sanitation for All?" Thematic Overview Paper No. 20, IRC International Water and Sanitation Centre. http://www.irc.nl/page/36722 (accessed January 2011).

de Graaf, Rutger, Nick van de Giesen, and Frans van de Ven. 2007. "Alternative Water Management Options to Reduce Vulnerability for Climate Change in the Netherlands." *Natural Hazards* 51 (3): 407–422.

Deoreo, W. D., A. Dietemann, T. Skeel, and P. W. Mayer. 2001. "Retrofit Realities." *Journal of the American Water Works Association* 93 (3): 58–72.

Dessler, Andrew, and Edward A. Parson. 2006. *The Science and Politics of Global Climate Change: A Guide to the Debate*. Cambridge: Cambridge University Press.

DFID (Department for International Development). 2001. "Addressing the Water Crisis: Healthier and More Productive Lives for Poor People." UK Department for International Development. http://www.bvsde.ops-oms.org/bvsacd/milenio/tspwater.pdf (accessed January 2011).

Dimitrov, Radoslav S. 2010. "Inside UN Climate Change Negotiations: The Copenhagen Conference." *Review of Policy Research* 27 (6): 795–821.

Dow, Kirsten, and Thomas E. Downing. 2006. *The Atlas of Climate Change: Mapping the World's Greatest Challenge*. London: Earthscan.

Dungumaro, Esther W. 2007. "Socioeconomic Differentials and Availability of Domestic Water in South Africa." *Physics and Chemistry of the Earth* 32: 1141–1147.

Fenemor, A., N. Deans, T. Davie, W. Allen, J. Dymond, M. Kilvington, C. Phillips, L. Basher, P. Gillespie, R. Yong, J. Sinner, G. Harmsworth, M. Atkinson, and R. Smith. 2008. "Collaboration and Modelling: Tools for Integration in the Motueka Catchment, New Zealand." *Water South Africa* 34 (4): 448–455.

Fonjong, Lotsmart N. 2008. "Gender Roles and Practices in Natural Resource Management in the North West Province of Cameroon." *Local Environment* 13 (5): 461–475.

Franke, Richard W., and Bárbara H. Chasin. 1980. *Seeds of Famine*. Montclair, NJ: Allanheld, Osmun.

Gandy, Matthew. 2008. "Landscapes of Disaster: Water, Modernity, and Urban Fragmentation." *Mumbai Environment and Planning A* 40: 108–130.

Gill,Suzannah, John F. Handley, Roland Ennos, and Stephan Pauleit. 2007. "Adapting Cities for Climate Change: The Role of Green Infrastructure." *Built Environment* 33 (1): 97–115.

Gine, Ricard, and Augusti Perez-Foguet. 2008. "Sustainability Assessment of National Rural Water Supply Program in Tanzania." *Natural Resources Forum* 32: 327–342.

Glantz, Michael H. 1994. *Drought Follows the Plow: Cultivating Marginal Areas.* Cambridge: Cambridge University Press.

Goldsmith, Edward, and Nicholas Hildyard. 1984. *The Social and Environmental Effects of Large Dams.* Vol. 1: *Overview*. Cornwall, UK: Wadebridge Ecological Centre.

Gregory, J. M., and J. A. Lowe. 2000. "Predictions of Global and Regional Sea-Level Rise Using AOGCMs with and without Flux Adjustment." *Geophysical Research Letters* 27 (19): 3069–3072.

Griffith-Jones, Stephany, Merylyn Hedger, and Leah Stokes. 2009. "The Role of Private Investment in Increasing Climate Friendly Technologies in Developing Countries." Paper for UN DESA. http://policydialogue.org/files/events/Griffith-Jones_Hedger_Stokes_Role_of_Private_Investment.pdf (accessed January 2011).

Hewitson, B. C., and R. G. Crane. 2006. "Consensus between GCM Climate Change Projections with Empirical Downscaling: Precipitation Downscaling over South Africa." *International Journal of Climatology* 26: 1315–1337.

Holling, C. S. 1973. "Resilience and Stability of Ecological Systems." *Annual Review of Ecology and Systematics* 4: 1–23.

Howard, Guy, and Jamie Bartram. 2003. *Domestic Water Quantity, Service Level and Health.* WHO Report WHO/SDE/WSH/03.02. http://www.who.int/water_sanitation_health/diseases/WSH03.02.pdf (accessed January 2011).

Howe, Joe, and Iain White. 2004. "The Mismanagement of Surface Water." *Applied Geography* 24: 261–280.

Hulme, Mike, ed. 1996. *Climate Change and Southern Africa: An Exploration of Some Potential Impacts and Implications in the SADC Region.* Report commissioned by WWF International and coordinated by the Climate Research Unit, UEA, Norwich, UK.

________. 2009. "The Science and Politics of Climate Change." *Wall Street Journal*, 2 December.

Hundecha, Y., and A. Bardossy. 2008. "Statistical Downscaling of Extremes of Daily Precipitation and Temperature and Construction of Their Future Scenarios." *International Journal of Climatology* 28 (5): 589–610.

Hunter, L. 2006. "Household Strategies in the Face of Resource Scarcity in Coastal Ghana: Are They Associated with Development Priorities?" *Population Resources Policy Review* 25: 157–174.

IPCC (Intergovernmental Panel on Climate change). 1990. *First Assessment Report.* Geneva: IPCC.

______. 1995. *Second Assessment Report: Climate Change.* Geneva: IPCC.

______. 2001. *Third Assessment Report: Climate Change.* Geneva: IPCC.

______. 2007a: *Climate Change 2007: Impacts, Adaptation and Vulnerability. Contribution of Working Group II to the Fourth Assessment Report of the Intergovernmental Panel on Climate Change*, ed. M. L. Parry, O. F. Canziani, J. P. Palutikof, P. J. van der Linden, and C. E. Hanson. Cambridge: Cambridge University Press.

______. 2007b. *Fourth Assessment Report: Climate Change.* Geneva: IPCC.

Jansen, Ada, and Caril-Erik Schulz. 2006. "Water Demand and the Urban Poor: A Study of the Factors Influencing Water Consumption among Households in Cape Town, South Africa." *South African Journal of Economics* 74 (3): 593–609.

Jeffrey, Paul, and Mary Geary. 2005. "Consumer Reactions to Water Conservation Policy Instruments." Pp. 305–330 in *Water Demand Management*, ed. D. Butler and F. A. Memon. London: IWA.

Johnson, Clare, Edmund Penning-Rowsell, and Dennis Parker. 2007. "Natural and Imposed Injustices: The Challenges in Implementing Fair Flood Risk Management Policy in England." *Geographical Journal* 173 (4): 374–390.

Jowit, Juliette. 2008. "Environment: Huge Increase in Spending on Water Urged to Avert Global Catastrophe." *Guardian*, 11 September. http://www.guardian.co.uk/environment/2008/sep/11/water.climatechange (accessed January 2011).

Kay, A. L., and H. N. Davies. 2008. "Calculating Potential Evaporation from Climate Model Data: A Source of Uncertainty for Hydrological Climate Change Impacts." *Journal of Hydrology* 358: 221–229.

Kenny, Douglas S., Christopher Goemans, Roberta Klein, Jessica Lowrey, and Kevin Reidy. 2008. "Residential Water Demand Management: Lessons from Aurora, Colorado." *Journal of the American Water Resources Association* 44 (1): 192–207.
Kerr, Richard A. 2007. "Global Warming Is Changing the World." *Science* 316: 188–190.
Kim, J.-W., J.-T. Chang, N. L. Baker, D. S. Wilks, and W. L. Gates. 1984. "The Statistical Problem of Climate Inversion: Determination of the Relationship between Local and Large-Scale Climate." *Monthly Weather Review* 112: 2069–2077.
Kingston, Daniel G., Martin C. Todd, Richard G. Taylor, Julian R. Thompson, and Nigel W. Arnell. 2009. "Uncertainty in Estimates of Potential Evapotranspiration under Climate Change." *Geophysical Research Letters* 36, L20403, doi:10.1029/2009GL040267.
Kolokytha, E. G., and N. A. Mylopoulos. 2004. "Evaluating Demand Management Aspects of Urban Water Policy: The City of Volos Case, Greece." *Water, Air, and Soil Pollution* 4 (4–5): 263–277.
Kruger, A. C. 2006. "Observed Trends in Daily Precipitation Indices in South Africa: 1910–2004." *International Journal of Climatology* 26: 2275–2285.
Lahlou, Zacharia M. 2001. "Leak Detection and Water Loss Control." Tech Brief: National Drinking Water Clearinghouse Fact Sheet, May.
Lemos, Maria-Carmen. 2003. "A Tale of Two Policies: The Politics of Climate Forecasting and Drought Relief in Ceara, Brazil." *Policy Sciences* 36 (2): 101–123.
Lenton, Roberto, and Mike Muller, eds. 2009. *Integrated Water Resources Management in Practice*. London: Earthscan.
Lloyd-Hughes, Benjamin, and Mark A. Saunders. 2002. "A Drought Climatology for Europe." *International Journal of Climatology* 22: 1571–1592.
Lundqvist, Jan. 2000. "A Global Perspective on Water and the Environment." *Physics and Chemistry of the Earth* 25 (3): 259–264.
Magiera, Philipp, Suzan Taha, and Lothar Nolte. 2006. "Water Demand Management in the Middle East and North Africa." *Management of Environmental Quality* 17 (3): 289–298.
Maier, H., A. Jain, G. C. Dandy, and K. P. Sudheer. 2010. "Methods Used for the Development of Neural Networks for the Prediction of Water Resource Variables in River Systems: Current Status and Future Directions." *Environmental Modelling and Software* 25 (8): 891–909.
Marsalek, J., B. E. Jiménez-Cisneros, P.-A. Malmquist, M. Karamouz, J. Goldenfum, and B. Chocat. 2006. "Urban Water Cycle Processes and Interactions." International Hydrological Programme Technical Documents in Hydrology No. 78, UNESCO, Paris.
Martinez-Espineira, R. 2002. "Residential Water Demand in Northwest Spain." *Environmental and Resource Economics* 21: 161–187.
McGuire, Michael J. 2006. "Eight Revolutions in the History of US Drinking Water Disinfection." *Journal of the American Water Works Association* 98 (3): 123–149.
Meir, M. F., M. B. Dyurgerov, U. K. Rick, S. O'Neel, W. T. Pfeffer, R. S. Anderson, S. P. Anderson, and A. F. Glazovsky. 2007. "Glaciers Dominate Eustatic Sea-Level Rise in the 21st Century." *Science* 317: 1064–1067.
Mollinga, Peter P., Ruth S. Meinzen-Dick, and Douglas J. Merrey. 2007. "Politics, Plurality and Problemsheds: A Strategic Approach for Reform of Agricultural Water Resources Management." *Development Policy Review* 25 (6): 699–719.
Moriarty, Patrick, and John Butterworth. 2003. "The Productive Use of Domestic Water Supplies." IRC International Water and Sanitation Centre. http://www.irc.nl/page/3733 (accessed January 2011).
Moser, Caroline, and Anis Dani, eds. 2008. *Livelihoods, Assets and Social Policy*. Washington, DC: World Bank.
Moser, Caroline, and David Satterthwaite. 2008. "Towards Pro-poor Adaptation to Climate Change in the Urban Centres of Low- and Middle-Income Countries." Human Settlements Working Paper Series Climate Change and Cities, No. 2. London: IIED.
Nicol, Alan. 2000. "Adopting a Sustainable Livelihoods Approach to Water Projects: Implications for Policy and Practice." Working Paper No. 133, Overseas Development Institute (ODI), London.
Nyong, A.O., and P. S. Kanaroglou. 2001. "A Survey of Household Domestic Water-Use Patterns in Rural Semi-arid Nigeria." *Journal of Arid Environments* 49: 387–400.

OECD (Organisation for Economic Co-operation and Development). 2003. *Social Issues in the Provision and Pricing of Water Services*. Paris: OECD.

Ofwat (Office of Water Supplies). 2007. "Household Water Consumption for 2007." UK Water Services Regulation Authority. http://www.ofwat.gov.uk/ (accessed January 2011).

Ogodo, Ochieng'. 2009. "Africa Facing Climate Data Shortage." Science and Development Network, 11 November. http://www.scidev.net/en/news/africa-facing-climate-data-shortage.html (accessed January 2011).

Oldfield, Frank. 2005. *Environmental Change: Key Issues and Alternative Approaches*. Cambridge: Cambridge University Press.

Oppenheimer, Michael, Brian C. O'Neill, Mort Webster, and Shardul Agrawala. 2007. "Climate Change: The Limits of Consensus." *Science* 317: 1505–1506.

Ortolano, Leonard, and Katherine K. Cushing. 2002. "Grand Coulee Dam 70 Years Later: What Can We Learn?" *International Journal of Water Resources Development* 18 (3): 373–390.

Page, Ben. 2005. "Paying for Water and the Geography of Commodities." *Transactions of the Institute of British Geographers* 30 (3): 293–306.

Parry, M., N. Arnell, P. Berry, D. Dodman, S. Fankhauser, C. Hope, S. Kovats, R. Nicholls, D. Satterthwaite, R. Tiffin, and T. Wheeler. 2009. *Assessing the Costs of Adaptation to Climate Change: A Review of the UNFCCC and Other Recent Estimates*. London: IIED.

Parry, Martin, Cynthia Rosenzweig, and Matthew Livermore. 2005. "Climate Change, Global Food Supply and Risk of Hunger." *Philosophical Transactions of the Royal Society B* 360: 2125–2138.

Pearce, Fred. 2006. "Mega-dams Back on the Agenda." *New Scientist* 191 (2569): 10.

Pelling, Mark. 2011. *Adaptation to Climate Change: From Resilience to Transformation*. London: Routledge.

Pereira, L. S. 2007. "Drought Impacts in Agriculture: Water Conservation and Water Saving Practices and Management." Pp. 349–373 in *Methods and Tools for Drought Analysis and Management*, ed. G. Rossi, T. Vega, and B. Bonaccorso. Berlin: Springer-Verlag.

Peter, G. 2006. "Gender Roles and Relationships: Implications for Water Management." *Physics and Chemistry of the Earth* 31: 723–730.

Plummer, Janelle, and Tom Slaymaker. 2007. "Rethinking Governance in Water Services." ODI Working Paper No. 284. Overseas Development Institute, London.

Ponte, Lowell. 1976. *The Cooling: Has the Next Ice Age Already Begun? Can We Survive It?* Englewood Cliffs, NJ: Prentice-Hall.

Postel, Sandra. 1992. *The Last Oasis: Facing Water Scarcity*. London: Earthscan

Potts, Deborah. 2009. "The Slowing of Sub-Saharan Africa's Urbanization: Evidence and Implications for Urban Livelihoods." *Environment and Urbanization* 21 (1): 253–259.

Prudhomme, Christel, Nick Reynard, and Sue Crooks. 2002. "Downscaling of Global Climate Models for Flood Frequency Analysis: Where Are They Now?" *Hydrological Processes* 16: 1137–1150.

Reed, Maureen G., and Shannon Bruyneel. 2010. "Rescaling Environmental Governance, Rethinking the State: A Three-Dimensional View." *Progress in Human Geography* 34 (5): 646–653.

Renzetti, Steven. 2002. *The Economics of Water Demands*. Dordrecht: Kluwer.

Rietveld, L. C., J. Haarhoff, and P. Jagals. 2008. "A Tool for Technical Assessment of Rural Water Supply Systems in South Africa." *Physics and Chemistry of the Earth* 34: 43–49.

Rockstrom, Johan. 2003. "Resilience Building and Water Demand Management for Drought Mitigation." *Physics and Chemistry of the Earth* 28: 869–877.

Rogers, Peter P., Manuel Ramón Llamas, and Luis Martínez-Cortina, eds. 2006. *Water Crisis: Myth or Reality?* London: Routledge.

Roth, Matthias. 2007. "Review of Urban Climate Research in (Sub)tropical Regions." *International Journal of Climatology* 27: 1859–1873.

Royal Society. 2008. "Climate Change Controversies: A Simple Guide." http://royalsociety.org/uploadedFiles/Royal_Society_Content/News_and_Issues/Science_Issues/Climate_change/Climate_booklet_RS1420_reprint_Dec08.pdf (accessed January 2011).

Ryan, Peter. 2004. *Scaling Up—A Literature Review*. Delft, Netherlands: IRC International Water and Sanitation Centre.

Sánchez, Luis D., Arlex Sánchez, Gerardo Galvis, and Jorge Latorre. 2006. "Multi-stage Filtration." Thematic Overview Paper No. 15, IRC International Water and Sanitation Centre, Delft, Netherlands. http://www.irc.nl/redir/content/download/28202/297431/.../TOP15_MFS_06.pdf (accessed January 2011).

Schiermeier, Quirin. 2010. "The Real Holes in Climate Science." *Nature* 463: 284–287.

Sharp, Liz. 2006. "Water Demand Management in England and Wales: Constructions of the Domestic Water User." *Journal of Environment Planning and Management* 49 (6): 869–889.

Silverman, Bernard A. 2003. "A Critical Assessment of Hygroscopic Seeding of Convective Clouds for Rainfall Enhancement." *Bulletin of the American Meteorological Society* 84: 1219–1230.

Solomon, Susan, Richard Alley, Jonathan Gregory, Peter Lemke, and Martin Manning. 2008. "A Closer Look at the IPCC Report." *Science* 319: 409–410.

Soussana, Jean-François, Anne-Isabelle Graux, and Francesco N. Tubiello. 2010. "Improving the Use of Modeling for Projections of Climate Change Impacts on Crops and Pastures." *Journal of Experimental Botany* 61 (8): 2217–2228.

Sprague, Lori A., David K. Mueller, Gregory E. Schwarz, and David L. Lorenz. 2009. *Nutrient Trends in Streams and Rivers of the United States, 1993–2003*. U.S. Geological Survey Scientific Investigations Report No. 2008–5202.

Stern, Nicholas. 2006. *Stern Review of the Economics of Climate Change*. Cambridge: Cambridge University Press.

Sultana, Farhana. 2009. "Fluid Lives: Subjectivities, Gender and Water in Rural Bangladesh." Special issue on "Gender Geographies of Water." *Gender, Place and Culture* 16 (4): 427–444.

Swyngedouw, Erik. 2006. "Power, Water and Money: Exploring the Nexus." Human Development Report 2006, Occasional Paper No. 14. New York: UNDP.

Symons, George E. 2006. "Water Treatment through the Ages." *Journal of the American Water Works Association* 98 (3): 87–98.

Tebbutt, T. H. Y. 1998. *Principles of Water Quality Control*. Oxford: Butterworth-Heinemann.

TERI (The Energy and Resources Institute). 2003. *Coping with Global Change: Vulnerability and Adaptation in Indian Agriculture*. New Delhi: TERI. http://www.iisd.org/pdf/2004/climate_coping_global_change.pdf (accessed January 2011).

Therkildsen, Ole. 1988. *Watering White Elephants? Lessons from Donor Funded Planning and Implementation of Rural Water Supplies in Tanzania*. Uppsala: Nordic Africa Institute.

Thompson, John, Ina Porras, James Tumwine, Mark R. Mujwahuzi, Munguti Katui-Katua, Nick Johnstone, and Libby Wood. 2001. *Drawers of Water II*. London: IIED.

Thornton, P. K., P. G. Jones, T. Owiyo, R. L. Kruska, M. Herrero, P. Kristjanson, A. Notenbaert, N. Bekele, and A. Omolo, with contributions from V. Orindi, B. Otiende, A. Ochieng, S. Bhadwal, K. Anantram, S. Nair, V. Kumar, and U. Kulkar. 2006. *Mapping Climate Vulnerability and Poverty in Africa*. Report to the Department for International Development, International Livestock Research Institute, Nairobi, Kenya.

Twort, Alan C., Don D. Ratnayaka, and Malcolm J. Brandt. 2000. *Water Supply*. 5th ed. Oxford: Butterworth-Heinemann.

UNCED (United Nations Conference on Environment and Development). 1992. "The Earth Summit, Rio de Janeiro." UNCED. http://www.un.org/geninfo/bp/enviro.html (accessed January 2011).

UNDP (United Nations Development Programme). 2006. *Beyond Scarcity: Power, Poverty and the Global Water Crisis*. Human Development Report 2006. http://hdr.undp.org/en/reports/global/hdr2006/ (accessed January 2011).

UNESCO. 2006. *The 2nd United Nations World Water Development Report: Water, a Shared Responsibility*. London: Earthscan.

______. 2009. *The 3rd United Nations World Water Development Report: Water in a Changing World*. London: Earthscan.

van Slobbe, Erik, Richard Morris, Neils Roling, Reinder Torenbeek, Koen Broker, and Hero Heering. 2008. "Social Learning in Wetland Development." Pp. 270–283 in *Wetlands: Functioning, Biodiversity*

Conservation, and Restoration, vol. 191 of *Ecological Studies*, ed. R. Bobbink, B. Beltman, J. T. A. Verhoeven, and D. F. Whigham. Berlin: Springer-Verlag.

Vidal, John, and Jonathan Watts. 2009. "Copenhagen Closes with Weak Deal That Poor Threaten to Reject." *Guardian*, 19 December. http://www.guardian.co.uk/environment/2009/dec/19/copenhagen-closes-weak-deal (accessed January 2011).

Walski, Thomas A. 2006. "A History of Water Distribution." *Journal of the American Water Works Association* 98 (3): 110–121.

Wamsler, Christine. 2007. "Bridging the Gaps: Stakeholder-Based Strategies for Risk Reduction and Financing for the Urban Poor." *Environment and Urbanization* 19 (1): 115–142.

Ward, Roy, and Mark Robinson. 1999. *Principles of Hydrology*. 4th ed. London: McGraw Hill.

WCD (World Commission on Dams). 2000. *Dams and Development: A New Framework for Decision-Making*. London: Earthscan. http://www.dams.org/report/earthscan.htm (accessed January 2011).

Were, Elizabeth, Jessica Roy, and Brent Swallow. 2008. "Local Organisation and Gender in Water Management: A Case Study from the Kenya Highlands." *Journal of International Development* 20: 69–81.

White, Iain. 2008. "The Absorbent City: Urban Form and Flood Risk Management." *Proceedings of the Institution of Civil Engineers: Urban Design and Planning* 161: 151–161.

Wilhite, Donald A., ed. 2000. *Drought: A Global Assessment*. Natural Hazards and Disasters Series. London: Routledge.

Wirfa, Linda, April Campbell, and N. Rea. 2008. "Implications of Gendered Environmental Knowledge in Water Allocation Processes in Central Australia." *Gender, Place and Culture* 15 (5): 505–518.

World Bank. 2009. *Water Data and Statistics*. Washington, DC: World Bank.

WWC (World Water Council). 2000. "2nd World Water Forum." The Hague, Netherlands, March. http://www.worldwatercouncil.org/index.php?id=16&L=0%2Fst...usuarios.arnet.com.ar%252 (accessed January 2011).

———. 2006. "4th World Water Forum." Mexico City, March. http://www.worldwatercouncil.org/index.php?id=1386&L=0targ (accessed January 2011).

———. 2010. "Water Crisis." http://www.worldwatercouncil.org/index.php?id=25 (accessed January 2011).

BOOK REVIEWS

DOWIE, Mark, *Conservation Refugees: The Hundred-Year Conflict between Global Conservation and Native Peoples*, 336 pp. Cambridge, MA: MIT Press, 2009. ISBN 978-0-262-01261-4.

The publication of Mark Dowie's *Conservation Refugees* provides a landmark in debates about the social impacts of conservation and protected areas—the point where it shifted from being a discussion among academics, conservationists, and indigenous peoples' organizations into the realm of popular writings. Academic studies of the relationship between protected areas and local peoples have expanded and developed considerably since the publication of West and Brechin's (1991) *Resident Peoples and National Parks*, which could be considered the first book on the subject. Evictions, exclusions, and injustices resulting from conservation actions, both historical and current, have been recorded throughout the world. Detailed analyses have revealed the wide range of power relations involved, their efficacy, and the implications for both biodiversity and local residents. Conservationists too have recently been debating dynamically, prompted in part by critiques from academics and indigenous activists.

Variously, past and ongoing injustices have been acknowledged, new commitments have been made, and new forms of action and partnership have been undertaken, although there remains considerable differences of opinion among conservationists about what the obligations and strategies of conservation should be with regard to local peoples. While in recent years there have been a few articles in environmentalist and conservation magazines and newspapers about evictions and exclusions involving protected areas (some authored by Dowie), *Conservation Refugees* represents the first time that this topic has been the subject of a book intended for the lay public. Aimed at exploring how indigenous peoples have long suffered as a result of conservation and how they can, and must, be part of the preservation of biodiversity, the volume also happens to be a well-researched, insightful, and pretty comprehensive read by a veteran and respected investigative reporter.

Starting with the case of Yosemite National Park in California, the book chronicles the evictions of indigenous peoples since the inception of modern protected areas. Subsequent chapters explore examples of evictions and exclusions in Ethiopia, the Congo, India, the Amazon, East Africa, and other places, examining the reasoning behind these actions and their consequences. Mixed in throughout is an exploration of the organizations and the conservation practices, concepts, and ideologies that underpin certain forms of protected areas. These observations enrich the analysis and help to explain what has led to the evictions. Also included are a few examples in which indigenous peoples have benefited from, or at least have not been harmed by, conservation. The book is clearly the result of a thorough review of the current academic literature on indigenous peoples and conservation, covering many well-known case studies, as well as key arguments on wilderness, the positive role of disturbances, and traditional ecological knowledge. However, the extensive

doi:10.3167/ares.2010.010109

research is not adequately represented in the slim documentation.

The opening line of the preface, which refers to the "good guy vs. good guy" (p. ix) struggle between conservationists and indigenous peoples, reveals the author's intent not to expose the evils of conservation but rather to promote forms that are good for both biodiversity and local people. Dowie considers that conservation efforts can succeed only when the latter are encouraged to become part of saving biodiversity, rather than being opposed to it. Overall, the book's treatment of large NGOs is less harsh than in Dowie's (1995) earlier work. The analysis of the big international conservation NGOs, who generally bear the brunt of criticism for their role in evictions, is rather nuanced, with a recognition that they can also treat indigenous peoples with respect, involving them in conservation. It explores how NGOs' attitudes toward indigenous peoples have changed (in Dowie's view, improved) in recent years—albeit with some distance left to travel—and how contradicting ideas can co-exist within the same NGO.

There are minor problems regarding some matters that are left out, which perhaps reflects a desire to have a cohesive book that is accessible to the non-specialist reader. Much blame is put on the largest international NGOs rather than on smaller organizations or states, which is of particular importance in countries, such as India, where the presence of large NGOs is thin. The writings of well-known conservationists are quoted, but the analysis could have benefited from personal interviews, particularly since some conservationists have rethought their attitudes toward indigenous peoples since the cited pieces were written. While Dowie rightly criticizes the perceptions of many conservationists regarding indigenous peoples' destructiveness, he could also explore in more detail situations in which these views are correct and the circumstances that might encourage local peoples to take destructive actions toward biodiversity. Likewise, Dowie's coverage of the ills that conservation NGOs can bring to rural peoples greatly outweighs that on potential benefits. The exclusive focus on indigenous issues ignores the impacts of conservation on other vulnerable groups who have been involved in long-standing stewardship of the environment but who lack or do not lay claim to ethnic indigeneity. Making it an issue solely of indigeneity excludes these other victims of conservation from our analysis and our sympathy, and possibly from better, more compassionate conservation practices. Despite this, *Conservation Refugees* should be read by anyone with an interest in conservation, indigenous issues, or the complexities of sustainability and development.

George Holmes
Sustainability Research Institute, School of Earth and Environment, University of Leeds

References

Dowie, Mark. 1995. *Losing Ground: American Environmentalism at the Close of the Twentieth Century*. Cambridge, MA: MIT Press.

West, Patrick C., and Steven R. Brechin. 1991. *Resident Peoples and National Parks: Social Dilemmas and Strategies in International Conservation*. Tucson: University of Arizona Press.

ESCOBAR, Arturo, *Territories of Difference: Place, Movements, Life, Redes*, 437 pp. Durham, NC: Duke University Press, 2008. ISBN 978-0-8223-4344-8.

The anthropologist Arturo Escobar is well-known for his work examining new social movements, 'post-development', and post-structural political ecology. Those who have found inspiration from these earlier works will enjoy his latest book, *Territories of Difference*, a lengthy monograph that draws together more than a decade's work with disparate literatures and social movements (especially in his native Colombia). Any attempt to state what the book accomplishes runs into difficulties. In the first place, the text's problematic is vast and poorly defined. As the title indicates, the volume is concerned with territory, difference, place, social movements, life, and networks (or *redes*). The one-word chapter titles do little to indicate the content of the distinct chapters, which in turn deliver few straightforward arguments. The same could be said of the book as a whole.

Consider two introductory passages in which Escobar defines the book's aims. The first comes immediately after a description of his collaborations with the Proceso de Comunidades Negras (PCN): "This book can be seen … as an ethnography of the practices, strategies, and visions of [a] particular group of activists [with whom Escobar has collaborated], including their own knowledge production. While the book is largely conceived from this perspective [i.e., as a study that emerges at the intersection of political and intellectual practice—JDW], however, it is more than that … [T]he book proposes a way of analyzing some of the most salient social, cultural, and ecological issues of the present day" (p. x). We are not told what this "way of analyzing … issues" is, but readers will gather that it involves examining the socio-natural networks within which political-economic issues arise (more on this later). Here is a second key framing passage (pp. 4–5):

> [This] book is about the incredibly complex intersections of nature and culture, space and place, landscape and human action, culture and identity, knowledge and power, economy and politics, modernity and globalization, and difference and sameness associated with imperial globality and global coloniality in a particular corner of the world; it is also about what has been called uneven geographies of poverty and livelihoods, and how they are related to historical political economies and culturally inflected patterns of development … I render these geographies manageable by a particular design in term of six basic concepts: place, capital, nature, development, identity, and networks. These concepts are both chapter titles and notions that articulate my argument throughout the book.

The reader might then expect the "argument [articulated] throughout the book" to be stated, but it is not. This points to the central weakness of this original and creative book. It sags under the weight of its diverse topoi, never uniting around a coherent theme or argument—perhaps by design. Two of the book's central tropes are weaving and networks, and the text is nothing if not an interweaving of threads of text and social life. The best way to enjoy it is to dip in, follow a thread, skip to another section, draw some connections, and move on. The book is an often frustrating mishmash, but that is not to say that it has nothing to teach. For while Escobar lacks a clear argument, there is no shortage of new ideas. While the book could be fairly described as an ambitious compilation of readings, the key question, then, is, how does Arturo Escobar read?

I will offer two brief observations to this question. The first concerns Escobar's approach to theory. He reads widely and with a penchant for upending every metaphysical pairing he finds (e.g., 'nature and culture') in post-structuralist fashion, thus producing a narrative that is productively fragmented, layered with bits and pieces of concepts that are neither wholly embraced nor discarded. Escobar seems less motivated to solder analytical arguments carefully together than to mix social theory with cultural-political experiences. His is a practiced blending, and much of the appeal of the text comes from the moments when it works. His reading is also practiced in the sense that it has an intended effect. As Escobar explains, one of his paramount aims is to "demystify theory that ignores subaltern experiences and knowledge … *in order to relocate their politics of place as key to our understanding of globalization*" (p. 15). This position could be characterized as populist insofar as it calls for an explicit return to the representation of subaltern experience by anti-colonial intellectuals like

Escobar. We should register the considerable distance between this ethnographic strategy and the post-colonial criticism of Gayatri Spivak, Partha Chatterjee, and Qadri Ismail.

My second observation is that Escobar's readings are not Marxist, at least not in any sense that the word is normally understood. Escobar posits that "Marxism was not very good at dealing with nature" (p. 8), and I get the impression that he thinks it was not very good at much else. Consider chapter 2, 'Capital', which prosecutes one of the most coherent sub-arguments in the book, concerning the status of the relations of production and consumption of the black communities under study. Escobar's opening salvo is to ask "whether these communal forms [of economic life] could be considered noncapitalist forms of economy" (p. 72). Citing the work of J. K. Gibson-Graham (a major influence on Escobar), he claims that "what is at stake is the possibility of finding noncapitalist forms of economy at play in concrete situations" (p. 74). Reviewing familiar histories of livelihood transformation wrought by African oil palm, "global shrimp" (p. 85), and their attendant market relations, Escobar explains that he will "bracket the assumption that these enterprises can be described in purely capitalist terms" (p. 83)—a reasonable strategy, particularly since arguing against their seemingly obvious capitalist character would "require more detailed study," as he rightly notes (ibid.). But within 20 pages, his questioning and 'bracketing' have become displaced by positive assertions, cast as research findings. We find the communities that he studied being compared against "Gibson-Graham's criteria for noncapitalism" (p. 100), and then, in the chapter's conclusion, the decisive claim that "in this region there exists a complex dynamic of various types of economies, economies one needs to understand without subjecting them to a single capitalist determination" (pp. 106–107). Translation: the Marxists are wrong; capital is not dominant here. Thus, his question is answered, albeit without ever making the case. (For a different reading of black practices, see Asher 2009).

I am not sure what motivates this rhetorical strategy, but I suspect that it derives from a desire to show that these communities reproduce non-capitalist spaces within which to resist coloniality. Or that Escobar can help the black community find itself in such spaces. In either case, his logic seems to be that if we can find socio-natural networks that are relatively autonomous from capital and bring them to light, alternatives will grow. Reasonable thinkers may disagree as to whether this political theory is tractable, or whether Escobar's research successfully demonstrates the existence of such non-capitalist economic networks. What is certain is that many readers will find this an ambitious and creative book.

Joel Wainwright
Department of Geography, Ohio State University

References

Asher, Kiran. 2009. *Black and Green: Afro-Colombians, Development, and Nature in the Pacific Lowlands*. Durham, NC: Duke University Press.

FAGAN, Brian, *The Great Warming: Climate Change and the Rise and Fall of Civilizations*, 304 pp. New York: Bloomsbury Press, 2008. ISBN 978-1-59691-392-9.

We are often told that those who do not know history are doomed to repeat it. As societies around the world come to terms with the complexities of global warming, historians and archaeologists are increasingly turning to the past, arguing that droughts, rising temperatures, and environmental degradation caused the collapse of various ancient civilizations. For these scholars, abandoned cities and the toppled monuments of their kings sound a clarion call, providing a preview of our own destiny if we fail to take global warming seriously. Such studies often have a moralizing quality, as clear as the verses of Shelley's *Ozymandias*. For example, in his widely read book *Collapse: How Societies Choose to Fail or Succeed*, Jared Diamond (2005) argues that the short-sightedness and avarice of rulers and leading elites repeatedly led societies to collapse, a clear indictment of our present condition.

Many scholars—anthropologists in particular—have reacted to Diamond's book and similar studies that give environmental and climate change starring causal roles in the collapse of ancient civilizations. Without downplaying the urgency of addressing global warming, they argue that these studies underestimate the ability of societies to respond to environmental and climatic changes; that they focus on the political dissolution of kingdoms and empires, while ignoring the resilience of past societies and the cultural continuities that persist in the wake of political collapses; and that they oversimplify the interaction between people, their environment, and climate change. Like the case studies brought together by McAnany and Yoffee (2010), Brian Fagan's *The Great Warming* serves as an important counterweight to simplistic environmental explanations of ancient social collapses. Fagan's book appeals to a wider audience, however, as it is written to be accessible to scholars in other disciplines and the general public as well.

The Great Warming is impressive in its comparative scope, analyzing case studies that show how climate change affected ancient societies around the world during the Medieval Warm Period, which he dates to about AD 800–1300. Two critical observations emerge from Fagan's synthesis, both of which are pertinent to the ways we think about global warming and make plans for the future. First, global climate trends affect local climate regimes in quite variable ways. To explain why this is so, Fagan nicely synthesizes recent advances in our understanding of global climate forces and processes, such as El Niño and the Intertropical Convergence Zone, while pointing out that there is much yet to learn. He also masterfully takes the reader through the various proxy records used to reconstruct climate histories (e.g., sediment records from lakes, ice cores from glaciers and ice sheets, historical documents, etc.), pointing out the strengths and weaknesses of each. One comes away with an awareness of how much our models of global climate and our reconstructions of climate history have advanced in the last 25 years, and with renewed appreciation for the fact that global warming will have many diverse local manifestations.

The second observation is equally critical: no matter how detailed our models of past and present climate change are, understanding its role in shaping human history requires an equally detailed study of people and culture. In this respect, Fagan provides a welcome nuanced treatment of the ways in which ancient people responded to changes in the natural world around them, and, in doing so, he demonstrates the limitations of Western constructs of collapse. He argues convincingly that the migration undertaken by the ancestral Puebloan peoples of the Southwest was an effective response to a long period of drying and drought. Thus, he recasts the abandonment of the great houses of Chaco Canyon and Mesa Verde from catastrophic collapse to wise long-term strategy.

Fagan's comparative analysis also demonstrates that drier conditions do not inevitably lead to collapse. For example, his discussion of the Chimú civilization of Peru's arid Pacific Coast illustrates a successful response to heightened desiccation. It is also refreshing to see a scholar address the ways that climate change has affected realms of culture beyond agriculture. For instance, Fagan shows how wind conditions during the Medieval Warm Period favored long-distance voyages by Polynesian sailors that led to the arrival of the Rapa Nui on Easter Island.

As an anthropologist, Fagan puts people and culture front and center in his analyses. He emphasizes the rich and detailed environmental knowledge of agrarian societies, past and present. Across the globe, people have developed effective methods of weather prediction on weekly to annual scales (see, e.g., Orlove et al. 2000), and thus they could pro-actively respond to changing weather. Fagan also highlights the role of technology in shaping the way that people interact with the environment. He discusses how technological innovations such as the moldboard plow, camel saddles, and toggle harpoons opened up new possibilities, allowing people to interact with both the environment and one another. By including these important observations about culture and technology, Fagan focuses the readers' attention squarely on the intersection between social and cultural history, on the one hand, and environmental and climate history, on the other.

Occasionally, one detects a whiff of moralizing in Fagan's analyses. For example, he suggests that the Maya collapse was due in part to an "inflexible ideology" and a "constant preoccupation with warfare" (p. 171), while the Chimú prospered during a period of drought because the increasingly dry conditions on the Peruvian coast had forced them to adopt careful conservation measures (pp. 171–172). I would suggest that the important differences lie in the fact that the Maya lacked the rich fisheries and the larger-scale imperial political economy of the Chimú, which, as Fagan duly notes, provided the latter with a buffer against agricultural shortfalls and spread the risk of drought across a broader geographic region.

In any broad, comparative synthesis, experts may take issue with the presentation of their area of expertise. *The Great Warming* is not immune, and I take some exceptions to Fagan's presentations of Tiwanaku and Classic Maya civilizations. In the case of the latter, in particular, Fagan overstates the Maya 'obsession' with water and overemphasizes the importance of water control to the authority of Maya rulers. He revives an old chestnut of Maya exceptionalism—that they developed in a particularly inhospitable environment—but underscores the long dry season and unpredictable rainfall instead of the impenetrable jungle.

These minor quibbles do not change the fact that *The Great Warming* is an engaging and well-written presentation of a remarkably large body of information that is accessible to experts and non-experts alike. They also do not detract from the larger, important contributions that Fagan's book makes to ongoing debates about the role of climate changes in shaping our past and our future. Anthropology and archaeology have demonstrated that human responses to changes in climate, no matter how severe, are always conditioned by culture and mediated by a society's institutions and technologies. Anthropologists and archaeologists are fortunate to have in Brian Fagan a gifted and committed intellectual ambassador who can convincingly articulate this critical point to a broader audience.

Jason Yaeger
Department of Anthropology, University of Texas at San Antonio

References

Diamond, Jared M. 2005. *Collapse: How Societies Choose to Fail or Succeed.* New York: Viking Books.

McAnany, Patricia A., and Norman Yoffee, eds. 2010. *Questioning Collapse: Human Resilience, Ecological Vulnerability, and the Aftermath of Empire.* Cambridge: Cambridge University Press.

Orlove, Benjamin S., John C. H. Chiang, and Mark A. Cane. 2000. "Forecasting Andean Rainfall and Crop Yield from the Influence of El Niño on Pleiades Visibility." *Nature* 403: 68–71.

HORNBORG, Alf, J. R. McNEILL, and Joan MARTINEZ-ALIER, eds., *Rethinking Environmental History: World-System History and Global Environmental Change*, 207 pp. Lanham, MD: AltaMira Press, 2007. ISBN 978-0759110281.

During an environmental history class, I once discovered that my students did not believe in capitalism. More precisely, they believed that capitalism differed in degree but not in kind from other economic systems, since everyone, at all times, was basically greedy and interested only in maximizing profit. What differentiated the capitalist mode of agriculture (if indeed such a thing existed) was not the new kinds of social relations, market structures, land tenure regimes, or debt, credit, and investment cycles that it entailed, but rather technology, which the students also saw—along with population growth, of course—as the great driver of environmental change.

After recovering from my initial shock, I realized that my students lacked exposure to a systematic theory of capitalism and its relationship to global environmental change. Thus, I welcomed the opportunity to review *Rethinking Environmental History*, which promised to fill this need. Containing 20 chapters by scholars from history, anthropology, geography, and ecological economics, among other fields, this edited volume seeks to integrate a structural political economy approach—particularly, Immanuel Wallerstein's world-system theory (WST)—with empirical analyses of international flows of materials, energy, wastes, capital, and people, from ancient times to the present. Yet, as is the case with many edited volumes, the end result is uneven, unwieldy, and disjointed. While certain chapters stand out, the analytical incoherence of the book as a whole may not advance the cause of using WST as a lens for a systematic exploration of global environmental change.

The argumentative thread that runs through the book, albeit somewhat sporadically, is that our global ecological crisis stems from the spatially uneven economic relations between core and periphery in the world-system. The story of modern times is thus the gradual expansion of the political influence and market territory of the core, which turns external arenas into peripheral units of the system, immiserating them and stripping them of natural resources in the process. This line of argument should be abundantly familiar to anyone with a passing interest in international development, colonialism, or global environmental change. Yet, as my classroom anecdote suggests, big theories require constant retesting, modification, and propagation through diligent empirical research. One major problem with *Rethinking Environmental History* is that only some of the contributors use WST as their primary analytical lens (Alf Hornborg, one of the editors, admits as much in his introduction). For example, Mats Widgren offers an illuminating chapter on the concept of landesque capital, and J. R. McNeill makes a compelling case for yellow fever's determining role in the imperial geopolitics of the New World, but their connections to WST are unclear. Perhaps fewer contributions, more tightly bound to the WST framework, would have generated greater analytical coherence and precision.

Yet, taken together, even the more WST-oriented chapters left me skeptical about the theory's usefulness for understanding environmental change. In part, this is due to the weakness or inconsistency of empirical claims, exacerbated by the editors' evident lack of interest in providing synthesis across chapters. In a rhetorically overheated chapter, Jason W. Moore proposes that a "metallurgical revolution," fired by wood fuel, drove Europe's early modern crisis of deforestation. But how does this jibe with the preceding chapter by the late Michael Williams, covering roughly the same period of Europe's forest history, which demotes metallurgy to a minor cause of forest clearance? In fairness, a much more coherent and innovative empirical program is presented in four chapters that attempt quantitative analyses of "unequal ecological exchange" and global-scale "social metabolism." This kind of historical accounting is extremely complex and necessary for refining increasingly popular concepts, such as the "ecological footprint," and for weighing intergenerational distributive justice claims (as highlighted by Wallerstein himself, in an ominous concluding chapter).

At a more theoretical level, the volume fails to resolve one of the more persistent critiques of WST: how countries advance from periphery to semi-periphery to core. Although Williams categorizes the world-system as a "highly dynamic arrangement" (p. 106), this assessment is not borne out much in his chapter or elsewhere. In fact, the spatial, economic, and geopolitical arrangement between core and periphery, as portrayed in this book, seems to crystallize irreversibly by the 1800s. The core hosts the 'productive' economies, with the periphery bound to focus on 'extractive' industries, leading to ever-diminishing terms of trade and spatially uneven development. This structure captures the world economy circa 1914, but today the lines between core and periphery (production and extraction) seem quite blurry. For example, how do we explain the transformation of the Powder River Basin of Wyoming, one of the world's richest sites of resource extraction (coal), being nestled in the heart of the global 'core'? Or the rise of China? Or the industrialization of Brazil?

Even more unnerving, the concept of 'ecologically uneven exchange', as elucidated by Joan Martinez-Alier and the late Stephen Bunker, seems, in the final analysis, to verge toward environmental determinism. At first glance, environmental determinism would seem to find little purchase in a critical, structural, Marxian political-economic analysis. Yet Martinez-Alier and Bunker, in their respective chapters, suggest that the unevenness of exchange and core-periphery relations are essentially structured by the heterogeneous nature of the spatial distribution of natural resources on earth—indeed, to use Bunker's phrase, it makes uneven development "physically inevitable" (p. 241). If that's the message, it is conveyed in much clearer prose by Jared Diamond in *Guns, Germs, and Steel.*

Ultimately, *Rethinking Environmental History*, by dint of its inconsistent analysis and confusing message, fails to displace Diamond and other popularizers of global environmental history, who succeed because of the simplicity of their mega-scale narratives and the primacy given to natural and technological, rather than political and economic, determinants. Such apolitical and triumphalist tales certainly deserve 'rethinking'. In the end, this book will be judged a success by a cadre of true believers in Marxian ecological economics and world-system theory, but—save for a few outstanding chapters—it is unlikely to convince those who are so fully immersed in capitalism that they misperceive it as the natural order of things.

Eric D. Carter
Grinnell College

JONES, Eric C., and Arthur D. MURPHY, eds., *The Political Economy of Hazards and Disasters*, 351 pp. Lanham, MD: AltaMira Press, 2009. ISBN 978-0-7591-1309-1 (cloth).

Eric C. Jones and Arthur D. Murphy's edited volume *The Political Economy of Hazards and Disasters* represents a valuable contribution to cross-cultural scholarship on the political-economic and historical dynamics that figure into the creation of and the response to disastrous situations. By synthesizing a diverse body of ethnographic and archaeological material, the book offers valuable insights into the vulnerability profiles and other determining factors that structure the experience of such events. As a key text in the anthropology of hazards and disasters, it provides an important perspective on the ways in which communities and governments manage the calamitous situations that are constant factors in human history.

Jones and Murphy state that their main interest is the social distribution of vulnerability and the relation between economic patterns and the occurrence of hazards and disasters. They define disaster as "a tragedy experienced by a human group at the hands of an identifiable event, mitigated by local capacity and broader intervention, and measured in terms of economic, spiritual, psychological, biological, political, or social impact" (p. 5). As a concept, 'vulnerability' ties together the book's 14 chapters. The ethnographic and archaeological analyses describe how disasters affect populations in a multiplicity of ways. In turn, the accounts portray local peoples' agency as they confront a cascade of crises.

The volume begins with two sections that focus on economic dynamics. Part 1 ("Economic Parameters of Disasters") sets the tone by discussing historical and contemporary trends in the examination of disastrous situations. The contributing authors strive to create a balanced perspective on the importance of economic parameters among the multitude of complex variables that analysts must confront when examining disaster and recovery. Part 2 ("Class-Based Vulnerability in Disaster Exposure, Impact, and Recovery") examines how class structures vulnerability within a population's social groupings. The section does a superb job explaining the socio-economic dynamics that structure the unequal experience of and recovery from disastrous situations. Moreover, it shows that little has changed for the most vulnerable populations and that major adjustments of recovery efforts are necessary.

Part 3 ("The Line between Hazard and Disaster for Primary Producers) and part 4 ("Product Distribution in Hazardous Settings") review variations in risk strategies and hazardous product distributions at a global level. Part 3 details how production strategies are adapted in Central and South America in times of crisis, from shifts in stone tool manufacturing in ancient Mexico to the accommodation of modern Ecuadorian life to the volatility of a volcanic environment. Part 4 follows this lead by examining disaster-related adaptations in the distribution patterns of various goods in the Philippines, the Indian Ocean, and Madagascar. Parts 3 and 4 use ethnographic and archaeological material to show how populations at high risk of repeated disasters must be able to adapt their production and distribution techniques if they hope to survive and recover as quickly and efficiently as possible.

Part 5 ("Political Economic Mitigation of Disasters") concludes the volume by focusing on the role of the state in assisting (or neglecting) populations in times of crisis. The chapters describe how governments both exacerbate and mitigate disasters and their local consequences. Analyses revolve around themes of industrial recovery, the use of knowledge from past catastrophes, and the idea of an effective 'culture of response'. The section does an excellent job tying together the topics from the rest of the book in relation to discussions of both effective and inadequate responses to calamitous events.

Overall, the volume presents a wonderfully detailed exploration of the different forms of theory, data, and methodology that can ground convincing analyses of disastrous situations. In a collaborative spirit, the authors provide a well-rounded perspective on all of the variables that contribute to the formation of hazards and disasters, as well as the experiences and responses of local populations. The book would be a useful addition to courses in economic anthropology and the anthropology of disasters and hazards, and in business or economics classes that seek to develop a more nuanced and political-economic understanding of disaster mediation.

Kelley L. Denham
University of Texas at San Antonio

LANGSTON, Nancy, *Toxic Bodies: Hormone Disruptors and the Legacy of DES*, 256 pp. New Haven, CT: Yale University Press, 2010. ISBN 978-0-300-13607-4.

In *Toxic Bodies*, environmental historian Nancy Langston traces the cultural history of DES, the synthetic estrogen diethylstilbestrol. DES acts as a hormone disruptor. It was used as medication for women and girls (1950s–1970s) and as growth supplements for poultry (1947–1953) and cattle (1950s–1970s). The dangerous potential of hormone disruptors to promote reproductive cancers and interfere with sexual development was recognized as early as the 1940s. How then, Langston asks, did DES use in humans and their food become so widespread?

Langston contends that specific institutional and cultural practices kept DES in circulation.

In the early 1940s, the Food and Drug Administration (FDA) made a historic shift in perspective from a 'precautionary principle', whereby drug companies had to demonstrate that a chemical was safe, to a more industry-friendly approach, whereby it was the responsibility of a chemical's critics to demonstrate harm. Moreover, disagreement over the relevance of animal studies to human health, confusion over the relationship between natural and synthetic chemicals, and outdated risk and dose/threshold models for establishing cause and effect tragically compounded the situation. Finally, political pressure from pharmaceutical, chemical, and agricultural industries, alongside cultural beliefs in the technocratic authority of biomedicine over women's bodies, kept DES circulating for decades as either a medication for women and, later, girls or as a growth supplement for poultry and cattle.

The book is part history and part anthropology. Throughout, it is infused with activism (and rightly so). Langston takes us back to the beginnings of the FDA and its early administrators to highlight how the FDA has had a variable but consistently close relationship with industry. She tells disturbing stories of how pharmaceutical companies sent out free samples of DES and thalidomide (later withdrawn in the early 1960s when found to cause congenital deformities), in order to get 'data' to submit for new drug applications. The process obviously violated informed consent and in no way followed a scientific approach to demonstrating effect. Langston ties these actions into wider cultural beliefs in the primacy of technology over the unruly female body. From there, she uses gender theorists, such as Anne Fausto-Sterling and Judith Butler, to unpack implicit biases of a male-female dichotomy, a genetic determinist model of gender, and a Western split between body and environment. She shows how each bias shaped the use of samples, the motivation for the development of the drugs, and their ultimate uses.

Langston ends the book with a call to restructure our understandings of the impact of hormone disruptors. Borrowing from Steingraber's (2003) ecological reconceptualization of pregnancy, Langston advocates for a new ecology of health, one in which our bodies are semi-permeable entities in a web of interconnectedness with surrounding physical and socio-cultural worlds. Our bodies—specifically, our immune systems—are in constant negotiation with these worlds. Health is the result of resiliency to past and present threats, not simply the absence of immediate threats. In Langston's refashioning, models of risk would thus include ongoing epigenetic effects that can encompass the often late manifestations of low-dose exposure to hormone disruptors. These models of risk challenge the FDA's current quantitative risk assessment that ostensibly measures the potential harm of a new drug. Langston then calls for the return of the precautionary principle and a new dialogue that would bring diverse voices to the research and (re)assessment of chemicals in our medications, foods, and environments.

Certainly, Langston's proscriptions are timely, and they give us a frame by which to assess current regulatory institutions' positions and policies on chemicals used in food production and medication. For example, on 27 February 2010, *Science News* reported on new Environmental Protection Agency (EPA) findings for atrazine, a popular commercial pesticide and hormone disruptor (Raloff 2010). Recent analyses demonstrate a persuasive correlation between atrazine and birth defects/reproductive disorders. As Langston forecast in the last chapter of her book, the EPA admitted that pressure from environmental advocates—not new findings from a disembodied group of scientists—forced the re-examination of its earlier approval of atrazine. And, just as with DES, available data are problematic because they come from animal studies and necessarily inexact epidemiological research. Human data are compounded by a lack of control, the delayed observable impact of hormone disruptors, numerous other chemical confounders, and unknown exposure timing, duration, and amount. Previous atrazine research was funded in part by industry and (not surprisingly) failed to find correlations. On the other hand, the EPA had not released research that had noted atrazine concentrations in drinking water that exceeded its standards. Indeed, Langston's analysis of DES gives us the ability to assess critically the multiple layers of influence on regulatory agencies' decisions.

Perhaps the major drawback to Langston's book is that it is almost too persuasive. The DES story she tells us is convincing. At the least, the permitted use of DES was an abrogation of the fiduciary responsibility of regulatory agencies. At the worst, it was intentional malfeasance. We get little in the way of industry documentation and perspective on DES, although, understandably, industry information is proprietary. One is left wondering about the individual actions and institutional processes that make up this thing we recognize as 'the bottom line' (i.e., profit). Moreover, how do cultural notions of 'risk' impact public response to industry and FDA findings? How do ideas of 'natural' impact the public's willingness to ingest certain chemicals? Indeed, how is 'chemical' defined in different contexts? To better regulate, as Langston suggests, we will need a more intimate historical and cultural understanding of the dynamics involved in the relationship between industries, regulatory agencies, and the public.

In sum, Langston deftly demonstrates that we need a closer and more critical look at the individuals and processes that have shaped the FDA and its response to DES *Toxic Bodies* would be a powerful learning tool in upper-level undergraduate and graduate courses in history, anthropology, environmental studies, and sociology. It would also be of significant interest to environmental advocates and health professionals who regularly deal with the after-effects of hormone disruptors.

K. Jill Fleuriet
University of Texas at San Antonio

References

Raloff, Janet. 2010. "Weed Killer in the Crosshairs: Concerns Prompt Reexamination of Atrazine's Safety." *Science News* 177 (5): 18.

Steingraber, Sandra. 2003. *Having Faith: An Ecologist's Journey to Motherhood*. New York: Perseus Books Group.

LI, Tania Murray, *The Will to Improve: Governmentality, Development, and the Practice of Politics*, 392 pp. Durham, NC: Duke University Press, 2007. ISBN 978-0-8223-4027-0 (paperback).

The Will to Improve is an ethnographically rich study of development and conservation in the highlands of Central Sulawesi, Indonesia. Li places this project in the context of the history of civilizing missions, migration and cash cropping, indigenous rights, and the cultural assumptions of the environmentalists who were active in establishing the Lore Lindu National Park. Li's close focus on the social positionality of the actors within the fields of power and her ability to expand the concept of governmentality toward the cultural nexus of the actors who carry it out make this essential reading for environmental activists and development agents, as well as anthropologists.

The ethnography and analysis of chapters 3 through 6 will be of particular interest to environmental activists, anthropologists, and sociologists. The ethnography is set up in chapters 1 and 2 with a rich social history of plans and policies for development in Indonesian history. Li concludes chapter 2 with a masterful discussion of the 'emergent positionings' of the subjects of these interventions, indigenous peoples who have been 'civilized' by forced resettlement out of the forests and hills. Chapter 3 traces the effects of global capital, stimulated by national development policies, which transformed land and labor relations. Growing cacao for global markets and accumulating land, migrants excluded indigenous populations, who were increasingly constructed as poor and unworthy of development. Resulting conflict over land was classified by government officials as anarchic rather than class-based. Into these sets of social relationships came a plan to protect the forests—the Lore Lindu National Park. The vision of improvement underlying the establishment of the park was that of saving biodiversity. High value was placed on the landscape of the rainforest. The defined problem was the "destabilizing practices" of highlanders (p. 125), their destruction of biodiversity. Huge amounts of money were committed on the basis of this connection that was drawn between these problems and potential solutions.

Methodologically, chapter 4 is interesting. Through a close reading of documents, Li analyzes conservation projects associated with the national park and assesses the effect of the plans on villagers who would be excluded from the park. Through this, she discerns the progressive stripping away of caveats and specific local information until, regardless of the actualities, the proposed project comes to look like the already existing vision of conservation and development agencies. This is 'rendering technical', a process that reduces planning to techniques, making politics invisible. Furthermore, at no point in these visions of 'integrated development' did project plans realistically calculate the actual cost of conservation that would be borne by the villagers. Their lives were simply to be improved.

To the dismay of conservationists, these plans resulted in organized resistance to the park by indigenous villagers. In essence, landless indigenous villagers asserted the right to be their own trustees, pro-park activists attempted to reassert their own position as trustees, and a leading conversation organization, the Nature Conservancy, lost an opportunity to integrate villagers into conservation plans. Li contends that "[t]he objective of trusteeship is not to dominate others [at least not openly, I might add]—it is to enhance their capacity for action, and to direct it" (p. 5). The power of trustees is often legitimated by science, and 'rendering technical' works alongside problematization to define problems and solutions. The project plan "constructed a boundary around a knowable, improvable, technical domain" (p. 154), from which villagers' knowledge was excluded.

Li's work is particularly powerful because it highlights villagers' agency within the construction of the capacity for that agency through their particular history (p. 228). She positions their agency not as a simple opposition to power, but as agency within multiple fields of power—intersecting matrices that had been brought to life during the processes of development. Similarly, Li refuses to examine development merely in terms of hidden motives of profit and domination; rather, she maintains that the trustees are sincere and try to manage the balance between orderly rule, capitalist profit, and the good of the natives (see, e.g., the conclusion of chapter 6). Nevertheless, I found her discussion of the pro-park activists far less empathetic than that of the villagers. Li does not take us into the world of the development actors—their historically constructed social positionality—nearly as effectively as she engages us in the life of the indigenous villagers.

Nevertheless, I found this volume to be a rich and theoretically significant contribution. Concepts such as 'the will to improve', 'trustees', and 'rendering technical', as well as Li's approach of situating agency within governmentality and her careful textual analysis of project documents, make a strong addition to the literature on environment and development.

Kathleen Gillogly
Department of Sociology/Anthropology, University of Wisconsin-Parkside

RADKAU, Joachim, *Nature and Power: A Global History of the Environment*, 448 pp. Cambridge: Cambridge University Press, 2008. ISBN 978-0-521-85129-9 (hardback).

In *Nature and Power: A Global History of the Environment*, German historian Joachim Radkau examines how human actions have affected the environment and addresses how these actions should be understood in the context of the ongoing, reciprocal relations between humanity and non-human nature. Originally published in German in 2002, first released in a revised English translation in 2008, republished in English in 2009, and widely reviewed in multiple languages, this is a work of global environmental history that has received something approaching a global readership, at least among scholars at work in environmental history and adjacent fields.

This volume is notable for tackling not only the global scale but also the *longue durée* of its topic. The author addresses the significant organizational challenges of such a broad work by structuring his account according to "the geographic reach of environmental problems, the level of social authority that deals with them, and the type of knowledge that is employed in the process" (p. 36). This approach allows Radkau to integrate an analysis of water and forests with attention to colonialism and globalization, resulting in a work that is as historiographically ambitious as it is geographically and temporally so. For Radkau, the key to all of this is power. It is the central issue, regardless of whether the matter at hand is the local management of hunting rights or the European exploitation of overseas colonies. This book makes clear that, around the world and over a broad swath of time, the management of nature has often been about the management of humanity.

A self-described reluctant global historian (p. xi), Radkau is concerned with the relations between seemingly local developments and evolving global processes. The cumulative effects of the daily practices of the world's population receive a great deal of attention in this work. Indeed, some of Radkau's most persistent concerns are intensely local, even intimate: dietary changes, the uses of excrement, and the effect on populations of early forms of birth control are all important and recurrent themes in *Nature and Power*. Ultimately, it is the national scale that receives limited attention in this rendering. While Radkau recognizes the significance of the nation-state (along with other institutions and organizations) in the establishment and enforcement of environmental regulations, the structure of the book implies that the national scale does not provide an especially useful frame through which to analyze the character of relations between humanity and non-human nature.

Particularly in the English edition that was revised to accommodate readers most familiar with environmental history as practiced in English-speaking regions, Radkau is concerned to bring the insights of what he calls Old World environmental history to bear on the field as practiced in North America. While acknowledging his debt to the US field in particular, Radkau hopes that the resulting shift away from debates over ideas of wilderness and toward considerations of sustainability in long-inhabited landscapes might provide insights of value in contemporary efforts at environmental administration. Such concerns come to the fore in the latter third of *Nature and Power*, reflecting a shift in emphasis over the course of the book from historiographical critique of the environmental history field to policy prescriptions for those interested in improving environmental management. Notably, however, in both cases Radkau's advice seems fairly similar. He emphasizes the need for greater complexity in understandings of relations between human and non-human nature, arguing that both environmental historians and those involved in contemporary environmental discourse have often failed to accommodate the variety of forms that these relations have taken or may take.

Nature and Power is intended to demonstrate the utility of an environmental perspective to an understanding of global history, which implies an audience that includes those outside the field of environmental history. However, it is professionals or senior students working in environmental history and closely allied fields that are most likely to appreciate Radkau's combination of history and historiography. For the English-speaking reader who is not fluent in German, this work has the advantage of making available some of the insights produced by so-called Old World environmental historians, although frustration is a likely outcome when the linguistic barriers to tracking down Radkau's source material become apparent. *Nature and Power* certainly demonstrates the value of closer dialogue among those practicing environmental history in different linguistic and cultural contexts, however difficult that may be to facilitate.

Shannon Stunden Bower
University of Alberta

ROBBINS, *Paul, Lawn People: How Grasses, Weeds, and Chemicals Make Us Who We Are,* 186 pp. Philadelphia: Temple University Press, 2007. ISBN 978-1-59213-579-0.

In her 2008 review of Paul Robbins's *Lawn People,* Julie Guthman celebrates this book as "a classic" that marks "political ecology's long overdue entry into American suburbia" (Guthman 2008: 425). Guthman's praise is not hyperbole. Robbins has produced a work of extraordinary range and sophistication. He has uncovered a hidden factor in human-environment relations and has brought a novel reading of social theory to bear on a remarkable set of empirical data. *Lawn People* is indeed a classic.

What makes the book so important? It is the first to provide scholars of environmental issues with a coherent theory of 'socio-natural' subject production. In the field of political ecology (where Robbins is unquestionably a central figure), a roiling debate has frothed about how geographers should theorize, and interpret, nature and society. One leading position within this debate argues that the very distinction between nature and society must be rethought by critically examining socio-natural phenomena, that is, things that are simultaneously natural and social. Meanwhile, many social theorists have sought to explain particular forms of subjectivity. *Lawn People* is the first book to bring together these two strands of inquiry effectively and to produce something genuinely new. Like a mad scientist zapping a biochemical brew with positive and negative electrodes, Robbins wields these two cutting-edge lines of inquiry to bring new life to an old problem: the lawn.

Space does not permit me to elaborate on the book's strengths, so I will focus only on its most novel quality: its thesis about nature and subjection. The central argument of the book is that the lawn—understood not as a mere natural thing but as a socio-natural assemblage, connecting grass, soil, and insects, as well as human labor-power, industrial chemicals, and so forth—produces a subject. We read that the "lawn is a system that produces a certain kind of … turfgrass subject" (p. xvi). But who or what is this subject? It is the very lawn person that gives this book its title. The term 'lawn person' is not hyphenated in the book, but it could be, the hyphen indicating the interconnection forged through the mutual adequation of lawn work: people work on lawns, and lawns work on people. How? Lawns, Robbins demonstrates, condition our behaviors and thoughts, particularly about ourselves and our neighbors. In conforming to the demands of the lawn, the lawn worker reproduces capitalist social relations. Chapter 5 documents how lawn care functions through a roughly $10 billion industry (dominated by a few chemical companies). This industry requires constant inputs of unpaid labor. Lawns and lawn people therefore cannot be understood outside of capitalist social relations.

Conceptually, *Lawn People* departs from others in its emphasis on Louis Althusser's (1971) concept of interpellation (to date, nearly all political ecologists have approached subjectivity through Foucault). Yet Althusser's essay says nothing about nature, let alone lawns. It offers a sonorous theory of subjection, one that emphasizes the ways that subjects are interpellated, or hailed, by ideology.[1] Yet a lawn cannot hail us like a cop. Therefore, Robbins's focus on Althusser is curious and, some have said, a failure. I do not think that these criticisms hold.

On its face, the turn to Althusser may seem unsuccessful for two reasons. First, Robbins does not offer much of an explanation for how he uses Althusser. Second, *Lawn People* departs from Althusser's account on several key points. To weigh these limitations, we need to return briefly to Althusser's theses on ideological state apparatuses (ISAs). What are ISAs? Althusser writes, "They must not be confused with the (repressive) State apparatus … Repressive suggests that the State Apparatus in question 'functions by violence.'"[2] ISAs rather function through "distinct and specialized institutions" that stand outside the state. Althusser gives many examples, including the "religious ISA," "the educational ISA," and "the family ISA." Notwithstanding this diversity, all ISAs "contribute to the same result: the reproduction of the relations of production, i.e., of capitalist relations of exploitation." And how? ISAs "function by ideology." What is ideology? Althusser contends that "[i]deology represents the imaginary relationship of individuals to their real conditions of existence."[3] Crucially, ideology is neither mere ideas nor propaganda.[4] Ideology, Althusser insists, "has a material existence." It is the lived, practical relation between humans and their world.[5] This brings us to Althusser's central thesis and the clearest link to *Lawn People*, which is that "*all ideology hails or interpellates concrete individuals as concrete subjects*, by the functioning of the category of the subject … [I]deology 'acts' or 'functions' in such a way that it 'recruits' subjects among the individuals (it recruits them all), or 'transforms' the individuals into subjects (it transforms them all) by that very precise operation which I have called *interpellation* or hailing, and which can be imagined along the lines of the most commonplace everyday police (or other) hailing: 'Hey, you there!'"

Part of the importance of this account of subjection stems from, in Mcwitt's (2002: 46–47) words, the fact that "interpellation at once presupposes and settles a certain ambiguity in the boundaries of subjective identity." This is because interpellation "relies on an individual's ability to perceive themselves addressed" and "the signal is received regardless of whether one is paying attention" (ibid.: 47). The hailing was not the result of attentive listening; on the contrary, it simply happened. It was an 'event' that productively ruptured the subject. As Butler (1997: 114–115) elaborates: "For Althusser, the efficacy of ideology consists in part in the formation of *conscience*, where the notion 'conscience' is understood to place restrictions on what is … representable … [I]t designates a kind of turning back—a reflexivity—which constitutes the condition of possibility for the subject to form." Crucially for the account of the 'lawn ISA' that is implicitly advanced in *Lawn People*, Butler (ibid.: 118) explains: "For Althusser, to perform tasks 'conscientiously' is to perform them … again and again, to reproduce those skills and, in reproducing them, to acquire mastery. Althusser places 'conscientiously' in quotation marks …, thus bringing into relief the way in which labor is moralized … To become a 'subject' is thus to have been presumed guilty, then tried and declared innocent. Because this declaration is not a single act but a status incessantly *reproduced*, to become a 'subject' is to … have become an emblem of lawfulness, a citizen in good standing, but one for whom that status is tenuous." The key point is that the iterative, practiced quality of subjection is closely tied up with the challenge of producing a certain status—for instance, the status of suburban citizenship that is earned through conformity to lawn norms.

We can now return to *Lawn People* and a criticism of the work, which is that the text does not elaborate how the lawn works ideologically in the respects that are key for Althusser. The ties between the lawn and ideology are not demonstrated, so it is hard to see what the lawn ISA does. In addition, the aurality of the hailing is unclear. How should we weigh this? First, we should acknowledge that Althusser never elaborated on the practical ties between the church, the school, and ISAs. (Perhaps part of what made his essay so fruitful was this very sketchiness.) In any event, *Lawn People* does not challenge Althusser's account of interpellation with regard to the state, churches, and schools as much as it asks, but what about the non-human? Herein lies *Lawn People*'s ambition. It is perhaps the first attempt to read the interpellative scene onto the natural spaces not mentioned by Althusser and, in so doing, to critique socio-natural subjection.

To his credit, Robbins recognizes that this task requires us to rethink the work of environmentalism as an ideology along Althusser's guidelines (e.g., we have to stop thinking of lawns as 'natural' and humans as 'social'). Taking his cue from Donna Haraway, Robbins (2007: 16) insists that this critique of environmentalist ideology must involve bringing in the non-human: "[Althusser's account of interpellation] really says very little about the daily interactions that actually dominate people's lives and human behaviors in nature, economy, and community. In the case of vast ecologies, what does the interpellating? Not the Church, nor the police. Who calls to the lawn chemical user so that they consistently respond as lawn workers? Whose voice does the lawn owner hear as they open the door and look out on the grass …? We hope to demonstrate here that it may be the lawn itself." Note that this claim is hedged: it 'may be' the lawn itself. And it turns out that there really is no such a thing as the lawn 'itself', because it already involves us and other things that are not usually considered part of the lawn. So, no, grass does not hail. The lawn hails us as a socio-natural ensemble.

But can the lawn act as an ISA in Althusser's terms? On this point, I think *Lawn People* is on strong ground, since Robbins demonstrates that, for many suburbanites at least, the lawn acts in a way that is analogous to the church and school in France of the 1960s: it conditions behaviors and thoughts. Consider one of the most important findings of the study: people who tend to know more about the hazards of lawn chemicals—that is, relatively well-educated consumers—are *more likely* to use chemicals. This upends the nostrum that the key to environmentalism is 'consciousness raising'—that if people only knew the real costs of their actions, their behavior would change. It turns out that 'they know not what they do' is wrong. They do know, yet still they do.

Yet again, what of the hailing? If interpellation is supposed to operate through the possibility of being hailed, then *Lawn People* forces the question about how we hear nature. Who or what is the parallel here to the friend knocking on the door? *Lawn People* asks that we accept that it is the lawn: as the text heading of one chapter declares, "The Lawn Speaks." The serious point to this joke is that hailing is not sonorous. It involves passive human listening. But where listening is understood broadly, hearing nature involves resonance, an openness to being affected by nature. And in hearing nature, we resonate with things that are not limited to what we narrowly call 'nature' (Nancy 2007). To take one example from Robbins (2007), 'Suzanne' persists in spraying her yard and strapping dog booties on her pet. She explains: "You try to make [the lawn] look as nice as you can, without offending other people" (ibid.: 113). This is socio-natural conformism of the first order. And how did it come to be? We read that people "have become 'responsible' lawn managers because a number of things are simultaneously at work on them—including their communities, their families, and their property values—all mediated by an aesthetic designed far away … They [have] become anxious" (ibid.: 130). Rather than a knock on the door, we resonate with the dry, bald patches in the grass and perceived glances over the fence (whether real or not). Thus, the character of interpellation is more than social, but it is emphatically *not* nature or grass that hails us. The nature of interpellation is socio-natural.

Lawn People does not teach us that the lawn hails us. Instead, it shows that who we are is formed through socio-natural relations that take place in our lives in mundane and demanding spaces, including, pre-eminently, lawn spaces. To say this differently: no, the lawn as such does not knock on the door. Rather, the lawn is a becoming-space of subjection, a space that exists and surrounds lawn people. It is that which surrounds us and thus opens up to our being what we are. The lawn is not Althusser's policeman who hails, "Hey, you there!" It is the 'there' at the end of that hail where we find ourselves today, as lawn people, such that we may be hailed.

Joel Wainwright
Department of Geography, Ohio State University

Notes

1. My emphasis on interpellation as a "conspicuously sonoric event" is inspired by John Mowitt (2002: 45–47).
2. All quotations in this paragraph are from Althusser (1971).
3. Althusser warns that ideology as such "is nothing insofar as it is a pure dream (manufactured by who knows what power …) … [I]deology has no history, which emphatically does not mean that there is no history in it … but that it has no history of its *own*."
4. As Mowitt (2002: 45) writes in his discussion of interpellation: "Prior to the Althusserian intervention, ideology was still essentially confused with propaganda."
5. This formulation draws on Brewster ([1969] 1997). In his glossary of Althusserian terms, Brewster states: "Ideology is *the 'lived' relation between men and their world*, or a reflected form of this unconscious relation, for instance, a 'philosophy' … It is distinguished from a science not by its falsity, for it can be coherent and logical … but by the fact that *the practico-social predominates* in it over the theoretical, over knowledge" (ibid.: 314, italics added). Here Brewster is summarizing Althusser, who is drawing from Marx.

References

Althusser, Louis. 1971. "Ideology and Ideological State Apparatuses (Notes towards an Investigation)." Pp. 127–186 in *Lenin and Philosophy and Other Essays*. New York and London: Monthly Review Press. (First published in 1970 in *La Pensée*.) http://www.marxists.org/reference/archive/althusser/1970/ideology.htm.

Brewster, Ben. [1969] 1997. "Glossary." Pp. 309–324 in Louis Althusser and Etienne Balibar, *Reading Capital*, trans. B. Brewster. New York: Verso.

Butler, Judith. 1997. *The Psychic Life of Power: Theories in Subjection*. Stanford, CA: Stanford University Press.

Guthman, Julie. 2008. "Review of *Lawn People*." *Professional Geographer* 60 (3): 425–426.

Mowitt, John. 2002. *Percussion: Drumming, Beating, Striking*. Durham, NC: Duke University Press.

Nancy, Jean-Luc. [2002] 2007. *Listening*. Trans. Charlotte Mandell. New York: Fordham University Press.

SHERIDAN, Michael J., and Celia NYAMWERU, eds., *African Sacred Groves: Ecological Dynamics and Social Change*, 240 pp. Athens: Ohio University Press, 2008. ISBN 978-0-8214-1789-8 (paperback).

Often overlooked in the literature is the importance of small, interstitial gallery forests in Africa. Much attention, especially with regard to conservation, human-wildlife conflict, and human livelihoods, has focused on larger forested tracts. The smaller forests, often completely surrounded by agriculture, are regarded as highly denuded, degraded, and of little conservation value. However, these smaller forests are essential for sustaining rural populations, providing fuel, income, food, and ecosystem services. Mysterious and often misunderstood, sacred groves are places whose existence and management is shaped by religious beliefs. This book is a valuable contribution to that discourse in its examination of the three-way importance—social, ecological, and spiritual—of gallery forests. Sheridan and Nyamweru bring together an impressive cadre of scholars whose interests in the sacred groves of Africa vary thematically and geographically.

The book is divided into four themes: human ecology, social organization, symbols, and the future of African sacred groves. The editors open with a comprehensive introduction that explores the symbolic and ideological aspects of these "power-laden landscape features" (p. 1). Sheridan continues in chapter 2 with a more in-depth analysis of sacred groves. Drawing from landscape ecology, he discusses the misplaced approach of many scholars to appreciate these groves as relics (i.e., relic theory) rather than purposeful manifestations of spiritual and livelihood needs. Their importance lies in what they represent more than a static physical entity. Sheridan claims that a proper evaluation of these sites for community-based resource management and the conservation of African culture and biodiversity must consider the social, ecological, and spiritual components that are involved and their interaction.

The first two sections mainly examine the social-ecological linkage between these forests and local communities with a diverse set of case studies. Alongside their historical, cultural, or religious significance, these groves, despite their island character, can concomitantly play important roles in conservation. These small sites (5–250 hectares) are often overlooked by international NGOs and conservationists in favor of larger tracts (e.g., Amazonian fragments of hundreds to thousands of hectares), but they are no less important. The last bastions of natural landscapes in the human-dominated mosaic, these floristically rich areas provide resources and buffers to reserves. In their chapters, co-authors Aiah Lebbie and Raymond Guries and author Tsehai Berhane-Selassie challenge the notion that these forests are relics (concurring with Fairhead and Leach [1996]). Rather, they are seen as purposeful establishments that evolve spatially and temporally, based on cultural, spiritual, and community relationships, and whose meanings and social values tend to shift according to the forests' changing symbolism and uses. In their chapter about kaya forests of coastal Kenya, Nyamweru et al. conclude that forests should be recognized more for their presence and how their dynamic (temporal and spatial) nature changes due to external influences, rather than for their diminishment. Much of Nadia Rabesahala Horning's chapter focuses on usufruct rights and the prohibitive and prescriptive rules that govern access to these groves.

The third section examines the symbolism associated with these forests, using examples from Côte d'Ivoire, Benin, and Ghana. Alma Gottlieb provides an insightful discussion about forests and the importance of local perception in modifying the behavior of actors. Many loggers in Côte d'Ivoire have become fearful of forests in the Beng region because of the powerful spirits who reside there and whose demands must be accommodated. Cultural patterns, traditions, and access privileges in communities neighboring these sacred spaces are often shaped, at least in part, by fear, respect, and worship for the powerful spirits that guard or reside in the groves in Benin. Ute Siebert stresses the importance of new 'integrative approaches' that account for cultural and religious values in local perceptions and their interaction with the environment. But while conservation initiatives can take advantage of 'sacred sites', this is a foreign designation, and these spaces are defined differently. Often the spiritual significance of these groves is entwined with daily life and exists as a by-product of human land use.

What does the future hold for these forests? Banana et al. and L. Alden Wily attempt to address this important question. Since informal tenure regimes and usufruct rights mainly dominate sacred groves, Banana et al. stress the need to strengthen existing and to create new pathways for legally recognized holdership and tenure. However, one of the difficulties with *de jure* designation of sacred groves is that the use and values bestowed on these groves is dynamic, and laws tend to capture these landscape elements as static spaces. Alden Wily argues that decentralized governance, improved legal status for customary land rights, and the involvement of communities in managing spaces and resources are all necessary. The human element (moderate land use and religious activities) should be included with the management of these groves, rather than excluded.

While the messages are clear and the writing is strong throughout, two aspects of this book give this reviewer pause. First, there is no distinction between sacred groves and multi-use forests managed by the community (with spiritual/religious purposes as one use, but not necessarily the most dominant one or even the motivating factor for conservation). Moreover, none of the authors define the term 'sacred grove', making it difficult to assess these groves as sacred per se. The absence of a definition aligns well with the general thesis of the book—that these places are fixed, but their meanings and the cultural values attached to them change over time. Perhaps the authors intend for the reader to wonder how a proper designation of 'sacred' is achieved when this definition is largely foreign. If a forest is designated as sacred, should these groves be viewed differently from other protected areas? And if so, how? Nonetheless, the sacred forest distinction is an important mechanism to unify different interests and rally community support and compliance. Sacred groves can function as spiritual and physical defense mechanisms against external threats. While the spiritual nature of a forest can enhance conservation objectives, it does not provide the sole motivation for those objectives to be met. I agree with Banana et al. and Alden Wily that local actors and communities should be engaged and involved, since designation has local implications and importance. Informal rules and customary regulations are no longer sufficient to mediate the effects of increased population, land shortage, and growing demands on resources.

Second, the book misses a golden opportunity at the end to synthesize these chapters and leave the reader with a clear way forward (e.g., regarding policy initiatives, research imperatives, etc.). The introduction provides a roadmap, but I felt that the individual chapters diverge thereafter, with the flow from one chapter to the next being lost. Chapters are loosely organized under four content areas, but in reading them, it did not seem that they all fit in this book. Further, each chapter did not build on the previous one. I recognize that a cohesive and fluid structure is difficult with edited volumes, but after the last chapter, the reader is left hanging. The editors could have added an afterword or epilogue to identify research and policy imperatives.

Joel Hartter
Department of Geography, University of New Hampshire

References

Fairhead, James, and Melissa Leach. 1996. *Misreading the African Landscape: Society and Ecology in a Forest-Savanna Mosaic*. African Studies Series. Cambridge: Cambridge University Press.

WALKER, Richard A., *The Country in the City: The Greening of the San Francisco Bay Area*, 378 pp. Seattle: University of Washington Press, 2008. ISBN 978-0-295-98701-9 (hardback).

One of the most iconic diagrams of the social sciences captures 'the growth of the city' (Burgess [1925] 1967: 55). In it, concentric rings of development radiate outward from an urban core, rapidly consuming, transforming, and ultimately downgrading everything in their wake. Chicago, the modern metropolis that this diagram idealized as 'the city', had in actuality two geographical features that challenged its growth—vast reaches of prairie and a great inland lake. Nevertheless, its master planner exhorted that both should be overcome to realize "a population capable of indefinite expansion" (Burnham and Bennett 1909: 80) and the commercial growth that such expansion would bring. In his environmental history of the San Francisco Bay Area, Richard Walker shows how the problem of urban growth took a profoundly different turn. This impressively comprehensive book details the emergence of a peculiar species of urban expansion that chose not to overcome nature by developing it into extractable and expendable forms of value. Walker narrates how, instead, locals built their nature and, in the process, achieved a distinctly green public culture whose repercussions have echoed far beyond the Bay Area.

Scholarship on contemporary urbanism in the United States does not suffer from a lack of work that condemns or celebrates the polycentric metropolis. Ever attune to these discussions, Walker nevertheless departs from them by asking what else a polycentric metropolis might wreak, beyond novel design aesthetics, lax zoning, or an automobile-oriented 'placelessness'. Walker focuses on the San Francisco Bay Area, a complex ecosystem that, by the turn of the last century, had already begun to bare the scars of feverish industrialization. Mining runoff had poisoned its rivers, railroad and construction booms had thinned its ancient redwood stands, and its prominent shallow bay choked on silt, infill, and refuse. By narrating how area locals and newcomers encountered and sought to hem in the rapid ecological degradation around them, Walker effectively deflates commonplace myths about the Bay Area's unusual 'natural endowments'. He demonstrates that intensive human engineering was instead responsible for forging these endowments. Walker thus persuasively argues that the way in which urbanites imagine and interface with the materiality of their landscapes steers the paths that urban expansion ultimately takes. In so doing, he challenges facile divides between 'the city' and 'the country', while also complicating analyses of sprawl as generalizable and random growth. More than that, he shows how the co-evolution of natural environments and asphalt jungles in the Bay Area gradually fashioned a distinct civic ethos. This is an ethos that anchors citizenly virtues and obligations in a regional consciousness—and conscience—that is underwritten by ecological stewardship.

The bulk of Walker's book charts a century of grassroots politics that produced a remarkable metropolis now consisting of 'ruralized cities' within an 'urbanized countryside'. Each of the volume's 10 chapters unpacks struggles that have yielded some of the most definitive characteristics of Bay Area urbanism. These include chapters on the development of robust greenswards, the pioneering of growth boundaries, the cultivation of rich and specialized agricultural zones, and finally early organizing efforts that fed the environmental justice movement. Walker investigates the intricacies of political maneuvering and coalition building from which a regional ecological conscience emerged. He also attends to the refraction of this conscience within broader domains by demonstrating how Bay Area actors incubated experiments that later inspired state and federal regulations. Most notably, these regulations include water conservation, toxic remediation, and ecological restoration policies. This book will therefore interest both academics and practitioners who seek to understand or affect environmental interventions at the complex interface of municipal, state, and federal governance.

Walker makes a pointed effort to broaden awareness about the Bay Area's role in state, national, and now international environmental politics beyond the work of a few elite men. He certainly does not overlook the contributions of those such as John Muir and David Brower. However, he shows that their grand gestures became possible only in a context where elite women—and later less elite people of color—did the work of rendering environmentalism into a matter of everyday politics. This is a politics preoccupied with smaller-scale but nevertheless high-stakes issues, for example, the dearth of adequate recreational space or the toxicity of fish caught to supplement meager diets. By attending to pragmatic efforts to resolve such concerns, Walker not only writes back into American environmental history those who have been eclipsed by its heavyweights, but also highlights the kind of painstaking organizing practices that are needed to shape broad-based and effective environmental coalitions. He thus tracks the evolution of an environmental conscience and the projects it engendered, from relatively narrow and elite preoccupations with conservation and consumption toward more inclusive questions about public well-being.

Walker's investment in the capaciousness of the Bay Area's ecological conscience and its record of achieving pragmatic solidarities across race, gender, and class divides makes it difficult for him to square the fissures that threaten environmental coalitions. He despairs especially over fragmentations that have plagued the environmental justice movement. In fact, after commenting on rapacious developers and the poor planning of San Jose, Walker reserves his sharpest words for environmental justice militants who sacrifice strategy by dwelling on essentially correct, yet also unrealistic and distracting, critiques of racism and capitalism (p. 248). This surprising tone stands out within an otherwise evenhanded narrative. It suggests the need to advance representation work—here, the important effort of inflecting the myriad voices that made the 'city in the country' thinkable and doable—toward a separate but not necessarily unrelated task.

In the spirit of this advancement, we might turn to the case of the Miwok, a people indigenous to northern California's coastal and mountain areas. The Miwok do not escape Walker's admirable representational commitments, and he notes briefly the stunning natural bounties that they must have enjoyed before the advent of Euro-American settlement. Yet these passing and past tense mentions are curious, as Miwoks living within the Yosemite Valley needled John Muir. They also seriously confounded subsequent national park boosters, officials, and tourists well into the twentieth century (Spence 1999). Exactly how did confusion surrounding the Miwok and efforts to reconcile their presence within settler anxieties about the nobility, purity, endangerment, and 'naturalness' of nature inform regional and national commitments to environmental stewardship? Rather than flagging the limits of representational work, to entertain such a question points toward a new task: clarifying how and with what effects public moral orders, including green or civic ones, distribute benefits, aspirations, and especially risks and demands differently across different populations.

Catherine Fennell
Columbia University

References

Burgess, Ernest W. [1925] 1967. "The Growth of the City: An Introduction to a Research Project." Pp. 47–62 in *The City*, ed. Robert E. Park, Ernest W. Burgess, and Roderick D. McKenzie. Chicago: University of Chicago Press.

Burnham, Daniel H., and Edward H. Bennett. 1909. *The Plan of Chicago*. Chicago: Commercial Club of Chicago.

Spence, Mark D. 1999. *Dispossessing the Wilderness: Indian Removal and the Making of the National Parks*. New York: Oxford University Press.

WRANGHAM, Richard, and Elizabeth ROSS, eds., *Science and Conservation in African Forests: The Benefits of Long-Term Research*, 280 pp. Cambridge: Cambridge University Press, 2008. ISBN 978-0-521-89601-6 (hardback).

Richard Wrangham and Elizabeth Ross's edited volume *Science and Conservation in African Forests: The Benefits of Long-Term Research* makes an important argument: the establishment of long-term research projects and field stations in African forests has been essential to successful wildlife conservation. Contributors from various backgrounds provide testimonies from seven of Africa's oldest research stations, which have played an important role in countering the disappearance of our closest living relatives, the great apes. Throughout the 20 chapters, the authors employ a myriad of methods to support the book's central argument, and the reader is left convinced that long-term conservation research gives us reason to be hopeful about the future of Africa's forests.

The book's chapters are divided into two sections. The first several chapters focus on Uganda's Kibale National Park and its main field station, Makerere University Biological Field Station, while the following chapters explore research and conservation in six other African forests. Each contribution is interesting and enjoyable, filled with engaging data on wildlife ecology and the social, political, and economic entities that influence each field site and its associated conservation activities. The editors arrange the collection in a fluid fashion, offering not only a detailed history of parks and the cultures that surround them but also an honest account of the successes and failures in each research situation, along with the hopes and dangers that confront protected areas.

Because the contributions make use of both the natural and the social sciences, *Science and Conservation in African Forests* represents an integrated resource for anyone interested in practicing conservation. The information offers a set of guidelines for establishing long-term research projects and successful conservation initiatives. Giving special attention to primatology and the great apes, the book features contributions from several renowned scholars, including Jane Goodall. For novice primatologists such as myself, this volume is a guiding light on a path to conserving the species that we so deeply respect.

Although the book focuses primarily on conservation of the African great apes, the insights are valuable for any practitioner or researcher, and they can be applied in almost any conservation situation. A major strength of the book is its clear statement that conservation does not happen by declaring an area a national park, evicting its human inhabitants, and ignoring their influence on natural processes within the region. Several authors adamantly stress the importance of working directly with local populations. According to them, long-term research and any associated conservation efforts cannot be realized without the acceptance, support, and participation of locals and nationals. This can be accomplished in many ways, including training nationals as scientists with legitimate advanced degrees, as well as providing health care, education, and community development to people who live in and around protected areas.

Unfortunately, there is little direct evidence to support the main argument of the book. As the editors state in the first chapter, this weakness results from the absence of studies that chart the actual effects of long-term research on forest ecology. All of the supposed positive outcomes—including higher reforestation rates, increasing animal populations, and intensifying community involvement—need to be studied in more detail to distinguish the genuine benefits of the proposed conservation and research strategies from chance occurrences, whether human or non-human in origin.

Nevertheless, the evidence of successful conservation of African forests is all but undeniable. As *Science and Conservation in African Forests* states, such positive findings can be attributed to the work of established field stations, their long-term research projects, and their associated conservation initiatives. Anyone interested in conservation will find this book a must-have resource that provides both concrete information and an inspiring sense of hope for a biologically diverse future.

Andrew Oberle
University of Texas at San Antonio

Berghahn Books

VIRTUALISM, GOVERNANCE AND PRACTICE

Vision and Execution in Environmental Conservation

Edited by James G. Carrier and Paige West

The chapters in the first section provide rich descriptions of environmental discourses, but the true contribution of Virtualism, governance and practice *is found in the second section, which explores the relationship between conservation design and conservation execution... [It] would be an excellent addition to courses on environmental discourses and community-based conservation. Even more, [it] would be useful reading for any individual who hopes to contribute to the policy and practice of conservation in the twenty-first century.* · Focaal

The breadth and detail of these studies, combined with the excellent conceptual framework provided by editors, make this a highly valuable collection for scholars, policy makers, and applied practitioners.
· James Igoe, Dartmouth College

Many people investigating the operation of large-scale environmentalist organizations see signs of power, knowledge and governance in their policies and projects. This collection indicates that such an analysis appears to be justified from one perspective, but not from another. The chapters in this collection show that the critics, concerned with the power of these organizations to impose their policies in different parts of the world, appear justified when we look at environmentalist visions and at organizational policies and programs. However, they are much less justified when we look at the practical operation of such organizations and their ability to generate and carry out projects intended to reshape the world.

James G. Carrier has taught and done research in Papua New Guinea, the United States and Great Britain. **Paige West** is Associate Professor of Anthropology at Barnard College, Columbia University.

Volume 13, Environmental Anthropology and Ethnobiology

208 pages, 7 ills, bibliog., index
ISBN 978-1-84545-619-1 Hardback

FISHERS AND SCIENTISTS IN MODERN TURKEY

The Management of Natural Resources, Knowledge and Identity on the Eastern Black Sea Coast

Ståle Knudsen

... Knudsen delivers an intriguing study about the meaning of the fish, the fishing activity as a way of life, and the culture of the fishermen. His ability to explain the culture and the life and hard work of Turkish fishers is outstanding." · Environmental History

Through the ethnography and history of fish production, seafood consumption, state modernizing policies and marine science, this book analyzes the role of local knowledge in the management of marine resources on the Eastern Black Sea coast of Turkey. Fishing, science and other ways of knowing and relating to fish and the sea are analyzed as particular ways of life conditioned by history, ideology and daily practice. The approach adopted here allows for a broader analysis of the role knowledge plays in the management of common pool resources (CPR) than is provided in much of the contemporary CPR debate that tends to have a somewhat narrow focus on institutions and rules. By contrast, the author argues that also local knowledge and the larger historical and ideological context of production, as manifest in state modernization policies and consumption patterns, should be taken into account when trying to explain the current management regime in Turkish Black Sea fisheries.

Ståle Knudsen is Associate Professor at the Department of Social Anthropology, University of Bergen.

Volume 8, Environmental Anthropology and Ethnobiology

304 pages, 23 ills, bibliog., index
ISBN 978-1-84545-440-1 Hardback
ISBN 978-1-84545-375-6 Paperback

www.berghahnbooks.com